AF581666

Novel Industry 4.0 Technologies and Applications

Novel Industry 4.0 Technologies and Applications

Editors

Nikolaos Papakostas
Carmen Constantinescu
Dimitris Mourtzis

MDPI • Basel • Beijing • Wuhan • Barcelona • Belgrade • Manchester • Tokyo • Cluj • Tianjin

Editors
Nikolaos Papakostas
University College Dublin
Ireland

Carmen Constantinescu
Fraunhofer Institute for
Industrial Engineering
Germany

Dimitris Mourtzis
University of Patras
Greece

Editorial Office
MDPI
St. Alban-Anlage 66
4052 Basel, Switzerland

This is a reprint of articles from the Special Issue published online in the open access journal *Applied Sciences* (ISSN 2076-3417) (available at: https://www.mdpi.com/journal/applsci/special_issues/Novel_Industry).

For citation purposes, cite each article independently as indicated on the article page online and as indicated below:

LastName, A.A.; LastName, B.B.; LastName, C.C. Article Title. *Journal Name* **Year**, *Article Number*, Page Range.

ISBN 978-3-03943-583-8 (Hbk)
ISBN 978-3-03943-584-5 (PDF)

Contents

About the Editors vii

Nikolaos Papakostas, Carmen Constantinescu and Dimitris Mourtzis
Novel Industry 4.0 Technologies and Applications
Reprinted from: *Appl. Sci.* **2020**, *10*, 6498, doi:10.3390/app10186498 1

Nikolaos Papakostas, Anthony Newell and Abraham George
An Agent-Based Decision Support Platform for Additive Manufacturing Applications
Reprinted from: *Appl. Sci.* **2020**, *10*, 4953, doi:10.3390/app10144953 3

Ci He, Shuyou Zhang, Lemiao Qiu, Xiaojian Liu and Zili Wang
Assembly Tolerance Design Based on Skin Model Shapes Considering Processing Feature Degradation
Reprinted from: *Appl. Sci.* **2019**, *9*, 3216, doi:10.3390/app9163216 25

José Luis Saorín, Jorge de la Torre-Cantero, Dámari Melián Díaz and Vicente López-Chao
Cloud-Based Collaborative 3D Modeling to Train Engineers for the Industry 4.0
Reprinted from: *Appl. Sci.* **2019**, *9*, 4559, doi:10.3390/app9214559 43

Alexios Papacharalampopoulos, Christos Giannoulis, Panos Stavropoulos and Dimitris Mourtzis
A Digital Twin for Automated Root-Cause Search of Production Alarms Based on KPIs Aggregated from IoT
Reprinted from: *Appl. Sci.* **2020**, *10*, 2377, doi:10.3390/app10072377 55

Tao Zan, Zhihao Liu, Zifeng Su, Min Wang, Xiangsheng Gao and Deyin Chen
Statistical Process Control with Intelligence Based on the Deep Learning Model
Reprinted from: *Appl. Sci.* **2020**, *10*, 308, doi:10.3390/app10010308 71

Christian Dahmen and Carmen Constantinescu
Methodology of Employing Exoskeleton Technology in Manufacturing by Considering Time-Related and Ergonomics Influences
Reprinted from: *Appl. Sci.* **2020**, *10*, 1591, doi:10.3390/app10051591 91

Xinda Wang, Xiao Luo, Baoling Han, Yuhan Chen, Guanhao Liang and Kailin Zheng
Collision-Free Path Planning Method for Robots Based on an Improved Rapidly-Exploring Random Tree Algorithm
Reprinted from: *Appl. Sci.* **2020**, *10*, 1381, doi:10.3390/app10041381 105

Susana Suarez-Fernandez de Miranda, Francisco Aguayo-González, Jorge Salguero-Gómez and María Jesús Ávila-Gutiérrez
Life Cycle Engineering 4.0: A Proposal to Conceive Manufacturing Systems for Industry 4.0 Centred on the Human Factor (DfHFinI4.0)
Reprinted from: *Appl. Sci.* **2020**, *10*, 4442, doi:10.3390/app10134442 119

Dimitris Mourtzis, Vasileios Siatras and John Angelopoulos
Real-Time Remote Maintenance Support Based on Augmented Reality (AR)
Reprinted from: *Appl. Sci.* **2020**, *10*, 1855, doi:10.3390/app10051855 149

Juanli Li, Yang Liu, Jiacheng Xie, Xuewen Wang and Xing Ge
Cutting Path Planning Technology of Shearer Based on Virtual Reality
Reprinted from: *Appl. Sci.* **2020**, *10*, 771, doi:10.3390/app10030771 165

Martin Pech and Jaroslav Vrchota
Classification of Small- and Medium-Sized Enterprises Based on the Level of Industry 4.0 Implementation
Reprinted from: *Appl. Sci.* **2020**, *10*, 5150, doi:10.3390/app10155150 . . . . **183**

Jaroslav Vrchota and Martin Pech
Readiness of Enterprises in Czech Republic to Implement Industry 4.0: Index of Industry 4.0
Reprinted from: *Appl. Sci.* **2019**, *9*, 5405, doi:10.3390/app9245405 . . . . **205**

Qiwei Zhang, Xiangdong Kong, Bin Yu, Kaixian Ba, Zhengguo Jin and Yan Kang
Review and Development Trend of Digital Hydraulic Technology
Reprinted from: *Appl. Sci.* **2020**, *10*, 579, doi:10.3390/app10020579 . . . . **231**

About the Editors

Nikolaos Papakostas, Associate Professor at University College Dublin. He is Research Director of the Laboratory for Advanced Manufacturing Simulation and Robotics (UCD-LAMS) at the School of Mechanical and Materials Engineering. His research areas include Robotics, Digital Manufacturing, and Optimization of Manufacturing Systems and Networks. He has cooperated with many leading companies in Ireland and Europe mainly from the bioprocessing, automotive, and aerospace industrial sectors as well as with top machine tool building and software firms. He has coordinated a significant number of European and national research projects, targeting the introduction of novel digital technologies in the manufacturing environment of the 21st century. He has authored over 100 scientific papers with more than 2500 citations. He has served as IT Manager of a large manufacturing company for several years. He is an Associate Member of the International Academy for Production Engineering (CIRP).

Carmen Constantinescu, Prof. Dr-Ing., MBA, at Fraunhofer Institute for Industrial Engineering IAO. She graduated as Mechanical Engineer at Technical University of Cluj-Napoca, Romania, and completed her PhD degree in the area of Industrial Engineering at the same university. Since 2001, Professor Constantinescu has been working at Fraunhofer Institute for Manufacturing Engineering and Fraunhofer Institute for Industrial Engineering in Stuttgart. In 2008, she founded the research group and department "Digital Factory" and is currently leading the research area "Digital Manufacturing 4.0". Prof. Constantinescu is involved in academic activities at University of Stuttgart; Royal Institute of Technology, Sweden; and Technical University of Cluj-Napoca, Romania, in the fields of Modeling, Simulation, and Digital/Virtual and Smart Factory. More than 30 of her research projects have been financed by the European Commission under FP6, FP7, Horizon2020, the German Ministry of Research and Education, the German National Science Foundation, and the German industry. She has co-authored more than 4 books and 30 scientific publications in the field of optimization of manufacturing processes through digitalization transformation technologies as part of her contribution to the academic and industrial communities.

Dimitris Mourtzis, Professor at University of Patras. He is Director of the Laboratory of Manufacturing Systems and Automation (LMS) at the Department of Mechanical Engineering and Aeronautics. His main research interests are focused on Manufacturing Systems, Robot Automation, and Virtual Reality in Manufacturing and Manufacturing Processes Modeling and Energy Efficiency. He is also actively involved in Digital Transformation and Implementation of Industry 4.0 in industry. He is an elected Fellow Member of the International Academy for Production Research (CIRP) and member of several International Federations, among others, IFAC, IFIP, and ASME. He is the coordinator of important projects of Manufacturing Innovation and Circular Economy in the framework of the European Factories of the Future Research Association—EFFRA, the European Institute of Innovation and Technology—EIT and, specifically, the project EIT Manufacturing—EIT M. He has served as Coordinator of the innovative educational activity for engineers "Teaching Factories—New training methods of the Engineer" regarding the education and training of future young engineers in collaboration with industry. Moreover, he is the Faculty Liaison representing University of Patras in the iPodia program. He is reviewer of more than eighty international scientific journals with a high Impact Factor and has published more than three hundred scientific papers with more than 6000 citations.

Editorial

Novel Industry 4.0 Technologies and Applications

Nikolaos Papakostas [1,*], Carmen Constantinescu [2] and Dimitris Mourtzis [3]

1 Laboratory for Advanced Manufacturing Simulation and Robotics, School of Mechanical and Materials Engineering, University College Dublin, D04 V1W8 Dublin, Ireland
2 Fraunhofer IAO, Nobelstraße 12, 70569 Stuttgart, Germany; carmen.constantinescu@iao.fraunhofer.de
3 Laboratory for Manufacturing Systems and Automation, Department of Mechanical Engineering and Aeronautics, University of Patras, 26504 Patras, Greece; mourtzis@lms.mech.upatras.gr
* Correspondence: nikolaos.papakostas@ucd.ie; Tel.: +353-1-716-1741

Received: 13 September 2020; Accepted: 16 September 2020; Published: 17 September 2020

The Industry 4.0 paradigm has led to the creation of new opportunities for taking advantage of a series of diverse technologies in the manufacturing domain, including Internet of Things, Augmented and Virtual Reality, Machine Learning, Advanced Robotics, Additive Manufacturing, System and Process Simulation, Computer-Aided Design/Engineering/Manufacturing/Process Planning systems as well as Product Lifecycle Management platforms.

The integration of such technologies, employing information that is generated during different phases of a product lifecycle, may lead to the better utilization and optimization of existing resources, such as labor, materials, energy, and equipment, as well as to the development of products of higher quality and performance in a sustainable manner.

Considering the continuous growth of available computational power, the proliferation of cloud-based platforms, the cost-efficient development and utilization of once prohibitively expensive equipment, such as robotic systems (stationary, mobile, collaborative, and wearable), advanced sensors, and 3D printers, there will be a time when engineers will be able to transform the requirements pertaining to a new product to detailed production, supply chain, and product lifecycle management configurations in a very accurate manner, exploring diverse demand and production scenarios. Engineers would at some point be capable of identifying very fast, perhaps in a fully automated and intuitive way, what the product design would look like, which resources would be needed for developing the product and how they should be configured, who would be supplying parts, equipment, and services, how the product could be repaired and updated, and how it could be recycled when reaching its end of life.

Although products and manufacturing processes are typically quite complex and are often associated with a high degree of uncertainty, it is expected that the availability of more information will lead to the generation of structured product development knowledge and models, which will make their way in tightly integrated digital manufacturing platforms, thus enabling the faster and overall more efficient development of products and services.

However, the first demonstrations of Industry 4.0 principles and technologies are already here and will pave the way towards further developments in manufacturing. This book includes 13 papers that discuss how the Industry 4.0 paradigm may be applied in real engineering and manufacturing cases. The topics covered span a series of diverse areas related to: product design and development [1–3], manufacturing systems and operations [4–8], process engineering [9,10], and Industry 4.0 technologies review and realization [11–13].

Author Contributions: All authors contributed equally to the preparation of this manuscript. All authors have read and agreed to the published version of the manuscript.

Funding: The partial financial support from a research grant from Science Foundation Ireland (SFI) under Grant Number 16/RC/3872, through the I-Form Advanced Manufacturing Research Centre, is gratefully appreciated.

Acknowledgments: This publication was only possible with the invaluable contributions from the authors, reviewers, and the editorial team of *Applied Sciences*. We would particularly like to thank our Managing Editor Melon Zhang.

Conflicts of Interest: The authors declare no conflict of interest.

References

1. Papakostas, N.; Newell, A.; George, A. An agent-based decision support platform for additive manufacturing applications. *Appl. Sci.* **2020**, *10*, 4953. [CrossRef]
2. He, C.; Zhang, S.-Y.; Qiu, L.-M.; Liu, X.; Wang, Z. Assembly tolerance design based on skin model shapes considering processing feature degradation. *Appl. Sci.* **2019**, *9*, 3216. [CrossRef]
3. Saorín, J.L.; De La Torre-Cantero, J.; Melian-Diaz, D.; López-Chao, V. Cloud-based collaborative 3D modeling to train engineers for the industry 4.0. *Appl. Sci.* **2019**, *9*, 4559. [CrossRef]
4. Papacharalampopoulos, A.; Giannoulis, C.; Stavropoulos, P.; Mourtzis, D. A digital twin for automated root-cause search of production alarms based on KPIs aggregated from IoT. *Appl. Sci.* **2020**, *10*, 2377. [CrossRef]
5. Zan, T.; Liu, Z.; Su, Z.; Wang, M.; Gao, X.; Chen, D. Statistical process control with intelligence based on the deep learning model. *Appl. Sci.* **2019**, *10*, 308. [CrossRef]
6. Dahmen, C.; Constantinescu, C. Methodology of employing exoskeleton technology in manufacturing by considering time-related and ergonomics influences. *Appl. Sci.* **2020**, *10*, 1591. [CrossRef]
7. Wang, X.; Luo, X.; Han, B.; Chen, Y.; Liang, G.; Zheng, K. Collision-free path planning method for robots based on an improved rapidly-exploring random tree algorithm. *Appl. Sci.* **2020**, *10*, 1381. [CrossRef]
8. De Miranda, S.S.-F.; Gonzalez, F.A.; Salguero, J.; Gutiérrez, M.J. Ávila life cycle engineering 4.0: A proposal to conceive manufacturing systems for industry 4.0 centred on the human factor (DfHFinI4.0). *Appl. Sci.* **2020**, *10*, 4442. [CrossRef]
9. Mourtzis, D.; Siatras, V.; Angelopoulos, J. Real-time remote maintenance support based on Augmented Reality (AR). *Appl. Sci.* **2020**, *10*, 1855. [CrossRef]
10. Li, J.; Liu, Y.; Xie, J.; Wang, X.; Ge, X. Cutting path planning technology of shearer based on virtual reality. *Appl. Sci.* **2020**, *10*, 771. [CrossRef]
11. Pech, M.; Vrchota, J. Classification of small and medium-sized enterprises based on the level of industry 4.0 implementation. *Appl. Sci.* **2020**, *10*, 5150. [CrossRef]
12. Vrchota, J.; Pech, M. Readiness of enterprises in Czech Republic to implement industry 4.0: Index of industry 4.0. *Appl. Sci.* **2019**, *9*, 5405. [CrossRef]
13. Zhang, Q.; Kong, X.-D.; Yu, B.; Ba, K.-X.; Jin, Z.-G.; Kang, Y. Review and development trend of digital hydraulic technology. *Appl. Sci.* **2020**, *10*, 579. [CrossRef]

Article

An Agent-Based Decision Support Platform for Additive Manufacturing Applications

Nikolaos Papakostas *, Anthony Newell and Abraham George

Laboratory for Advanced Manufacturing Simulation and Robotics, School of Mechanical and Materials Engineering, University College Dublin, D04 V1W8 Dublin 4, Ireland; anthony.newell@ucd.ie (A.N.); abraham.george@ucdconnect.ie (A.G.)

* Correspondence: nikolaos.papakostas@ucd.ie; Tel.: +353-1-716-1741

Received: 23 June 2020; Accepted: 14 July 2020; Published: 18 July 2020

Abstract: The effective estimation and consideration of process cost, time, and quality for additive manufacturing operations, when a series of suitable technologies and resources are available, is very important for making informed product design and development decisions. The main objective of this paper is to propose the design, deployment, and use of an agent-based decision support platform, which is capable of proposing alternative additive manufacturing resources and process configurations to design engineers while reducing the number of communication steps among engineering teams and organizations. Different computer-aided systems are utilised and interfaced for automating the information exchange as well as for accelerating the overall product development process.

Keywords: process planning; scheduling; design for additive manufacturing; multiple criteria

1. Introduction

Additive Manufacturing (AM) is characterized by the layer-wise production of a part as opposed to conventional manufacturing (CM) of subtractive or forming methods, such as computer numerical control (CNC) milling and injection moulding [1,2]. Additive manufacturing processes can allow for increased design freedom to produce functional freeform products with high geometric complexity, previously not possible with CM, and are further enabled by advances in design for additive manufacturing (DfAM) [3]. AM is also an attractive option as it can provide significant cost savings compared to CM, due to the abatement of tooling and the reduced material usage [4]. AM can complement CM when used to quickly produce injection moulds and tools with conformal cooling for enhanced performance [5]. Further to this, the on-demand production of spare parts using AM enhances supply chain management by lowering warehousing and transport costs as well as by reducing lead times [6–10]. Designs optimised for AM using topological optimisation have significantly reduced weights, leading to substantial reductions in lifecycle fuel costs and emissions, which is of particular importance to aircraft parts [11]. AM parts see a diverse range of applications in aerospace, automotive, defence, medical implants, as well as in toys, and jewellery industries [12].

Various AM technologies are available for producing parts, including stereolithography, powder bed fusion, and directed energy deposition, among others [1,2]. Each AM technology has its own associated material preparation requirements, material phases, and workable materials, from polymers to metals, composites and ceramics, and even biological materials [13]. Cost, quality, and time performance achieved in the production of AM parts are important considerations for the product development process [3] and vary depending on the technologies and the resources used. Indeed, two different machines using the same AM technology can produce parts of different mechanical properties and performance in terms of quality, cost, and process time, depending on a series of factors. These factors include the type of material used as well as process parameters, such as layer resolution, build rate, support design, build orientation, and scan strategy [14]. The implementation of

AM processes by companies, especially small and medium enterprises (SMEs), is not trivial and is fraught with several issues mainly due to engineers not always having enough experience in using these systems. The high investment and training costs, as well as the economic viability challenges associated with the deployment of these systems, raise concerns when a decision needs to be taken about whether a company would invest in AM or not [15–17]. Instead of purchasing AM equipment, companies or individuals may instead have their products produced by AM service providers. In this case, design engineers would send their product or part designs to AM service providers in the form of Computer-Aided Design (CAD) file formats, such as STL or 3MF files, together with a set of technical specifications, and then would ask for further information regarding delivery time, cost, and quality expectations. The broad range of machines, technologies, AM original equipment manufacturers (OEMs), and suppliers make it increasingly difficult for designers and engineers to understand which AM equipment or process configurations could be best used for producing the product or part that they design, utilising AM technologies. Consequently, this leads to a high degree of uncertainty regarding the estimation of process time, cost, and quality performance of the AM process itself. Furthermore, designers are typically not aware of the AM machines' availability and process capabilities. For this reason, they usually rely on the expertise of AM service suppliers or engineering bureaus, which often take advantage of Manufacturing Execution Systems (MES) for keeping track of their AM resources' operation and availability, while utilising Computer Aided Manufacturing (CAM) systems or slicing software that are associated with their AM equipment. Special machine profiles, comprising different materials and process configurations, are used by CAM software for generating the machine process instructions (G-code) and for estimating the time, materials usage, and cost of the AM production process [18]. Although interfacing different Computer Aided technologies is a fairly complex process, recent studies have shown that even SMEs have started addressing with moderate success the need for having integrated systems in different product development phases [19]. Regarding how the configuration of AM processes can be facilitated, a number of approaches have been used and reported and will be discussed in the following sections.

1.1. Decision Support Systems for Additive Manufacturing (AM) Process Selection

Decision Support Systems (DSS) are used to aid the decision-making process by generating alternatives, which then may be evaluated utilising a set of predefined criteria [20,21]. Different types of DSSs exist, including communication-driven, data-driven, document-driven, knowledge-driven, and model-driven DSSs [22,23]. DSSs and models for AM processes have been reported in the literature [24–35]. Park and Tran [24] used a knowledge-based DSS with separate decision tables for materials, quality, and overall cost associated with each individual AM method, for developing a rule-based decision support system. With this system the user could input specific material, method and accuracy requirements and the output report would recommend a single printing method. Limitations of this system are that just a single AM method is recommended, whereas multiple AM methods could achieve similar performance and could be considered by a designer or an engineer. The overall cost and build time were also not included in the output report.

Kretzschmar et al. [25] implemented a DSS for the selection of AM machines for metal powder bed fusion (PBF) processes based on cost (machine, material, and labour) and build time. In this system, the user inputs the part's STL file, then selects a machine, material, accuracy levels, production volumes and densities of support structures. Outputs to the user include itemised cost per part and build time. Limitations of this study are a large number of required user inputs, including machine selection.

Ghazy [26] developed an updateable knowledge-based DSS for selecting AM systems based on materials, finishing methods, and machines with a Simple Multiple Attribute Rating Technique (SMART). User inputs for this system included process dependent specifications, such as minimum wall thickness, surface finish, size, quantity and accuracy, material dependent specifications, such as strength, hardness, electrical and thermal properties, the output of both of these being a ranked table of suitable AM processes and materials for the build. Limitations of this DSS are the omission of consideration of cost and build speed parameters for process time estimation as well as the lack of STL file reading functionality, which would reduce any potential errors in the process dependent selection phase, where many part geometry parameters are input instead. Byun and Lee [29] used a modified TOPSIS method for analysing qualitative and quantitative data, using factors such as accuracy, roughness, strength, elongation, part cost and build time for ranking alternative AM processes in a pairwise comparison matrix. Borille and de Oliveira Gomes [27] used an analytic hierarchy process (AHP) and multiplicative analytic hierarchy process (MAHP) for comparing and ranking appropriate AM processes with consideration to cost, build time, accuracy, roughness, tensile strength, and elongation. Braglia and Petroni [28] used AHP for AM process selection based on cost, time, size, complexity, and surface texture. Zhang et al. [30] used a Multi-Attribute Decision-Making (MADM) approach, taking advantage of a knowledge value measuring method for making decisions involving production cost, time and quality, using a case study on AM system parameters from Rao and Padmanabhan [33]. These system parameters included mechanical quality properties, such as accuracy, surface roughness, tensile strength, and elongation, as well as part cost and build time. Meisel et al. [31] developed a DSS framework for the selection of suitable AM processes in remote or austere environments. Bikas et al. [32] presented a framework for facilitating process selection in AM. This framework included the evaluation of AM technologies, AM process selection, assessment of technical feasibility of AM processes, evaluation of the design, and finally process planning for hybrid manufacturing to reduce costs by combining CM and AM. Watson and Taminger developed a decision support framework for selecting AM vs CM methods, such as CNC milling, based on energy consumption indicators [34]. This framework accounted for the entire manufacturing lifecycle energy consumption required for both AM and CM and determined that there lies a critical value for the fraction of the bounding envelope that contains material whereby the energy consumption for AM and CM is equivalent. For volume fractions below this critical value, AM is more efficient and above this, CM is more efficient. Wang et al. [35] developed a DSS for AM process selection using a hybrid MADM with TOPSIS for ranking possible solutions. This DSS accounted for various performance parameters, including tensile strength, dimensional accuracy, surface finish, and material cost. A comprehensive review of the methods currently used for AM process selection was conducted by Wang et al. [36]. This review could support DfAM using knowledge-based DSS on top of examining and evaluating user preferences and AM process performance. Zhang et al. [37] developed a build orientation optimisation method for multi-part production in AM. They used a feature-based method to constrain the number of possible orientations for each part within a group of parts in a single build, which would ensure build quality was not compromised. Following this, a genetic algorithm was used for optimising the decision index of an integrated MADM model for each alternative orientation in order to minimize cost.

Agent-Based Decision Support Systems and Cloud Manufacturing

A software agent in the manufacturing domain is typically a computer programme that may act on behalf of an engineer or operator. Agent-based decision support systems (ABDSS) are typically autonomous computer or software systems, which communicate with the environment on behalf of a designer, an engineer or an operator to achieve a predefined goal [38]. Similar platforms have been developed previously for a plethora of applications. For example, an agent-based system (ABSTUR) capable of picking an optimal route to avoid overcrowded and non-profitable tourist routes was devised. The agents in ABSTUR represented the simulator, different categories of tourists and the route manager [39]. ABDSS can be adaptive and intelligent, tailoring their behaviour to environmental changes and can apply a fixed set of rules to enable reasoning, learning, and planning functionality. Multi-agent systems are a group of agents, which work baring similarity to a community of human workers, collaborating with predefined roles, towards a particular goal through effective communication and reasoning [38]. They enable decentralized problem solving [40] and can combine machine learning, simulation, and multi-criteria decision-making features. ABDSS have been used in areas, such as engineering design [41], process planning [42], production planning and resource allocation [43], production scheduling and control [44], process control monitoring and diagnosis [45], enterprise organisation and integration [46], networked production [47], assembly, and life-cycle management [48,49].

A Multi-Agent Systems (MAS) architecture was developed by Legien et al. [21] for supporting technology recommendations, specifically related to material choice support for casting with cost estimations. In the field of DfAM, Dhokia et al. proposed an agent-based generative design tool [50]. In the area of process planning, a number of multi-agent systems and platforms have been proposed for machining prismatic parts utilising STEP-NC [51,52]. MAS for modelling AM processes have also been proposed, without however elaborating on how different AM technologies and equipment may be utilised as a part of the same platform [53]. Recent research works have suggested the integration of MAS in cloud manufacturing platforms for AM [54]. Cloud manufacturing is a paradigm, which is enabled by cloud computing technologies as well as by the integration of networked manufacturing systems, including networked manufacturing, virtual manufacturing, internet of things (IoT), and agile manufacturing technologies [55,56].

In a previous work of this paper's authors, the use of blockchain technology principles was introduced for securely managing AM product development data with a MAS, mainly for storing product development information in a secure way [57]. In this paper, a novel MAS approach is presented for facilitating and accelerating the process of designing and manufacturing a product utilising AM technologies, by integrating CAD, CAM tools, and MES data, and by automatically selecting the most suitable AM process configuration, equipment, and service provider.

1.2. Comparison with Other AM Platforms

Several different AM platforms in the market offer manufacturing execution or AM production capabilities as a service. Manufacturing service providers such as Materialise, Stratasys, Formlabs, Shapeways, Protiq, 3D-Hubs, and others provide similar manufacturing options at different capacities. These companies usually have access to proprietary machines and software. These platforms come with their own features and require a significant amount of time for engineers and designers to understand how they work and how they may be adapted to their own needs.

When reviewing the services provided by the likes of Materialise, Protiq, Shapeways (Figure 1), or others, it is to be noted that what is offered to users is more of a marketplace, where the user defines the requirements to print parts and then place an order for having those parts printed. These platforms typically are also capable of providing a delivery date. However, being a web-based service, the downside with such services is that there is no direct integration with CAD software environments. As the designer must exit the design environment and move to a different software application, which requires the designer to upload the file in a specific CAD format, experimentation with a different version of the same design is more error-prone and time-consuming.

At the same time, whenever a user wishes to check what the consequences will be in case the quality parameters (such as finishing quality) are changed, the same process will have to be followed again. Additionally, the user typically receives no feedback about the process parameters, such as layer thickness, or the AM machine that will be used for printing the part. In case there are strict requirements regarding layer thickness, density or even the machine to be used, there is usually no straightforward way to input these requirements in these platforms. Furthermore, the cost and time estimations received by the user are typically based on part volume and support information and not on the output of a CAM/slicing software tool, which may mean that cost and time estimations are not as accurate as they could be.

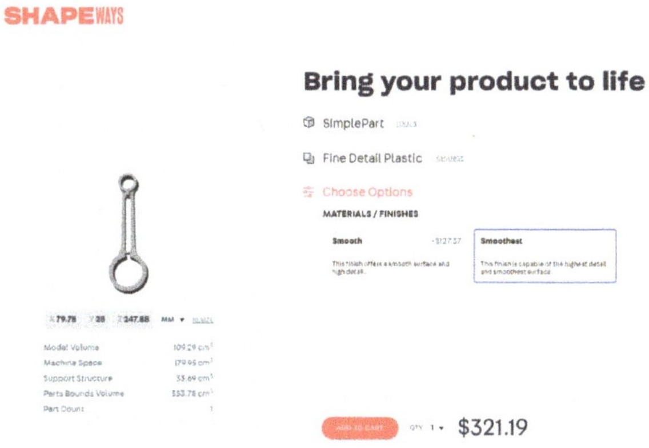

Figure 1. Shapeways platform estimating build cost for a part.

Features of these platforms include:

- Automated generation of support structures (not user viewable);
- Repair of STL model (not user viewable);
- Selection of process parameter (not user viewable);
- Quality selection (the user has limited options);
- They are oriented towards enterprise or early-adoption customers that have a finished design and need a prototype;
- Material selection is limited—they typically use medium to high-end materials.

Furthermore, there are other software packages like Netfabb, Repetier (Figure 2), or Simplify3D, that can cater to a wide range of 3D printing systems, or build platforms by importing machine/system settings directly from their website repository or by specifically feeding in different parameters for specific platforms and materials within the software to create profiles or printer configurations. Such software provides more flexibility to the designer by allowing for more customization options down to the minute details like support density, layer thickness, infill density, material options, extrusion options, and many others depending on the AM technology. The use of this software, however, is not an automated procedure since the designer is required to have intricate knowledge of the slicing/CAM tool software. Specific versions of software packages like Netfabb or Repetier are free-to-use software packages, whereas Simplify3D is a commercial software tool, providing the flexibility to import printers and configuration files through their website with a proprietary slicing engine.

The list of features of these platforms include:

- They are highly flexible—and therefore the setup with host and server is complex.
- The capability of repairing defective STL meshes.
- Control of layer thickness, material, density, and other process parameters is allowed.
- Custom generation of support structures, patterns, infill density for supports is allowed.
- Custom slicers can be imported—for example, newer versions of the slicing engines from Prusa, Cura, Netfabb, or others. This is a feature that is not typically addressed to average users, but it is still useful when special design and production requirements need to be considered.

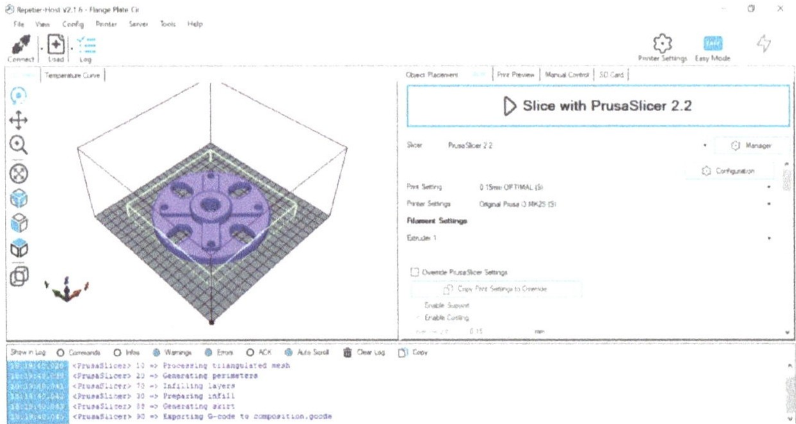

Figure 2. Repetier-Host Platform generating toolpath for a build.

In this paper a different approach is presented, whereas the selection of AM part process configuration options is invoked directly within a CAD environment. The proposed approach would allow for the accurate simulation of AM processes utilising different process configurations. The proposed software platform is designed to extract the required part information automatically from the CAD model design. The platform is capable of testing different process configurations on a diverse range of AM machines and then allows the consideration of multiple criteria, related to cost, time and to parameters affecting quality, for selecting a suitable AM machine and process configuration. The proposed platform is capable of considering custom machine profiles that have been developed by AM service providers over time, based on their experience, diverse material profiles, as well as vendors from different locations. Since the platform can directly be integrated into a CAD environment, the designer may experiment with different design versions or part features without exiting the CAD system.

2. Proposed Methodology

2.1. Scope

Designers would welcome receiving feedback regarding the impact of design features—typically modelled in CAD software—is on the production process, as there is no information on the printed product cost as well as on its delivery. A series of part characteristics need to be considered, including their geometry and technical specifications. The first features that need to be checked, before a specific AM is considered for producing a part, are its external dimensions and the minimum wall thickness. In particular, each part will have a maximum x, y, and z dimension, which will dictate the appropriate build volume requirement for that part. This is of importance to AM machine selection, as each machine will have a maximum build envelope and recommended minimum wall thickness.

Designers are also typically not aware of the capabilities and performance characteristics of diverse AM equipment and technologies. They usually do not have enough knowledge about the different process configurations of each available machine that are suitable for the part designs. Furthermore, they do not have access to information pertaining to the machines' availability as well as to their cost, time performance capabilities and to their process parameters options.

Although many different AM technologies and equipment are available from a number of third-party companies or cooperating suppliers, there is no way of comparing the high number of the resulting process configurations, comprising of different materials, specifications, and supplier/machine options, other than obtaining this information by receiving bids directly from the available suppliers.

However, this is a time-consuming process on which product or part designers have little or no control at all.

2.2. Approach

In this paper, an agent-based approach is proposed, whereby different agents undertake the role of representing diverse stakeholders and resources in the process of designing and manufacturing a product utilising AM technology. Design Agents, in the form of desktop software applications, represent product design engineers and their role is to automate the process of identifying and then submitting the technical specifications of a product or part, while Machine Agents represent the process engineers who would analyse the part geometry and technical specifications and would then return to the designer with information about the process cost and time performance as well as about the process configuration itself. Both Design and Machine Agents are in essence software modules that may potentially replace human operators in a series of activities related to the negotiation and handling of 3D printing orders as well as to the selection of a suitable machine and process configuration. The Design Agent may, in the end, select the AM service provider (represented by a Machine Software Agent) who would make the best offer, considering cost, time, and process criteria. The proposed approach allows Machine Agents to communicate with specific CAM systems that correspond to each 3D printer they represent. These CAM systems are typically associated with a number of process profiles that contain special parameters values for different combinations of materials, process accuracy, part density, and support options. In this paper, the Fused Deposition Modelling (FDM) AM processes have been modelled, but other AM technologies, such as Powder Bed Fusion or Direct Energy Deposition, could also be modelled, taking advantage of the same platform and design principles. Each Machine Agent may choose from a multitude of process configuration profiles. Specifically, each AM process profile associated with a specific AM machine and its CAM software contains the following information:

- Material to be used, including its properties, such as density, cost per weight unit, chamber, or bed temperature;
- Support options, such as support profiles, support on build plate or x-y basis, and parameters per support option;
- Part density;
- AM layer thickness, which is usually directly related to part quality (finish quality, tensile strength, impact strength, and hardness) [58,59].

Each of these profiles corresponds to a specific combination of material, support option, part density, and quality. The standard and most accurate way to check the print feasibility of a part, for a specific AM machine, is to simulate the execution of its G-code, that is generated by a CAM software programmed for that AM machine. This simulation could provide useful information, such as material usage, cost, and build times. It is assumed that the CAM software utilises the latest firmware update for the 3D printer, as well as the most updated information pertaining to the availability of the machines. This information is typically not known to the organisation that wishes to place an order for their parts. It can become quite challenging for a design engineer working for that organisation to explore possible combinations of available equipment, suitable material, process configurations, and service providers [60]. The proposed platform (Figure 3) is based on the Java Agent Development Framework (JADE) [61]. Design Agents are software modules representing design engineers and provide engineers with a Graphical User Interface for specifying the CAD file of the part and its technical specifications, including support options, materials that could be used, the minimum density, the maximum layer thickness (minimum 3D printing accuracy), quantity, and the order due date. One of the main reasons why an agent-based approach was selected is that this way it is easier to distribute the overall computational load to all Machine Agents that are available. Since the proposed approach is in principle simulation-based, the overall computational time required for simulating the 3d printing process for a wide range of machines would be quite long. Instead, by using agents, the designer will only have to wait for as long as it is required for the slowest Machine Agent to return its process alternatives. At the same time, the interaction of each agent with operators and other IT systems could also be done independently, i.e., each Agent may be interfaced to a different MES or CAM platform.

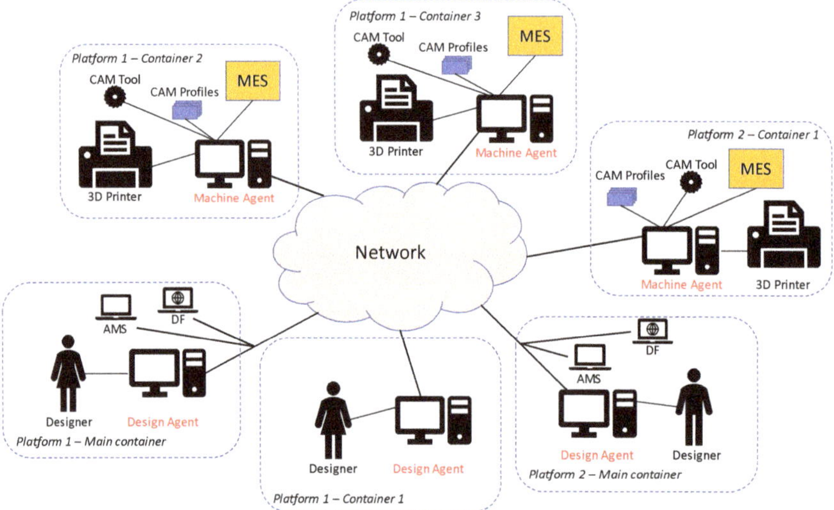

Figure 3. Agent-based platform.

A part geometry analysis tool has been developed and integrated with a commercial CAD platform for launching the Design Agent and for automatically passing to it basic design information, such as the CAD file and the maximum external dimensions and the minimum wall thickness of the part. The designer may edit these values, especially in cases where the geometry of the parts is complex and the calculation of the values of certain features, such as the wall thickness, is not straightforward.

Each platform (belonging, for instance, to a particular manufacturing OEM) will have to host the main container and can have as many other containers (each one belonging, for instance, to a particular AM service provider) as needed. Different platforms may also be connected, which in principle would allow different networks of service providers to be connected. An Agent Management System (AMS) is also part of the platform, providing an assigning service. The Directory Facilitator (DF) provides a Yellow Pages service, allowing agents to find other agents that are also part of the proposed JADE framework.

The sequence of steps followed, and the information contained in the messages exchanged among the Design and Machine Agents and the platform components, including the CAM software and the MES are depicted in the UML sequence diagram of Figure 4. The list below provides more details about each of these steps:

- Once a part is designed, the design engineer (Designer) uses the part geometry analysis tool, which launches the Design Agent (a software instance) who then uploads the CAD file to the cloud or to a remote repository and provides the minimum technical specifications (maximum layer thickness, minimum part density), the basic geometric information (external dimensions and minimum wall thickness), the list of alternative materials that can be used, the number of parts to be produced, the order due date, and decision criteria weights.
- The Design Agent finds all available Machine Agents (software instances) through the platform DF and then sends a message to all suitable Machine Agents, encapsulating all information provided

by the Designer, including the link to the repository containing the CAD file, while asking for a bid.

- The Machine Agent selects all CAM process profiles that fulfil all material, density specifications, and support options and calls the CAM software to generate the G-code and simulate the AM process for each one of these profiles.
- The output of the CAM software is then received by the corresponding Machine Agent and the processing time and cost per part are calculated.
- The Machine Agent then requests information regarding the availability of the machine from its MES and estimates the end date for the specific order. This is done by utilising all available idle time slots for producing the number of parts requested by the Design Agent.
- All alternative process configurations are sent by the Machine Agents back to the Design Agent who then estimates their utility by considering the relative importance of all criteria identified by the Designer.
- Then the best alternative is identified by the Design Agent, who sends a message with an order placement request to the corresponding Machine Agent.
- As soon as the Machine Agent confirms the order, the best alternative process configuration with all pertinent information, regarding the process parameters, the service provider and cost, tardiness performance is presented back to the Designer.
- The Machine Agents are then reset so that they can receive new requests from new instances of the same or other Design Agents from the same or another Agents' Platform.

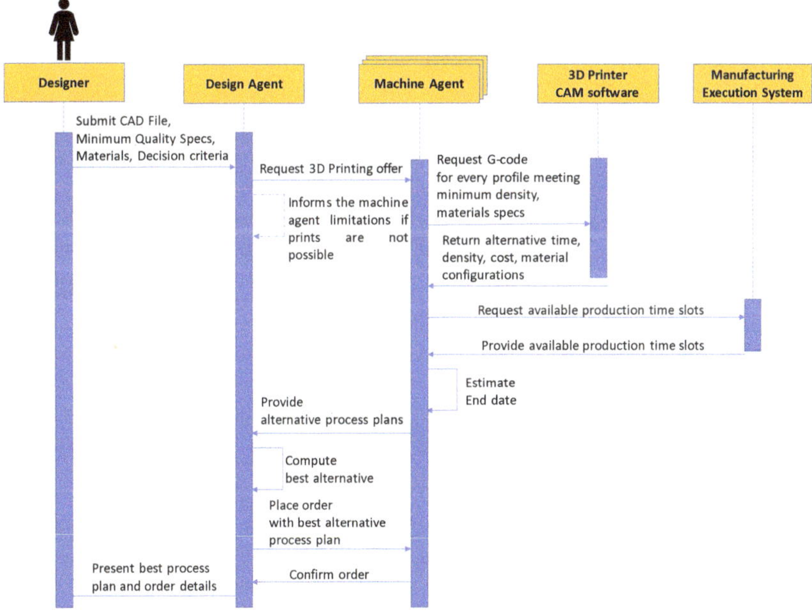

Figure 4. UML Sequence diagram showing the interaction among agents and platform components.

A simplified example of the overall information exchange and alternatives generation process is presented in Figure 5. The Designer submits a part design together with the associated technical specifications to the Design Agent, which are then sent, as part of a 'MessageA' type message, to 2 Machine Agents, representing two 3D printers. Each Machine Agent generates all alternatives that satisfy the technical specifications and then invoke their corresponding CAM tool, providing as input the process parameters for each alternative and the CAD file. As soon as the CAM process simulation (G-code generation) is completed, the output of the simulation together with planning information from the MES are used for calculating the performance indicators for each alternative. This information is sent back to the Design Agent as part of a 'MessageB' type message. The Design Agent ranks all received alternatives, by calculating the utility of each one of them after considering the relative importance of all decision criteria, using the Simple Additive Weighting method, and presents the best ones to the designer.

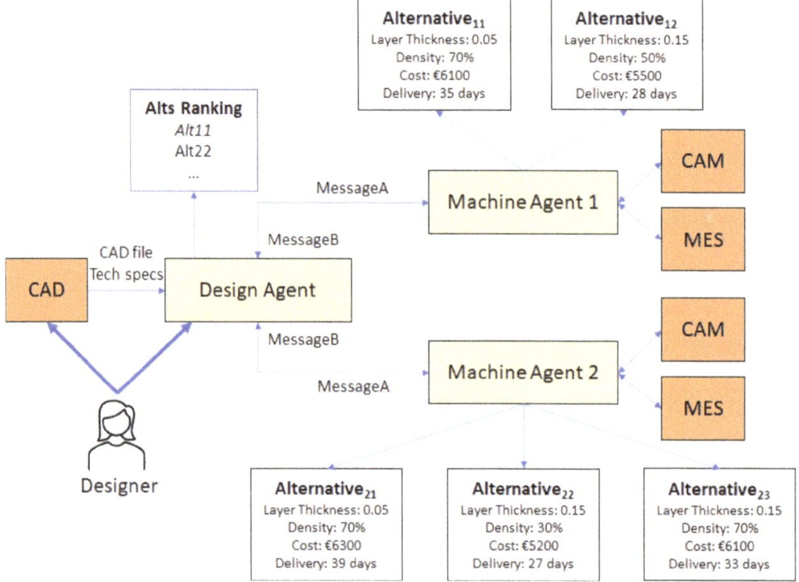

Figure 5. Example of the information exchange process.

2.3. *Software Design and Implementation*

The main components of the proposed platform are the Design and the Machine Agents both implemented in the 'DesignAgent' and 'MachineAgent' classes, respectively (Figure 6). Both components are implemented in Java, inheriting the initialisation, lifecycle, and communication functions from the JADE core Agent class. Each Agent has its own settings, such as the location of the Agents' facilities, the decision criteria weights for the Design Agent and the process profiles and cost parameters for the Machine Agent. The information sent from the Design Agent to Machine Agents is contained in a 'MessageA' object, while all alternatives generated by each Machine Agent and sent back to the Design Agent are part of a 'MessageB' object.

Each Agent can be instantiated in a computer that has a Java Runtime Environment. The platform has been tested and used in networked environments with multiple nodes (computers). A special interface has been built, allowing a commercial CAD system to launch a Design Agent.

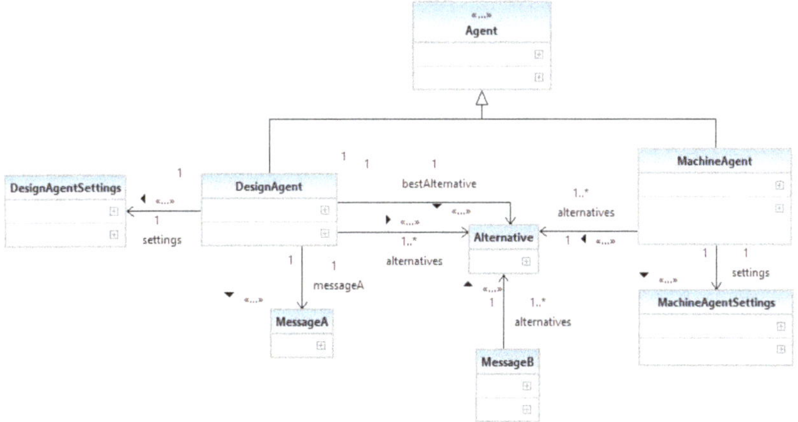

Figure 6. UML Class diagram of main platform components.

2.4. Cost Function

Significant consideration must be given while assessing the cost function for an AM process. The selection criteria of an AM technology largely depend on machines, location, materials, post-processing operations, and many other factors that are deployed in an AM production line. This must also be considered while deriving a cost function for an AM technology. This will determine the most viable AM technology to build a product. The calculation of a cost function will vary significantly in different AM technologies that are currently available in the market. This is because, depending on the AM technology deployed, the values of these factors will vary. For example, the cost factors of a material extrusion process are different when compared with the ones associated with a powder bed fusion process. This could be in the form of energy consumption, labour, overheads, materials cost, capital investment for an AM machine, support volume generated for a build, which in particular can affect the material cost depending on the chosen orientation, or it could be in the form of necessary post-processing operations required for a finished product.

The cost per order is calculated by each Machine Agent for every machine and process profile by considering process material and time requirements as well as shipping cost elements [62,63]. Equation (1) shows a generic cost function used for estimating the overall cost. This equation is derived based on existing techno-economical models used for conventional manufacturing processes [63].

$$C_{sc} = Q.\{A_{sc} \,.\, [1\text{-}d(Q,W_{sc})] \,.T_{sc} + M_{sc} \,.\, W_{sc} + P_{sc} + K_s \,.\, W_{sc} \,.\, D_s\} + F.\, [1 - d(Q,W_{sc})] \quad (1)$$

where:

- s denotes the Machine Agent representing a specific machine of a service provider,
- c denotes the process configuration index,
- C_{sc} is the overall cost for agent s and configuration c,
- Q is the order quantity,
- A_{sc} is the cost rate (€/h) for configuration c of the machine represented by agent s,
- T_{sc} represents the processing time per piece if configuration c of the machine represented by agent s is selected,
- M_{sc} is the total material cost (€/kg) for building the part using configuration c of the machine represented by agent s,
- W_{sc} is the overall weight (kg) of the piece if configuration c of the machine represented by agent s is used, including support material,
- P_{sc} represents the set-up and post-processing operations cost per part,

- K_s is the average shipping cost rate from the service provider represented by agent s [€/(km.kg)],
- D_s is the distance between the service provider represented by agent s and the Designer's location and
- F is the fixed cost per order,
- $d(Q,W_{sc})$ is the discount rate applied, based on the overall cost, excluding material costs, which is in turn a function of the part weight and ordered quantity.

Since the cost function is generic enough so that it can be applied across multiple machine profiles and configurations irrespective of the AM technology or machine used, machine and build plate utilisation were assumed to be at their maximum of 100%. Furthermore, the costs regarding labour or energy consumption would be categorized under the cost rate (A_{sc}). Fixed cost (F) includes overheads and the initial cost of the machines and infrastructure required for the production facility.

3. Test Cases

The test cases in this paper were devised for validating the proposed approach, utilising the latest stable version of a platform, which is being used by researchers of the I-Form Advanced Manufacturing Research Centre for designing and planning 3D printing experiments. Part of these experiments are conducted in cooperation with the industry. The cases are related to an injection moulding company in Ireland that wishes to produce several prototype plastic parts for allowing their client (OEM) and their sales representatives to review the part before committing to the final design that will lead to the development of the mould, which is an expensive process. It is assumed that six different service providers are available in five European countries, using three different types of FDM 3D printers.

3.1. Test Case 1

In the first test case, a relatively simple part was selected, where lower layer thickness, cost, and tardiness are favoured. The criteria weights for evaluating the alternative process configurations from all six service providers were: layer thickness (40%), density (0%), tardiness (30%), and cost (30%). The materials to be considered were Polylactic Acid (PLA) and Acrylonitrile Butadiene Styrene (ABS). Once the product design is completed, a design engineer has the option to use the part geometry analysis tool, which will automatically convert the native CAD file to an STL file and will calculate the external dimensions of the part (Figure 7). Then it will launch the Design Agent (Figure 8). For all 3D printers, the Prusa Slicer CAM tool [18] was interfaced with the corresponding Machine Agents for generating the G-code, while plain text files were used for representing the information related to the machines' availability as stored in standard MES platforms. The process profiles that are available for each machine correspond to 3 different layer thicknesses, i.e., 0.05 mm, 0.15 mm, and 0.30 mm, 2 materials (PLA, ABS), 4 infill densities (0%, 20%, 50%, 70%), and 3 support options. The overall number of profiles is therefore 72 per machine. More process configuration profiles could be prepared and used but for illustration purposes, the number of profiles in this paper was limited to 72. In this case, as per the user's layer thickness, density and material requirements, 2 layer thicknesses (0.05 mm, 0.15 mm), 2 materials (ABS, PLA), 2 infill densities (50%, 70%), and 1 support option were tested. The overall number of combinations is therefore 8, which is equal to the total number of profiles tested and simulated per agent and machine. Figure 9 presents the least and most expensive process alternatives, as well as the one that was selected as the best.

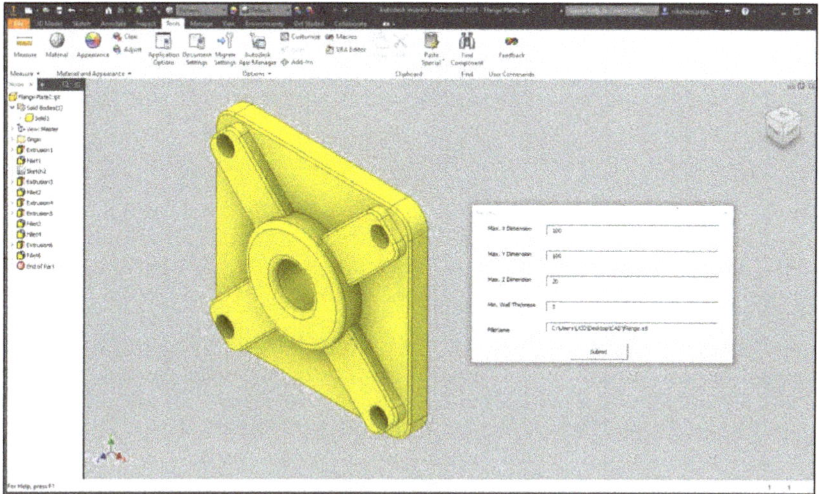

Figure 7. Test Case 1—Part geometry analysis tool integrated with a commercial Computer-Aided Design (CAD) system.

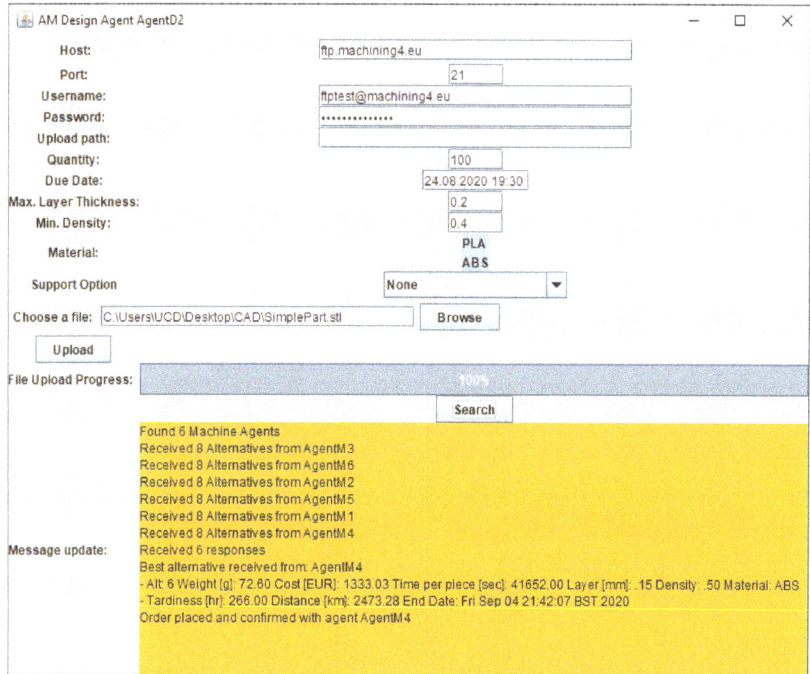

Figure 8. Test Case 1—Design Agent Graphical User Interface (GUI) with technical data and information from Machine Agents.

For the test case 1, agent MA6 provided the least expensive configuration. However, due to its lower degree of availability, this alternative was not the one selected. MA4 provided the best-balanced configuration, exhibiting low cost and tardiness. The most expensive process configuration was provided by agent MA2, since the AM service provider the agent represents is the most expensive one and this particular configuration is related to the lowest possible layer thickness and the highest infill density, leading also to very high tardiness.

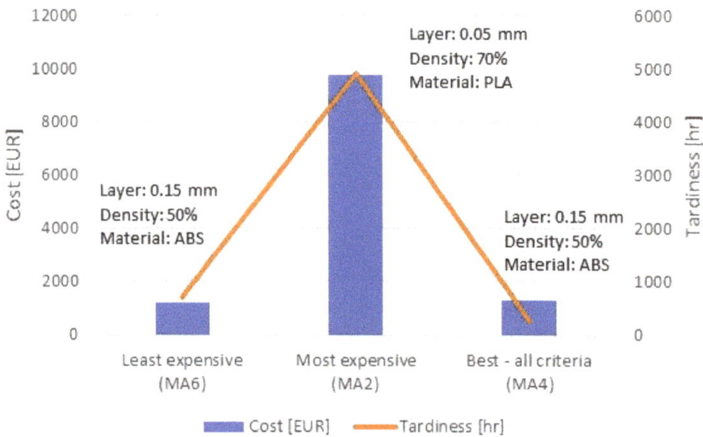

Figure 9. Test Case 1—Cost and tardiness of different configurations.

3.2. Test Case 2

In the second test case, a more complex part was selected (Figure 10). The company would prefer a part with a higher infill density, while cost would not be as important as density, delivery date, and layer thickness. The criteria weights for evaluating the alternative process configurations from all six service providers were: layer thickness (20%), density (50%), tardiness (20%), and cost (10%). The materials to be considered were PLA and ABS. All cost parameters and machines' availability have been assumed to be the same as in test case 1. The parameters used for launching the Design Agent and the information received from the Machine Agents are shown in Figure 11.

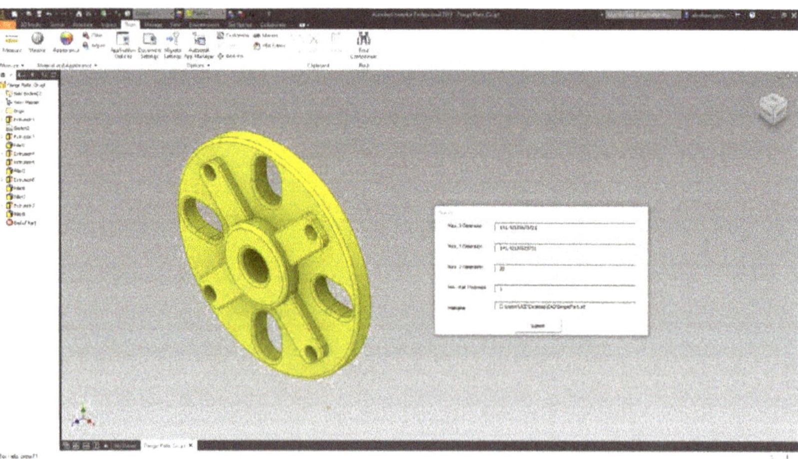

Figure 10. Test Case 2—Part geometry analysis tool integrated with a commercial CAD system.

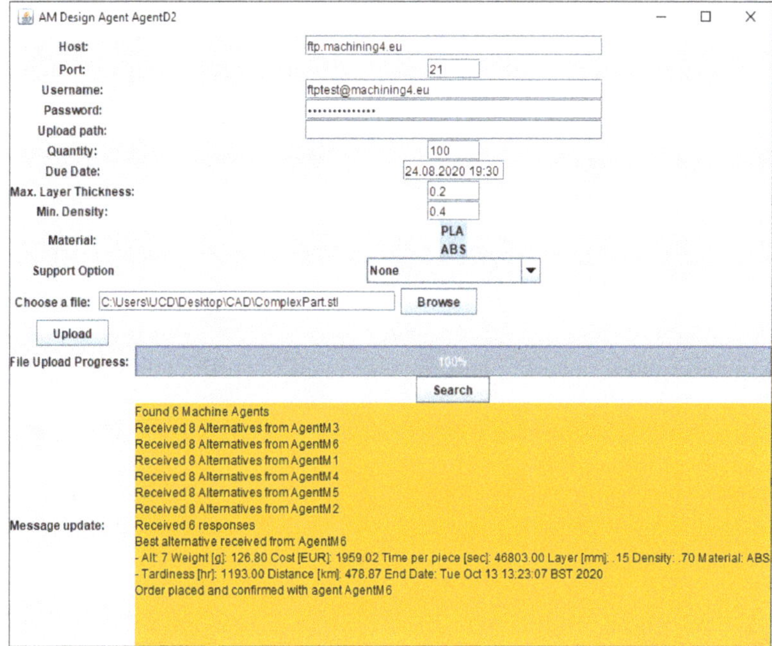

Figure 11. Test Case 2—Design Agent GUI with technical data and information from Machine Agents.

The best and least expensive alternatives were generated by Agent MA6 (Figure 12).

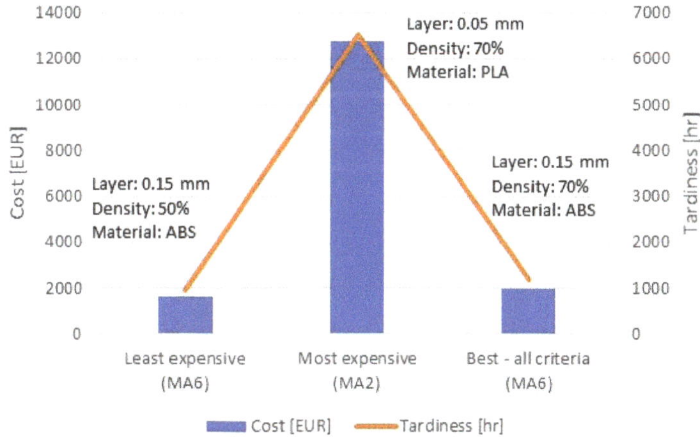

Figure 12. Test Case 2—Cost and tardiness of different configurations.

The best alternative, in particular, is a balanced solution, where a profile with the highest density and medium layer thickness was chosen so that cost and tardiness could be kept at low levels. The most expensive alternative was produced by Agent MA2, whose cost rates are the highest among the six AM service providers.

3.3. Cases Results Comparison and Simulation Validation

The results obtained from the two cases are summarised in Table 1. The 30% higher volume of the part corresponding to test case 2, together with the fact that a higher density was selected led to higher process times (and therefore tardiness) and cost.

Table 1. Comparison of test cases results.

Criterion	Weight/Performance—Case 1	Weight/Performance—Case 2
Layer thickness	40%/0.15 mm [ABS]	20%/0.15 mm [ABS]
Density	0%/50%	50%/70%
Tardiness	30%/266 h	20%/1193 h
Cost	30%/1333.03 EUR	10%/1959.02 EUR

For the validation of the test cases, two alternatives were chosen: the best alternative of test case 1 and a randomly selected alternative generated for test case 2. The corresponding parts were printed, using the process parameters suggested by the corresponding Machine Agents in the same 3D printer type, which is associated with these agents. The process time and the weight of the parts were measured. The variations observed from the values simulated versus actual part build time and weight are summarised in Table 2:

Table 2. Observations from the test cases for time and weight of the build.

Information	Test Case 1 (MA4)	Test Case 2 (MA1)
Time (simulation)	11 h 34 min (41,652 s)	15 h 19 min (55,114 s)
Time (actual)	11 h 29 min (41,340 s)	15 h 11 min (54,660 s)
Weight (simulation)	73.30 g	102.90 g
Weight (actual)	75.55 g	99.28 g

The differences between simulated and actual process times and part weights are:

- Time per part: 1–2%,
- Weight per part: 3–4%.

Assuming that a 95% accuracy is expected from the simulation process, these variations are well within range. The printed parts are shown in Figure 13.

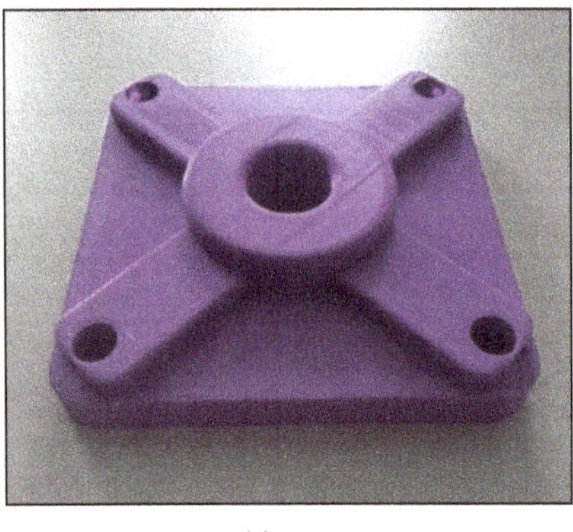

(a)

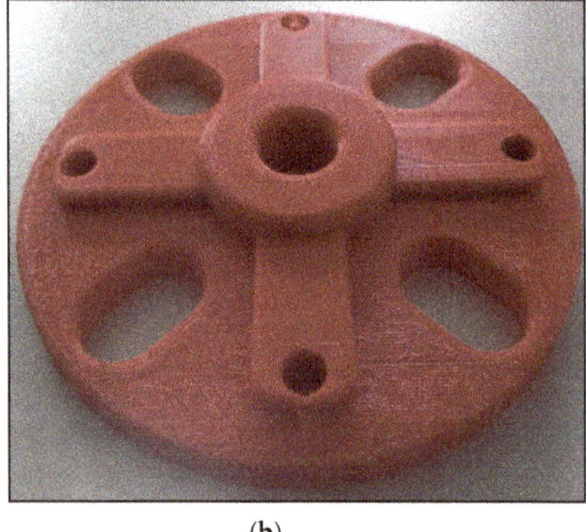

(b)

Figure 13. Builds for test cases 1 (**a**) and 2 (**b**).

4. Conclusions

This paper presents an agent-based approach for automating the process of selecting an AM service provider, the corresponding equipment and desired process configuration, while considering a set of often conflicting criteria.

The main goal of this approach is the implementation of a platform that is capable of supporting designers and engineers towards making informed product design and development decisions. The Machine Agents are in principle capable of interfacing open-ended CAM tools that are used with 3D Printers as well as of evaluating quite accurately the performance of several alternative process configurations. One of the main advantages of the proposed approach is that it can handle as many alternative AM service providers, equipment, and configurations as needed since the overall computing load is distributed evenly to the Machine Agents.

The proposed approach and platform could in principle be used with any kind of 3D equipment, given that the associated CAM software could be interfaced. However, this cannot always be the case, as, especially in the case of metal AM equipment, the CAM tools used by these machines are often proprietary and do not provide an Application Programming Interface (API) that would allow for the straightforward integration with the corresponding Machine Agents. Recent developments in the domain of robotic process automation (RPA) could provide an alternative way for interfacing and utilising the proprietary CAM systems in a near-automated way. With this technology, the interaction between a human operator and a software system may be replicated and executed on-demand in a fully parameterized manner. This opens possibilities for the integration of different CAM tools and platforms that could be interfaced with the proposed agent-based platform. The proposed approach could also complement existing platforms and approaches by, for instance, providing information regarding cost and time performance of diverse process configurations so that a limited number of configurations be reviewed in these platforms.

Further information, such as specific parts' geometry characteristics and feature sets could also be useful for identifying the most suitable process configurations and print profiles, based on the past performance of these profiles in the production runs with parts sharing similar features or characteristics. The platform is planned to be presented to the public utilising the central computer server of the Laboratory for Advanced Manufacturing Simulation and Robotics at UCD.

As part of the I-From Advanced Manufacturing Centre, it is also planned to provide further support for different AM manufacturing technologies that are of particular interest for research teams and the industry.

Author Contributions: Conceptualization, N.P.; methodology, N.P.; software, N.P.; validation, N.P., A.G.; formal analysis, N.P.; investigation, N.P.; resources, N.P.; data curation, N.P.; writing—original draft preparation, A.N. and N.P.; writing—review and editing, A.N. and N.P.; visualization, N.P., A.G.; supervision, N.P.; project administration, N.P.; funding acquisition, N.P. All authors have read and agreed to the published version of the manuscript.

Funding: This publication has been supported in part by a research grant from Science Foundation Ireland (SFI) under Grant Number 16/RC/3872 and is co-funded under the European Regional Development Fund.

Conflicts of Interest: The authors declare no conflict of interest.

Abbreviations

ABDSS	Agent-Based Decision Support System
ABS	Acrylonitrile Butadiene Styrene
AHP	Analytic Hierarchy Process
AM	Additive Manufacturing
AMS	Agent Management System
API	Application Programming Interface
CAD	Computer-Aided Design
CAM	Computer-Aided Manufacturing
CM	Conventional Manufacturing
CNC	Computer Numerical Control
DF	Directory Facilitator
DfAM	Design for Additive Manufacturing
DSS	Decision Support System
FDM	Fused Deposition Modelling
GUI	Graphical User Interface
IoT	Internet of Things
JADE	Java Agent Development Framework
MADM	Multi-Attribute Decision-Making
MAHP	Multiplicative Analytic Hierarchy Process
MAS	Multi-Agent Systems
MES	Manufacturing Execution System
OEM	Original Equipment Manufacturer
PBF	Powder Bed Fusion
PLA	Polylactic Acid
RPA	Robotic process automation
SMART	Simple Multiple Attribute Rating Technique
SME	Small and Medium Enterprise

References

1. ISO International; ASTM International. *ISO ASTM 52910: 2018 Additive Manufacturing—Design—Requirements, Guidelines and Recommendations*; ISO–International Organization for Standardization: Geneva, Switzerland, 2018.
2. Bikas, H.; Stavropoulos, P.; Chryssolouris, G. Additive manufacturing methods and modeling approaches: A critical review. *Int. J. Adv. Manuf. Technol.* **2016**, *83*, 389–405. [CrossRef]
3. Thompson, M.K.; Moroni, G.; Vaneker, T.; Fadel, G.; Campbell, R.I.; Gibson, I.; Bernard, A.; Schulz, J.; Graf, P.; Ahuja, B.; et al. Design for additive manufacturing: Trends, opportunities, considerations, and constraints. *CIRP Ann.* **2016**, *65*, 737–760. [CrossRef]
4. Mellor, S.; Hao, L.; Zhang, D. Additive manufacturing: A framework for implementation. *Int. J. Prod. Econ.* **2014**, *149*, 194–201. [CrossRef]
5. Brøtan, V.; Berg, O.Å.; Sørby, K. Additive manufacturing for enhanced performance of molds. *Procedia CIRP* **2016**, *54*, 186–190. [CrossRef]
6. Khajavi, S.H.; Partanen, J.; Holmström, J. Additive manufacturing in the spare parts supply chain. *Comput. Ind.* **2014**, *65*, 50–63. [CrossRef]
7. Li, Y.; Jia, G.; Cheng, Y.; Hu, Y. Additive manufacturing technology in spare parts supply chain: A comparative study. *Int. J. Prod. Res.* **2017**, *55*, 1498–1515. [CrossRef]
8. Hasan, S.; Rennie, A.E.W. The application of rapid manufacturing technologies in the spare parts industry. In Proceedings of the 19th Annual International Solid Freeform Fabrication Symposium, Austin, TX, USA, 4–8 August 2008; pp. 584–590.
9. Liu, P.; Huang, S.H.; Mokasdar, A.; Zhou, H.; Hou, L. The impact of additive manufacturing in the aircraft spare parts supply chain: Supply chain operation reference (scor) model based analysis. *Prod. Plan. Control* **2014**, *25*, 1169–1181. [CrossRef]

10. Beiderbeck, D.; Deradjat, D.; Minshall, T. *The Impact of Additive Manufacturing Technologies on Industrial Spare Parts Strategies*; University of Cambridge: Cambridge, UK, 2018. [CrossRef]
11. Gardan, N.; Schneider, A. Topological optimization of internal patterns and support in additive manufacturing. *J. Manuf. Syst.* **2015**, *37*, 417–425. [CrossRef]
12. Hague, R.; Reeves, P. *Additive Manufacturing and 3D Printing*; Wohlers Associates, Inc.: Fort Collins, CO, USA, 2013; ISBN 978-0-9913332-4-0.
13. Zadpoor, A.A.; Malda, J. Additive manufacturing of biomaterials, tissues, and organs. *Ann. Biomed. Eng.* **2017**, *45*, 1–11. [CrossRef] [PubMed]
14. Fayazfar, H.; Salarian, M.; Rogalsky, A.; Sarker, D.; Russo, P.; Paserin, V.; Toyserkani, E. A critical review of powder-based additive manufacturing of ferrous alloys: Process parameters, microstructure and mechanical properties. *Mater. Des.* **2018**, *144*, 98–128. [CrossRef]
15. Thomas, D.S.; Gilbert, S.W. *Costs and Cost Effectiveness of Additive Manufacturing*; NIST Special Publication: Gaithersburg, MD, USA, 2014.
16. Patalas-Maliszewska, J.; Topczak, M.; Kłos, S. The Level of the Additive Manufacturing Technology Use in Polish Metal and Automotive Manufacturing Enterprises. *Appl. Sci.* **2020**, *10*, 735. [CrossRef]
17. Huang, Y.; Leu, M.C.; Mazumder, J.; Donmez, A. Additive manufacturing: Current state, future potential, gaps and needs, and recommendations. *J. Manuf. Sci. Eng.* **2015**, *137*, 014001. [CrossRef]
18. Slic3r Slic3r-Open Source 3D Printing Toolbox. Available online: https://slic3r.org/ (accessed on 12 October 2019).
19. Koomen, B.; Hoogeboom, M.; Schellens, V. PLM Implementation Success Rate in SME. An Empirical Study of Implementation Projects, Preliminary Findings. In Proceedings of the IFIP Advances in Information and Communication Technology, Kyiv, Ukraine, 9–10 October 2019; Volume 565, pp. 47–57.
20. Sokolova, M.V.; Fernández-Caballero, A. Modeling and implementing an agent-based environmental health impact decision support system. *Expert Syst. Appl.* **2009**, *36*, 2603–2614. [CrossRef]
21. Legien, G.; Sniezynski, B.; Wilk-Kołodziejczyk, D.; Kluska-Nawarecka, S.; Nawarecki, E.; Jaśkowiec, K. Agent-based decision support system for technology recommendation. *Procedia Comput. Sci.* **2017**, *108*, 897–906. [CrossRef]
22. Alter, S. Taxonomy of Decision Support Systems. *Sloan Manag. Rev.* **1977**, *19*, 39–56.
23. DeSanctis, G.; Gallupe, R.B. *Decision Support Systems: Concepts and Resources for Managers*; Greenwood Publishing Group: Santa Barbara, CA, USA, 1987; Volume 33, ISBN 156720497X.
24. Park, H.S.; Tran, N.H. A decision support system for selecting additive manufacturing technologies. In Proceedings of the 2017 International Conference on Information System and Data Mining (ICISDM '17), New York, NY, USA, 1–3 April 2017; pp. 151–155.
25. Kretzschmar, N.; Ituarte, I.F.; Partanen, J. A decision support system for the validation of metal powder bed-based additive manufacturing applications. *Int. J. Adv. Manuf. Technol.* **2018**, *96*, 3679–3690. [CrossRef]
26. Ghazy, M.M. Development of an Additive Manufacturing Decision Support System (AMDSS). Ph.D. Thesis, Newcastle University, Newcastle, UK, 2012.
27. Borille, A.V.; Gomes, J.D.O. Selection of additive manufacturing technologies using decision methods. In *Rapid Prototyping Technology—Principles and Functional Requirements*; IntechOpen: London, UK, 2011.
28. Braglia, M.; Petroni, A. A management-support technique for the selection of rapid prototyping technologies. *J. Ind. Technol.* **1999**, *15*, 1–6. [CrossRef]
29. Byun, H.S.; Lee, K.H. A decision support system for the selection of a rapid prototyping process using the modified TOPSIS method. *Int. J. Adv. Manuf. Technol.* **2005**, *26*, 1338–1347. [CrossRef]
30. Zhang, Y.; Xu, Y.; Bernard, A. A new decision support method for the selection of RP process: Knowledge value measuring. *Int. J. Comput. Integr. Manuf.* **2014**, *27*, 747–758. [CrossRef]
31. Meisel, N.A.; Williams, C.B.; Ellis, K.P.; Taylor, D. Decision support for additive manufacturing deployment in remote or austere environments. *J. Manuf. Technol. Manag.* **2016**, *27*, 898–914. [CrossRef]
32. Bikas, H.; Koutsoukos, S.; Stavropoulos, P. A decision support method for evaluation and process selection of additive manufacturing. *Procedia CIRP* **2019**, *81*, 1107–1112. [CrossRef]
33. Rao, R.V.; Padmanabhan, K.K. Rapid prototyping process selection using graph theory and matrix approach. *J. Mater. Process. Technol.* **2007**, *194*, 81–88. [CrossRef]
34. Watson, J.K.; Taminger, K.M.B. A decision-support model for selecting additive manufacturing versus subtractive manufacturing based on energy consumption. *J. Clean. Prod.* **2018**, *176*, 1316–1322. [CrossRef]

35. Wang, Y.; Zhong, R.Y.; Xu, X. A decision support system for additive manufacturing process selection using a hybrid multiple criteria decision-making method. *Rapid Prototyp. J.* **2018**, *24*, 1544–1553. [CrossRef]
36. Wang, Y.; Blache, R.; Xu, X. Selection of additive manufacturing processes. *Rapid Prototyp. J.* **2017**, *23*, 434–447. [CrossRef]
37. Zhang, Y.; Bernard, A.; Harik, R.; Karunakaran, K.P. Build orientation optimization for multi-part production in additive manufacturing. *J. Intell. Manuf.* **2017**, *28*, 1393–1407. [CrossRef]
38. Monostori, L.; Váncza, J.; Kumara, S.R.T. Agent-based systems for manufacturing. *Cirp Ann.-Manuf. Technol.* **2006**, *55*, 697–720. [CrossRef]
39. García-Magariño, I. ABSTUR: An agent-based Simulator for Tourist Urban Routes. *Expert Syst. Appl.* **2015**, *42*, 5287–5302. [CrossRef]
40. Ricci, A.; Santi, A. Agent-Oriented Computing: Agents as a Paradigm for Computer Programming and Software Development. In Proceedings of the Future Computing 2011: The Third International Conference on Future Computational Technologies and Application, Rome, Italy, 25–30 September 2011; pp. 42–51.
41. Jin, Y.; Lu, S.C.Y. An agent-supported approach to collaborative design. *CIRP Ann.* **1998**, *47*, 107–110. [CrossRef]
42. Amara, H.; Dépincé, P.; Hascoët, J.Y. A human-centered architecture for process planning. *CIRP J. Manuf. Syst.* **2004**, *33*, 363–372.
43. Maropoulos, P.G.; McKay, K.R.; Bramall, D.G. Resource-aware aggregate planning for the distributed manufacturing enterprise. *CIRP Ann.* **2002**, *51*, 363–366. [CrossRef]
44. Zweben, M.; Fox, M.S.; Francisco, S.; Kaufmann, M.; Arbor, A. *Intelligent Scheduling*; Morgan Kaufmann Publishers Inc.: Burlington, MA, USA, 1994; Volume 6, ISBN 1558602607.
45. Monostori, L.; Prohaszka, J. A Step towards intelligent manufacturing: Modelling and monitoring of manufacturing processes through artificial neural networks. *CIRP Ann.* **1993**, *42*, 485–488. [CrossRef]
46. Balakrishnan, A.; Kumara, S.R.T.; Sundaresan, S. Manufacturing in the digital age: Exploiting information technologies for product realization. *Inf. Syst. Front.* **1999**, *1*, 25–50. [CrossRef]
47. Fox, M.S.; Barbuceanu, M.; Teigen, R. Agent-oriented supply-chain management. *Int. J. Flex. Manuf. Syst.* **2000**, *12*, 165–188. [CrossRef]
48. Bussmann, S.; Sieverding, J. Holonic control of an engine assembly plant an industrial evaluation. In Proceedings of the IEEE International Conference on Systems, Man and Cybernetics, Tucson, AZ, USA, 7–10 October 2001; Volume 1, pp. 169–174.
49. Niemann, J.; Ilie Zudor, E.; Monostori, L.; Westkämper, E. Agent-based product life cycle data support. *Manuf. Model. Manag. Control* **2005**, 105–110.
50. Dhokia, V.; Essink, W.P.; Flynn, J.M. A generative multi-agent design methodology for additively manufactured parts inspired by termite nest building. *CIRP Ann.* **2017**, *66*, 153–156. [CrossRef]
51. Allen, R.D.; Harding, J.A.; Newman, S.T. The application of STEP-NC using agent-based process planning. *Int. J. Prod. Res.* **2005**, *43*, 655–670. [CrossRef]
52. Mourad, M.; Nassehi, A.; Newman, S.; Schaefer, D. C-MARS-ABM: A Deployment Approach for Cloud Manufacturing. In Proceedings of the Advances in Transdisciplinary Engineering; IOS Press BV: Amsterdam, The Netherlands, 2017; Volume 6, pp. 213–218.
53. Dong, C.; Yuan, Y.; Lei, W. Additive manufacturing cloud based on multi agent systems and rule inference. In Proceedings of the 2016 IEEE Information Technology, Networking, Electronic and Automation Control Conference, Chongqing, China, 22–26 May 2016; pp. 45–50.
54. Wang, L.; Yao, Y.; Yang, X.; Chen, D. Multi agent based additive manufacturing cloud platform. In Proceedings of the 2016 IEEE International Conference on Cloud Computing and Big Data Analysis (ICCCBDA), Chengdu, China, 5–7 July 2016; pp. 290–295. [CrossRef]
55. Mourtzis, D.; Vlachou, E.; Xanthopoulos, N.; Givehchi, M.; Wang, L. Cloud-based adaptive process planning considering availability and capabilities of machine tools. *J. Manuf. Syst.* **2016**, *39*, 1–8. [CrossRef]
56. Holligan, C.; Hargaden, V.; Papakostas, N. Product lifecycle management and digital manufacturing technologies in the era of cloud computing. In Proceedings of the 2017 International Conference on Engineering, Technology and Innovation: Engineering, Technology and Innovation Management Beyond 2020: New Challenges, New Approaches, Funchal, Portugal, 27–29 June 2017; ICE/ITMC 2017-Proceedings. 2018; Volume 2018-Janua, pp. 909–918.

57. Papakostas, N.; Newell, A.; Hargaden, V. A novel paradigm for managing the product development process utilising blockchain technology principles. *CIRP Ann.* **2019**, *68*, 137–140. [CrossRef]
58. Shubham, P.; Sikidar, A.; Chand, T. The influence of layer thickness on mechanical properties of the 3D printed ABS polymer by fused deposition modeling. In *Key Engineering Materials 2016*; Trans Tech Publications Ltd.: Lausanne, Switzerland, 2016; Volume 706, pp. 63–67.
59. Pérez, M.; Medina-Sánchez, G.; García-Collado, A.; Gupta, M.; Carou, D. Surface quality enhancement of fused deposition modeling (FDM) printed samples based on the selection of critical printing parameters. *Materials.* **2018**, *11*, 1382. [CrossRef]
60. Terzi, S.; Bouras, A.; Dutta, D.; Garetti, M.; Kiritsis, D. Product lifecycle management-From its history to its new role. *Int. J. Prod. Lifecycle Manag.* **2010**, *4*, 360–389. [CrossRef]
61. Bellifemine, F.; Bergenti, F.; Caire, G.; Poggi, A. Jade—A Java Agent Development Framework. In *Multi-Agent Programming*; Springer: Boston, MA, USA, 2005; pp. 125–147.
62. Costabile, G.; Fera, M.; Fruggiero, F.; Lambiase, A.; Pham, D. Cost models of additive manufacturing: A literature review. *Int. J. Ind. Eng. Comput.* **2016**, *8*, 263–282. [CrossRef]
63. Chryssolouris, G. *Manufacturing Systems: Theory and Practice*, 2nd ed.; Springer: New York, NY, USA, 2006.

Article

Assembly Tolerance Design Based on Skin Model Shapes Considering Processing Feature Degradation

Ci He, Shuyou Zhang, Lemiao Qiu *, Xiaojian Liu and Zili Wang

State Key Laboratory of Fluid Power and Mechatronic Systems, Zhejiang University, Hangzhou 310027, China
* Correspondence: qiulm@zju.edu.cn; Tel.: +86-1385-800-2332

Received: 23 May 2019; Accepted: 5 August 2019; Published: 7 August 2019

Abstract: To increase the reliability and accuracy of tolerance design, more and more research works are considering not only orientation and position deviations; they are also forming errors in tolerance modeling. As a direct cause of form errors in industrial mass production, the processing features of the machining system degrade over time. Under the Industry 4.0 paradigm, an assembly tolerance design method based on Skin Model Shape is proposed to take the effect of degrading processing features into consideration. A continuous-time multi-dimensional Markov process is trained through maximum likelihood estimation based on the nodal sampling point set on the machined surface. Degradation of the machined surface is modeled based on the joint probability distribution of nodal displacements. Assembly force constraints and assembly entity constraints are applied to spatial assembly simulations. Tolerance synthesis takes the manufacturing cost and assembling probability as design objectives. A design example of the rotary feed component in a five-axis machine tool is proposed for explanation and verification.

Keywords: computer-aided tolerance; processing features degradation; skin model shape; statistical tolerance analysis; tolerance allocation

---

1. Introduction

Under the innovative concept of Industry 4.0, automated and digitized systems in smart factories could enable the real-time integration and analysis of massive amounts of data by the use of electronics and information technologies. Moreover, this would finally result in a more flexible and optimized manufacturing process [1–3]. This theory points to the improvement of intelligent solutions in the tolerance design of mechanical products, including the replacement of empiricism with knowledge-intensive and data-based processes. The general practice of mechanical tolerance design is to treat tolerance design as a combination of tolerance modeling, assembly simulating, tolerance allocation, and optimization. The best combination of part tolerance is assigned through multi-dimensional objective constraints in order to meet environmental clearance, conform to functional requirements, improve product quality, and lower manufacturing costs in the late design stage [4–6]. To increase the accuracy and robustness of the designed tolerance scheme, the influence of processing feature degradation caused by machining precision deterioration is taken into consideration.

In the past few decades, researchers have presented several efficient computer-aided methods for estimating geometry variation and product tolerance. Many models have been developed for geometrical feature representation, deviation accumulation, and tolerance zone estimation [7]. Conventional tolerance modeling methods are mainly categorized into those based on vector loops, Tolerance-Map (T-Map), small displacement torsor (SDT), a Homogeneous Transformation Matrix (HTM), and some other innovative models. Vector loops provide vectorial tolerance representation and analysis for the surfaces of different components. However, only five standard surfaces are typically considered (plane, cylinder, sphere, cone, and torus) [8–11]. Consisting of a hypothetical volume of points, a T-Map contains all the possible locations for geometrical features, and is used to illustrate the accumulation of product variations [12–14]. The SDT model uses screws and constraints to establish the extreme limits of part tolerance zones [15–17]. To overcome the lack of analytical models between the target feature and each feature on the dimensional chain, HTM was combined with T-Map to obtain an expression of the decoupling pose of tolerance features [18]. Since the Jacobian matrices are quite suitable for deviation propagation [19], the unified Jacobian–Torsor model was developed for product precision performance estimation [20–22]. Furthermore, some innovative modeling methods, such as neural network [23], volumetric envelope [24], modal analysis [25,26] and graph theory [27,28] have also been applied to model the mating tolerance of both parts and assemblies. Some of the aforementioned models have contributed to the mainstream commercial computer-aided tolerancing software, such as 3DCS Variation Analyst, eM-TolMate, VisVSA, and CETOL 6 Sigma [29]. However, most of these tolerancing models excel at estimating the precision of a mechanical product at one state, while they cannot adapt to simulate a sequence of degradation trajectories. Moreover, the simplification of form deviation into a series of tiny dimensional, rotational, and translational deviations of geometric features in some of these models brings about inaccurate virtual representation in the modeling stage and unneglectable simulation errors in the synthesizing and optimizing stages.

As a response, the Skin Model Shape theory was introduced as a new paradigm for the modeling of product geometry considering shape variability. The theory of Skin Model Shapes is an integration of international geometric standards and the concept of the Skin Model [30–33]. Within the theoretical framework of GeoSpelling [34], the Skin Model Shape is commonly formed by discrete geometry schemes such as point clouds and surface meshes. It provides an approach for the employment of computational techniques on the Skin Model. Research studies have shown that the Skin Model Shape has great efficiency for the representation of product geometry considering geometrical deviations and form errors [29,35,36]. Besides, since deviations in the Skin Model Shape are generated mainly by duplicating a sample pattern from the non-ideal manufacturing process, real part measurement, shape defect simulation, and statistical shape analysis are employed on Skin Model Shapes for convenience. For example, finite element analysis is introduced to help reduce local modeling drawbacks and improve model quality [37]. In addition, the discrete form of the Skin Model Shape has been proven to have great potential in further assembly simulation, performance evaluation, and tolerance analysis [38–40].

Significant efforts have been made to improve the accuracy of product models when processing features are taken into consideration. Processing features, including systematic geometric deviations and microscopic form patterns, are essentially affected by and have been employed as an estimation for the performance of machine tools. The pattern of machining defects is detected in order to provide a comparison between the ideal model and the actual machining result, which is usually obtained by a coordinate measuring machine [41]. Most of the previous research has identified the machining state as a Boolean attribute: precision reliable or precision failure. If the measured degradation reaches a threshold, precision failure occurs, and the product fault time is defined. However—mainly due to the wear of the driving system, transmission gear, and cutting tool—a stochastic process, rather than binary classification, is more suitable for simulating and predicting the deterioration of machining precision and degradation of processing features. Sun et al. [42] emphasized that tool wear is a dynamic process extending from sharp to worn and possibly to breakage, and that multi-state classification could provide a more timely and accurate estimation. Dai et al. [43] pointed out that the degradation paths would be different between one product and another. They conducted a comparison of Gaussian and logarithmic distribution in order to establish the degradation model. Ozcelik and Bayramoglu [44] verified the effect of tool wear on machining surface roughness and established a statistical model for prediction under various cutting conditions. Shu et al. [45] made the assumption that the next state of machine tool wear only depends on its current wear state. So, a non-homogeneous continuous-time Markov process model was used for modeling the total experience over the target life. After that, a linear mixed-effects model and maximum likelihood estimation was implemented to assess the wear evolution and lifetime of the tool [46]. Distribution of the residual life up to the wear threshold and estimation of the wear level have also been researched. Moghaddass and Zuo [47] modeled the gradual degradation of a mechanical device as a continuous-time degradation process with multiple discrete states. A condition-monitored device based on an unsupervised parameter estimation method was developed with only incomplete information observable. A homogeneous continuous-time, finite-state semi-Markov model was established based on the past history of components by Cannarile et al. [48]. It was verified to be of great help in improving the diagnostic performance of an empirical classification system involving the degradation of mechanical systems. These applications of stochastic processes in manufacturing modeling have shown great potential in simulating the tolerance of a large branch of mechanical products in mass production based on measuring data.

The intention of this paper is to propose a solution to assembly tolerance design problems considering processing feature degradation. These problems are commonly encountered in the practical production of equipment manufacturing in industries such as high-precision computer numerical control machining, aeronautics, and astronautics. This paper focuses on assemblies containing observable parts (mainly outsourced parts, such as standard components or ultra high-precision parts) and predictable parts (mainly self-made parts, such as lathe beds, columns, and pillars). These two kinds of parts are treated in different ways—sampling and measuring are usually implemented on the observable parts for precise and detailed modeling, while the predictable parts are modeled through mathematical simulations of the geometric tolerance, form errors, and assembly clearance. Some of the tolerance indices are given through measurements, while the to-be-designed tolerance indices are assumed and experimented repeatedly. Feature precision degradation is taken into consideration in tolerance modeling to enhance the reliability and robustness of a designed tolerance scheme. As a solution, a continuous-time variational multi-dimensional Markov process is introduced for modeling the degradation process, and the degraded surfaces are used in assembly simulations. On that basis, target assembly tolerance indices are synthesized through a series of numerical experiments. As a goal, a reduction in manufacturing cost and a guarantee in assembling probability is expected through the integration of these methods to the existing tolerance design framework, especially for products in mass production.

To achieve this, a tolerance design method considering processing feature degradation is proposed. The Skin Model Shapes form is adapted as a geometric basis. The nodal sampling point set on the machined surface is obtained by means of a high-precision coordinate measuring machine. A continuous-time multi-dimensional Markov process is trained to calculate the nodal displacement joint probability distribution on the machined surfaces. The degraded machined surfaces are predicted and applied in subsequent numerical experiments. To provide the precise assembly tolerance indices, assembly simulations and data analysis are conducted, and the assembly force constraints and assembly entity constraints are taken into consideration. Then, these tolerance indices are synthesized to provide tolerance schemes with low manufacturing costs and high assembly probabilities. Estimation and verification of the efficiency of the proposed method are illustrated through an example of the transmission shaft on a five-axis high-precision machine tool (VTM200F).

2. Predictive Machined Surface Modeling

In this paper, machined surfaces are regarded as an integration of processing features and random geometric deviation. To a specific machine surface, the processing feature is determined by the basic geometric shape (plane, cylinder, sphere, etc.), the application and sequence of processing techniques (turning, milling, grinding, etc.), and the corresponding operating parameters (cutting speed, feed rate, cutting depth, etc.). Random geometric deviation depicts unpredictable and inevitable machining errors. Previous research has shown the rationality behind the idea that a stochastic process could represent the precision deterioration of a machining system. In the condition of mass production, the processing features are degrading due to the deterioration of the machining tool. Therefore, a multi-dimensional Markov model based on data training is introduced for modeling and prediction. The pattern of degradation becomes more explicit if the machined parts are from different branches.

2.1. Modeling of the Multi-Dimensional Markov Process

The multi-dimensional Markov model is presented in this section, and a machined surface is used as an example. A uniform grid with a sampling interval Δ is placed on the nominal machined surface. The measuring process is calibrated through the grid nodes. A high-precision coordinate measuring machine is used to obtain the actual coordinates of sampling points on the real surface. The coordinate set is called the nodal sampling point set Ω. To a plane, the sampling interval Δ is a set distance along the x-direction and y-direction in the local machining coordinate system. To a spherical or cylindrical surface, the sampling interval Δ is a set angle around the central point or central axis.

First, we use $P(X_k^t = L_k^t), i \in \Omega$ to describe the probability that the actual coordinate of node k on the real surface is L_k^t at time t. L_k^t represents a discrete precision state by an interval of coordinate variation, as shown in Figure 1c. In a maintenance cycle of the machine tool, the shape error, dimension error, and location error of geometric features on real surfaces deteriorate over time. Therefore, the actual precision state of nodal points gradually moves away from the nominal positions:

$$P(X_k^0 = X_{k\_norm}) > P(X_k^1 = X_{k\_norm}) > \cdots > P(X_k^n = X_{k\_norm}) > \cdots \tag{1}$$

Obviously, the actual precision state L_k^{t+1} of node k at time $t+1$ is directly related to L_k^t, and has no relation to the precision states at and before time $t-1$. Treated as a sampling on the time-axis of a continuous-time stochastic model of the machine surface, the coordinate change of a single nodal point over a period of time conforms to the discrete-time Markov property:

$$\begin{aligned} P(X_k^t = L_k^t) &= P(X_k^t = L_k^t | X_k^{t-1} = L_k^{t-1}) \times P(X_k^{t-1} = L_k^{t-1} | X_k^{t-2} = L_k^{t-2}) \times \cdots \\ &\times P(X_k^1 = L_k^1 | X_k^0 = L_k^0) \times P(X_k^0 = L_k^0) \end{aligned} \tag{2}$$

In the machining process, a continuous cutting motion on the machining surface causes the coordinate change of relevant nodal points to conform to the Markov property. With the same form and parameters of the driving system and gear motion, the combination of motion between adjacent nodal points is fixed and constant. These nodal points are called relevant points. Inside the machining system, the same motion combination is conducted between pairs of relevant points. L_k^t relates to the coordinates of all the neighboring points , and has no relation to those points $\{k_1, k_2, \cdots, k_m\}$, and has no relation to those points outside the neighborhood, as shown in Equation (3):

$$P(X_k^t = L_k^t) = P(X_k^t = L_k^t | X_{k_1}^t = L_{k_1}^t, X_{k_2}^t = L_{k_2}^t, \cdots, X_{k_m}^t = L_{k_m}^t),\ k_1, k_2, \cdots, k_m \in S_k \tag{3}$$

Combining Equations (1) and (2), the actual precision state of nodal points on the machined surface conform to the temporal and spatial Markov properties at the same time, as shown in Figure 1a.

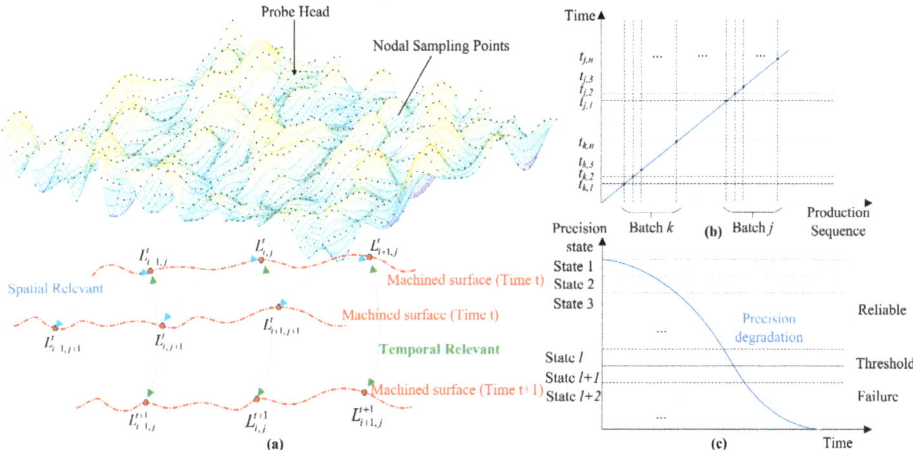

Figure 1. Temporal and spatial Markov properties of planar machining surface (**a**), and an illustration of the precision state degradation in the production sequence (**b**,**c**).

2.2. Calculation of Model Weight Parameters

The degradation of the feature precision state of parts from the same batch $\{T_0, T_1, \cdots, T_n\}$ is a continuous-time process if the parts are manufactured on the same machine tool. Condition characteristics m and transmission characteristics n are unique properties on or between relevant points. Therefore, the weight parameters set $\{w\} = \{w_1, w_2, \cdots, w_n\}$ is applied to the nodal sampling point set.

We use x to describe the condition of the rough surface before machining, and y to describe the machined surface at time t. As a description of feature precision degradation, the continuous-time nodal displacement joint probability distribution $P(y|x,t)$ in the form of the temporal Markov process is formulated in Equation (4a), while $P(y|x,t)$ in the form of the spatial Markov process is formulated in Equation (4b), as a description of feature precision specification at time t:

$$\begin{cases} P(y(t)|y(t-1),x,t) = \dfrac{\exp\{\sum\limits_{i,j}\lambda_i m_i(t_j,y,x,i)+\sum\limits_{l,j}\mu_{l,j}n_{l,j}(t_{j+1},t_j,y,x,l)\}}{\sum\limits_{t}\exp\{\sum\limits_{i,j}\lambda_i m_i(t_j,y,x,i)+\sum\limits_{i,j}\mu_j n_j(t_{j+1},t_j,y,x,j)\}}(a) \\ P(y|x,t) = \dfrac{\exp\{\sum\limits_{i,j}\lambda_i m_i(j,x_i,y,i)+\sum\limits_{h,j}\eta_{h,j}n_{h,j}(x_{h+1},x_h,y,h)\}}{\sum\limits_{x,y}\exp\{\sum\limits_{i,k}\lambda_h m_k(x_{i-1},x_i,y(t),i)+\sum\limits_{i,l}\eta_l n_l(x_i,y,i)\}}(b) \end{cases} \tag{4}$$

where i is the number of nodal points, j represents the time sequence, and l and h represent the temporal and spatial relevant points. λ represents the weight parameter of m, and μ represents the weight parameter of n. The unified form of Equation (4) is shown in Equation (5):

$$P(y|x,t) = \frac{\exp\{\sum\limits_{k=1}^{K} w_k f_k(x,y,t)\}}{\sum\limits_{x,y}\{\exp\sum\limits_{k=1}^{K} w_k f_k(x,y,t)\}} \tag{5}$$

Through the measured data from coordinate machining machine and the model shown in Equation (5), maximum likelihood estimation is applied to calculate the weight parameters. The value of w_k indicates that the observed point set has the highest probability. We use a one-order step function as the object function. The overfitting problems in the training process caused by sampling data loss or over-training can possibly cause the verification error to grow larger as the calculation complexity grows and the training error decreases. To overcome this, an L_2 regularization term weighted as $\frac{1}{2\delta^2}$ is added to the object function, as shown in Equation (6):

$$g(w) = \frac{\partial \ln(p(y|x,t))}{\partial w_k} - \frac{\sum w^2}{2\delta^2}. \tag{6}$$

As a solution to Equation (6), a quasi-Newton iteration is implemented to calculate the parameter weights, which makes $g(w)$ converge to zero:

(1) Set the initial counter $k = 0$, initial positive definite matrix $M^{(0)}$, and initial parameter weight $w^{(0)}$. Calculate the initial object function value g^0.
(2) If $g(w^{(k)}) \neq 0$, set the current search direction $p_k = -(M^{(k)})^{-1} \cdot g(w)$.
(3) Update the positive definite matrix $M(k)$ based on the Armijo rule with calibration coefficients:

$$M^{(k+1)} = M^{(k)} + \frac{y_k {y_k}^T}{y_k^T \delta_k} - \frac{M^{(k)} \delta_k \delta_k^T M^{(k)}}{\delta_k^T M^{(k)} \delta_k} \tag{7}$$

(4) Use a line search function $h(\lambda) = f(w^{(k)} + \lambda_k p_k)$. Update $w^{(k)}$ when $h(\lambda)$ reaches the global minimum.
(5) Output $w^{(k)}$ if $g(w^{(k)}) = 0$. Otherwise, $k = k + 1$ and go back to step (2).

2.3. Prediction of Degraded Surface

When $P(y|x,t)$ (as calculated in Section 2.2) reaches its maximum, the predictive machined surface is equal to y^*, which means that the greatest possible location distribution of sampling points at time t is obtained as the predictive surface:

$$y^* = \underset{y}{\text{argmax}} P_w(y|x,t) = \underset{y}{\text{argmax}} \sum_{k=1}^{K} w_k f_k \tag{8}$$

Under a certain value of $P(y|x,t)$, a backward path searching iteration is implemented after obtaining a specific sampling point's initial location probability $\delta_1(l)$, so as to calculate the most probable location of the sampling point during the whole machining process:

$$\delta_i(t) = \max_{1 \leq j \leq m} \{\delta_i(t-1) + \sum_{k=1}^{K} w_k f_k | y_{i-1} = j, y_i = t\}, t = 1, 2, \ldots, N \tag{9}$$

The prediction of the feature precision state on the machined surface is achieved when the backward path searching reaches its end point, as shown in Equations (10) and (11):

$$\max_{y} \sum_{k=1}^{K} w_k f_k(y,x) = \max_{1 \leq j \leq m} \delta_n(j), \tag{10}$$

$$y_i^* = \underset{1 \leq j \leq m}{\text{argmax}} \delta_n(j). \tag{11}$$

The calculated $P(y|x,t)$ contains the information of crafts and steps. At time t, the predictive machined surface y^* (in Equation (12)) consists of the predicted coordinates of a nodal sampling point set under processing features. If time $t = t_f$ is large enough to cover the designated lifecycle of the to-be-designed assembly, the degradation of the part feature is considered by using the predictive machined surface y_f^*:

$$y_f{}^* = \{y_{f,1}^*, y_{f,2}^*, \cdots, y_{f,N}^*\} \tag{12}$$

3. Constrained 3D Assembly Simulation and Tolerance Synthesis

3.1. Constrained Assembly Simulation

Then, the tolerance models generated by the methods described in Section 2 are used in assembly simulations. Generally, the process of assembling two parts involves the transformation of one part from its initial condition to its assembling condition, while the other part is relatively static. In this paper, the former part is called assembly part Q, and the latter part is called reference part P. The assembling transformation ξ is in the form of a twist because of its splendid performance in representing the orientation, and its independence from the coordinate system transformation. Parts are considered under the geometric basis of Skin Model Shapes in this paper, which means that Q and P are in the form of meshes or point clouds. So, the assembly part Q is transformed through ξ from its initial condition Q to assembling condition Q', as shown in Equation (13):

$$Q' = \xi \cdot Q \tag{13}$$

Twist ξ is essentially an affine transformation. It can always be decomposed to a motion screw $\xi_i = (s_i, v_i)$ in a three-dimensional (3D) velocity vector field, containing a rotation around its axis and a translation along the axis, as shown in Equation (14):

$$\xi_i = s_i + p_i \times v_i \tag{14}$$

where p_i is the vector from the spatial origin to point i. v_i is the axial velocity vector at point i. s_i is the tangential velocity vector at point i. Assembly twists of all the points of the assembly part consist the assembly transformation: $\xi = \{\xi_1, \xi_2, \cdots, \xi_i, \cdots\}$.

Mathematically, Q' is the global extremum of an object function containing Q, P, and twist ξ under assembly constraints, as shown in Equation (15):

$$\xi = \operatorname{argmin}(f(P, Q')) \tag{15}$$

Choose the root mean square Euclidean distance function in Equation (16) of the closest point pairs in point clouds P and Q:

$$f = RMSED(P, Q') = \sqrt{\frac{1}{n} \sum_{p_i, q_i \in CPP(i)} \|p_i - q_i\|^2} \tag{16}$$

Establish the iteration object function based on Equations (13)–(16). The end condition of the iteration is the appearance of three pairs of contact points in the assembly. A pair of contact points contains a point in P and a point in Q. Furthermore, the Euclidean distance between the two points is smaller than a threshold value. Usually, using the iterative closest points (ICP) is a common method under the circumstances to find the target. However, assembly entity constraints and assembly force constraints have restricted the direct use of ICP.

Assembly entity constraints restrict the intersection of part entity in assembly simulation. As shown in Figure 2a, the direct use of ICP causes an unreasonable intersection of part materials. However, if the rule of assembly entity constraints is considered, ICP would give a result that is much more rational (in Figure 2b). A signed distance field is introduced to supervise the compliance of the assembly entity constraints. The positive direction of the signed distance field is always against the assembling direction. The signed distance vector $d_i^{sdv} (i = 1, 2, \cdots, n)$ is from the points in P to the corresponding closest point in Q. When the angle between d_i^{sdv} and the positive direction is under 90 degrees, d_i^{sdv} is labeled positive. Otherwise, it is labeled negative. As shown in Figure 2c, all the signed distance vectors $d_i^{sdv} (i = 1, 2, \cdots, n)$ are positive during iteration before assembling. However, negative signed distance vectors appear when the rule of assembly entity constraints is violated (in Figure 2d). So, the signals of $d_i^{sdv}(i = 1, 2, \cdots, n)$ are repeatedly checked during the iteration.

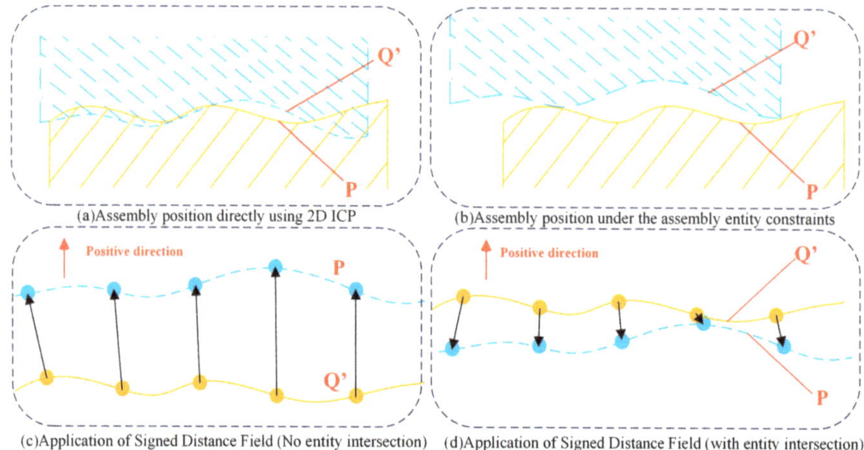

Figure 2. Illustration of assembly entity constraints and the application of a signed distance field in realizing the rule.

Assembly force constraints ensure the stability of assembling conditions through restricting the location of contact points according to the assembly force. Obviously, three points provide a stable support in 3D assembly simulation, so the three contact points should not be in a line. Additionally, the assembly force vector should cross the spatial triangle made up by the three contact points. Otherwise, the current condition is unstable, and additional rotations would happen. The compliance of assembly force constraints is only checked when the iteration ends, ensuring that the final assembling state is stable. The solution of the assembling condition is decomposed into four main steps:

(1) Compute any state of Q' with one contact point using ICP and ensure $d_i^{sdv} > 0, \forall i = 1, 2, \cdots, n$ (Figure 3a). The contact point is defined as P_1^{con}.
(2) Ensure $d_i^{sdv} > 0, \forall i = 1, 2, \cdots, n$ and search the second contact point P_2^{con} using twists whose direction of v_i is $\overrightarrow{P_{AF} - P_1^{con}} \times v_{AF}$ (Figure 3b), where P_{AF} is the point of intersection of assembly force vector v_{AF} and P.
(3) Ensure $d_i^{sdv} > 0, \forall i = 1, 2, \cdots, n$ and search the third contact point P_3^{con} using twists whose direction of v_i is $\overrightarrow{P_1^{con} - P_2^{con}}$, as shown in Figure 3c.
(4) Check if the current spatial contact triangle is qualified using the assembly force constraints rule: $(\overrightarrow{P_{AF} - P_1^{con}} \times \overrightarrow{P_2^{con} - P_1^{con}}) \cdot (\overrightarrow{P_{AF} - P_2^{con}} \times \overrightarrow{P_3^{con} - P_2^{con}}) > 0$ and $(\overrightarrow{P_{AF} - P_3^{con}} \times \overrightarrow{P_1^{con} - P_3^{con}}) \cdot (\overrightarrow{P_{AF} - P_2^{con}} \times \overrightarrow{P_3^{con} - P_2^{con}}) > 0$. If the conditions are not met, choose the two contact points that constitute a line closer to P_{AF} (in Figure 3d) and go back to step (3). Once the conditions are met, the current condition of Q' is the qualified assembling condition, as shown in Figure 3e.

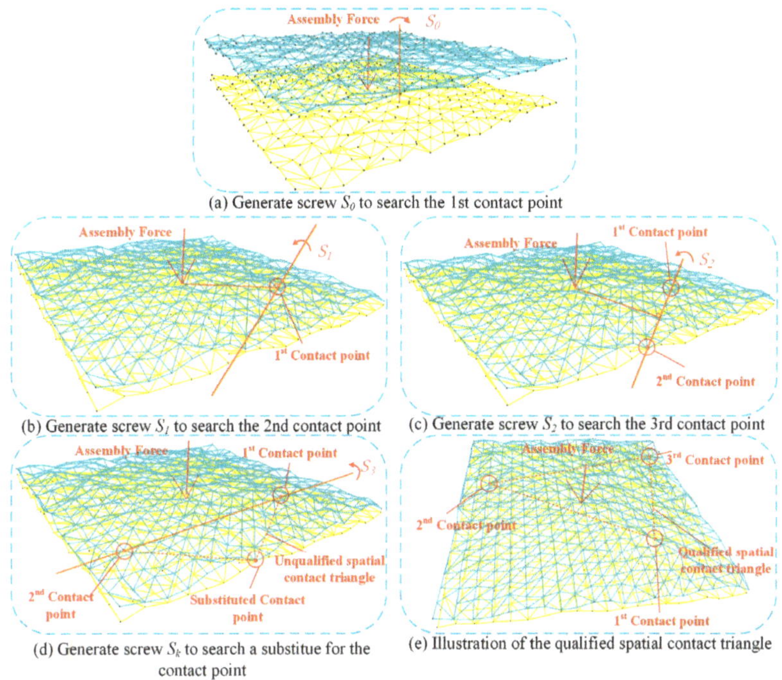

(a) Generate screw S_0 to search the 1st contact point

(b) Generate screw S_1 to search the 2nd contact point

(c) Generate screw S_2 to search the 3rd contact point

(d) Generate screw S_k to search a substitue for the contact point

(e) Illustration of the qualified spatial contact triangle

Figure 3. Main steps in the computation of assembling condition when considering assembly force constraints.

3.2. Static and Dynamic Tolerance Synthesis

In this paper, tolerance synthesis assigns tolerance indices X, minimizes manufacturing cost, maximizes assembling probability, and meets the assembly functional requirement Y at the same time. The X tolerance index comprises the given indices X_g (mainly in outsourced parts) and to-be-designed indices X_d (mainly in self-made parts). X_g and X_d are independent from the perspective of manufacturing, but they become relevant because of the assembly functional requirement Y. Additional objectives including the manufacturing cost and assembling probability turn this issue into an optimization problem.

To increase the robustness of the design scheme, feature degradation during the machining process is considered when generating observable parts. Predictions of part surfaces at any time during the designed life are made using the methods in Section 2. Generally, surfaces in their late design age with seriously deteriorated geometric features are used in tolerance synthesis. The surfaces of predictable parts are simulated by a multivariate Gaussian process (MGP) based on given tolerance indices. Repetitive experiments and variated models are applied in the design process due to the stochastic characteristic of the MGP. Numerical experiments of assembly simulation provide assembly tolerance indices, which are usually categorized into static tolerance indices (such as verticality and parallelism) and dynamic tolerance indices (such as circular runout and end face runout). Static tolerance indices are certain indices based on measurements and the analysis of assembled Skin Model Shapes. These only vary when the tolerance models are varied (in Figure 4a). However, dynamic tolerance indices are obtained through monitoring a series of static tolerance indices by analyzing the possible motion and translational/rotational samplings. Measuring positions are sampled for experiments through a Monte Carlo process. As shown in Figure 4b, the upper end face parallelism of a sliding table on a high-precision guide is a typical dynamic tolerance index, which is calculated through a series of assembly simulations after sampling measuring positions in its translational motion. Similarly, common runout errors of cylindrical parts (such as axle parts) can also be obtained by sampling measuring orientations in the rotational motion of those parts. A small sampling interval and large sampling quantity lead to a precise estimation.

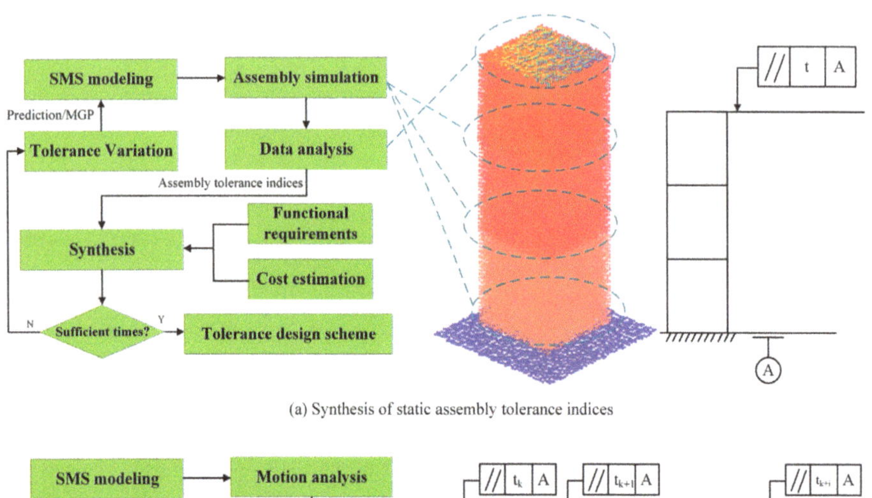

(a) Synthesis of static assembly tolerance indices

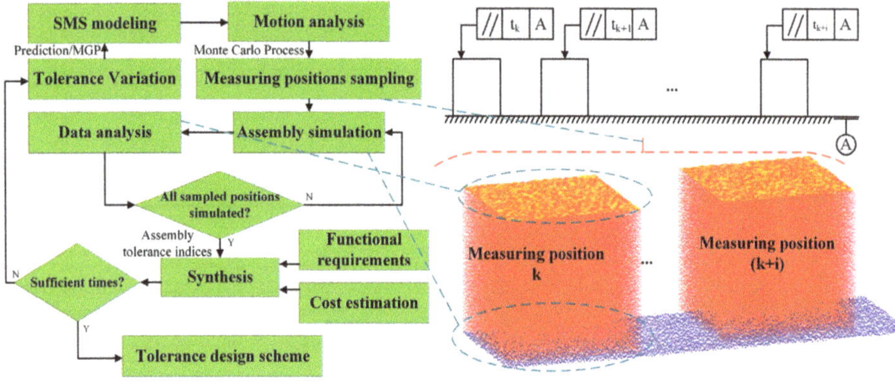

(b) Synthesis of dynamic assembly tolerance indieces

Figure 4. Synthesis of static and dynamic assembly tolerance indices based on assembly simulation.

As shown in Figure 4, assembly tolerance indices are estimated through data analysis in each measuring position/orientation. The distribution of estimated assembly tolerance indices to the to-be-designed tolerance indices is summarized in the form of tables or figures. Then, an integration of assembly functional requirements and manufacturing cost estimation is introduced to outline the objectives. Due to the stochastic characteristic of tolerance models, not all the examples with the same assigned tolerance scheme can meet the assembly functional requirements. Generally, if the ratio of conformation reaches a certain threshold (such as 95%), the tolerance scheme is treated as reliable. On that basis, the manufacturing costs of all the reliable tolerance schemes are computed and compared based on industrial statistics and empirical formulation. That with the lowest costs is selected as the designed tolerance scheme.

4. Case Study

4.1. Description of the Tolerance Allocation Problem

The proposed method is applied to solve an assembly tolerance design problem in the rotary feed system of a VTM200F5 five-axis turning and milling complex center. General tolerance indices of the rotary feed system have a significant influence on the cutting precision and are extremely sensitive to geometric deviation in the cutting process due to its direct link to the cutting tool. Considering assembly tolerance, the rotary feed system is simplified to two key parts: the hole part describing devices fixing rotary bearings, and a shaft part depicting the motor spindle component. The shaft part is an outsourced part made by Ti–Al thermostable alloy with its surface specially heat-treated. So, the key accuracy indices such as location accuracy, shape accuracy, and surface accuracy should not be modified in order to protecting the special surface coating. The hole part is a self-made part made by common bearing steel. Its key accuracy indices are to be designed, as shown in Figure 5. This is a typical component containing observable parts (with the same pattern of processing features and from different batches) and predictable parts (to-be-designed parts).

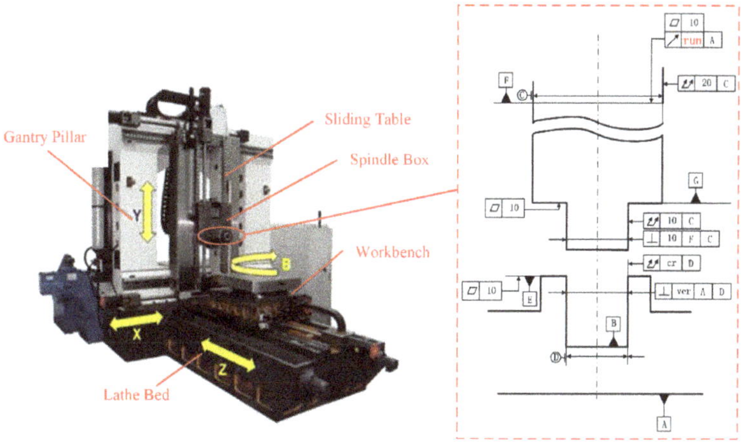

Figure 5. An assembly tolerance design example in designing the rotary feed system of a five-axis turning and milling complex center (VTM200F5).

In this component, the runout accuracy of the distal end of the shaft part is set as the target accuracy index due to its direct influence on the cutting accuracy. In the shaft part, the nominal flatness of the distal end surface is 0.010 mm. The nominal total runout of the distal end surface is 0.020 mm. The nominal total runout of the proximal end surface is 0.010 mm. The nominal verticality between the distal end surface and axis C is 0.010 mm. The nominal flatness of the shoulder surface is 0.010 mm. In the hole part, surface A is the assembly positioning surface. The key accuracy indices to be designed in this tolerance allocation problem are listed in Table 1. The flatness of the shoulder surface is FL_F. The total runout of inner hole is TR_D. The verticality between axis D and surface A is V_{DA}. The parallelism of the shoulder surface and surface A is PL_{EA}.

Table 1. List of the target accuracy term of the assembly, the nominal value of the accuracy terms of the shaft part, and the to-be-designed accuracy terms of the hole part.

Target Accuracy Term	Explanation	Symbol	Design Requirement
FR_F	Runout of surface F	e_f	0.020 mm
Accuracy Terms of the Shaft Part	**Explanation**	**Symbol**	**Nominal Value**
FL_F	Flatness of surface F	ε_1	0.010 mm
TR_{C1}	Total runout of the distal end surface	ε_2	0.020 mm
TR_{C2}	Total runout of the proximal end surface	ε_3	0.010 mm
FL_G	Flatness of shoulder surface G	ε_4	0.010 mm
VT_{CF}	Verticality between the distal end surface and axis C	ε_5	0.010 mm
Accuracy Terms of the Hole Part	**Explanation**	**Symbol**	**Nominal Value**
FL_E	Flatness of the shoulder surface E	e_1	0.010 mm
TR_D	Total runout of axis D	e_2	0.012 mm
VT_{DA}	Verticality between axis D and surface A	e_3	0.005 mm

4.2. Stochastic Process Training and Parameter Calculation

Twenty outsourced parts were randomly chosen. Parallel lines in two directions were drawn on assembly surfaces with a uniform interval of 0.01 mm. Sampling points are defined as the intersection points of these lines. We used a non-contact coordinate measuring machine with an accuracy of at least 0.001 mm to measure the actual location of sampling points. The measured coordinates amounted to 20-point cloud sets $\{T_0, T_1, \cdots, T_{20}\}$. Take surface F as an example: the measured point clouds sets are $\{T_{F0}, T_{F1}, \cdots, T_{F20}\}$. Two of the three coordinates are determined by sampling points. The third coordinate is called the feature parameter. A first-order gradient object function of maximum likelihood estimation with weight decay is shown in Equation (17):

$$g(w) = \sum_{i=1}^{N}\sum_{t=1}^{T} f_k(x_t, y_t, t) - \sum_{i=1}^{N}\sum_{y} [p(y|x,t) \times \sum_{t=1}^{T} f_k(x_t, y_t, t)] - \frac{\sum w^2}{2\delta^2} \quad (17)$$

Set the initial parameter weights as $w^{(0)} = \{0.1, 0.1, \cdots, 0.1\}$ and the initial positive definite matrix as $M^{(0)} = diag(0.1, 0.2, \cdots, 2.0)$. The coefficient of the correct term is set to 0.5. The maximum number of iterations is 1000. The feature weights of $P(y|x,t)$ corresponding to processing features are calculated according to Section 2.1. Some of the feature weights of surface F are shown in Table 2.

Table 2. Value of partial feature weights of surface F.

Feature Terms	Feature Weight Value	Feature Terms	Feature Weight Value
w_1	0.036	w_8	0.124
w_2	0.498	w_9	0.965
w_3	0.015	w_{10}	2.326
w_4	0.968	w_{11}	0.216
w_5	1.526	w_{12}	2.979
w_6	0.732	w_{13}	1.521
w_7	1.998	w_{14}	1.104

The prediction of degraded machined surface F is obtained using the methods in Section 2.3. Consider a particular sampling point (i, j), with the assumption that its feature parameter of the location on the predictive machined surface is $y^*_{i,j}$ and its feature parameter of the location on the real surface is measured as $y^{(1)}_{i,j}$. The geometric average of the overall location residuals E_r is set as an estimation of the precision of the predictive machined surface model, as shown in Equation (18):

$$E_r = \frac{\sum\limits_{N=1}^{20} \sqrt{\sum\limits_{i,j} \left(y^{(N)}_{i,j} - \overline{y}^{(N)}_{i,j}\right)^2}}{\sum\limits_{N=1}^{20} \sqrt{\sum\limits_{i,j} \left(y^*_{i,j} - y^{(N)}_{i,j}\right)^2}} \tag{18}$$

The value of E_r decreases when the difference between the predictive location and actual location of all the sampling points grows larger. Otherwise, the increase of E_r indicates that the predictive machined surface model is in good agreement with the real surface. Considering all the values of surface F in the 20-part samples, a comparison of the predictive machined surface using a multivariate Gaussian process and using the methodology in this paper is shown in Figure 6. Compared with the commonly used multivariate Gaussian process, the multi-dimensional Markov process provides a predictive model with higher accuracy. When time t is applied to the designed life of the assembly, the degraded machined surface near the precision failure threshold is generated and used in the assembly simulation.

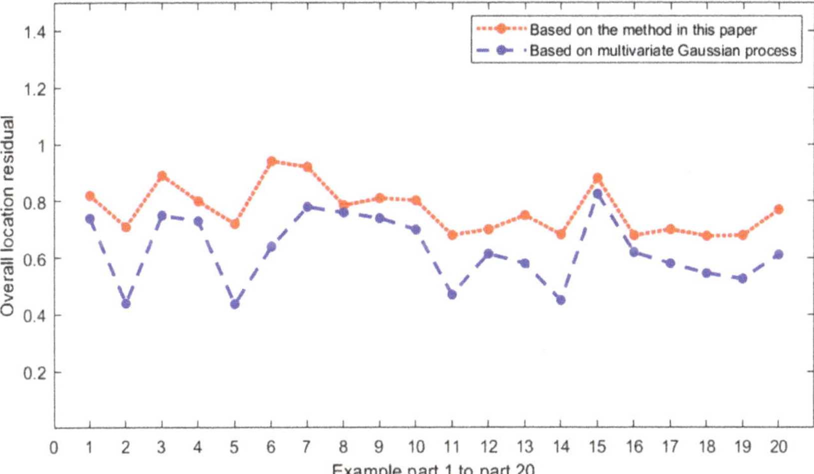

Figure 6. Comparison of overall location residuals of the predictive machined surface using the multivariate Gaussian process and using the method proposed in this paper.

4.3. Tolerance Synthesis of Example Rotary Feed Component

Numerical experiments were implemented with tolerance models of the shaft part and the hole part (in Figure 7a). Machined surface degradation prediction was combined as the model of the outsourced shaft part. Testing models for the hole part were generated based on MGP with given tolerance indices e_1, e_2, and e_3. Using the method in Section 3.1, assembling conditions and contact points were estimated based on ICP when the assembly entity constraints and assembly force constraints were checked during iterations. It is worth noting that the location of contact points varies according to the assembling contact. There are mainly three basic patterns from the perspective of the position distribution of contact points (illustrated in Figure 7a–c), which is the probability influenced by the scale of planar form errors, cylindrical form errors, and perpendicularity.

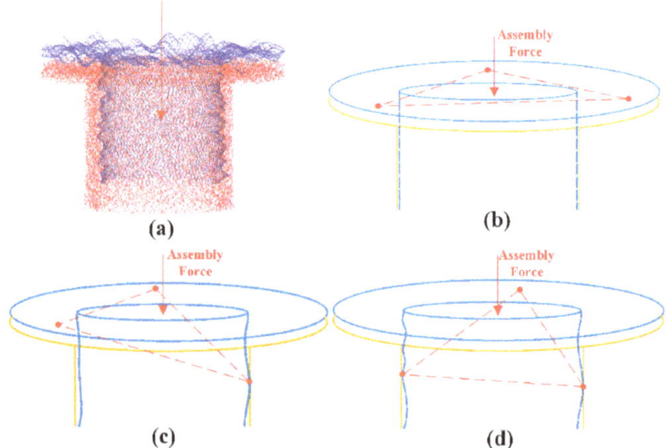

Figure 7. Numerical experiments (**a**) and three basic patterns of contact point positions (**b–d**).

The target assembly tolerance index, the runout error e_f of surface F, was estimated by analyzing the transformed point cloud in assembling condition. A four-dimensional solution space was generated as an explanation of tolerance synthesis, constructed with e_1, e_2, and e_3 as input and e_f as an output. According to the functional requirement of the assembly, the nominal value of e_f is 0.020 mm, which means the upper limit of e_f is 0.020 mm in all the adopted tolerance schemes. However, an unnecessarily narrow tolerance interval arrangement leads to extremely high costs in fabrication and maintenance, while a combination of wide tolerance intervals causes a high ratio of assembling failure. As a result, the influence of manufacturing cost and assembly probability based on an empirical formula and experimental statistics is considered in tolerance synthesis. Assembly examples with acquired functional requirements are called "successful" assemblies. Assembling probability is estimated by the proportion of successful assemblies of all the experimental assemblies. According to industrial practice, a yield rate of 95% is accepted, which means that the assembling probability should be more than 95%. The region of interest in the solution space is generated following this rule.

The empirical relationship between the tolerance interval and manufacturing cost based on historical industrial practice is shown in Equation (19), where α_1, α_2, and α_3 are manufacturing difficulty coefficients, and K is a scaling factor:

$$C = K(\frac{\alpha_1}{e_1^2} + \frac{\alpha_2}{e_2^2} + \frac{\alpha_3}{e_3^2}) \tag{19}$$

The influences of e_1, e_2, and e_3 on the assembly probability and manufacturing cost are shown in Figure 8a. A Monte Carlo sampling method was applied in the four-dimensional solution space to calculate the output of all the sampled inputs. High-density sampling was conducted in the region of interest to ensure searching accuracy; the density of sampling points was lower, but even outside the region of interest, it avoided falling into the local optimum. The top 1% with the lowest manufacturing cost was approximately fitted into an ellipsoid limit surface, as shown in Figure 8b, where the axes of e_1, e_2, and e_3 are even, and the axis of e_f is uneven.

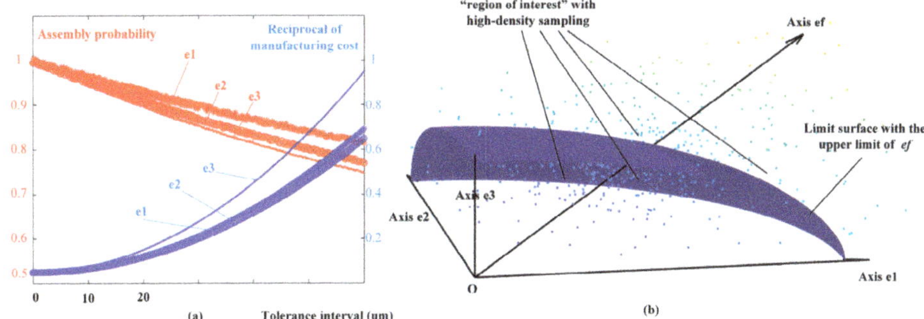

Figure 8. Distribution of assembly probability and manufacturing cost (**a**), and solution searching in a four-dimensional solution space constructed by even axes of e_1, e_2, and e_3, and the uneven axis of e_f (**b**).

Three examples with the lower manufacturing cost and satisfactory assembling probability constitute the designed tolerance scheme, as shown in Table 3. The designed scheme with an assembling probability larger than 95% and the lowest manufacturing cost is listed in bold. The designed value of all three to-be-designed tolerance terms e_1, e_2, and e_3 were enlarged by 0.003, 0.004, and 0.004, respectively. The objective functional requirement was still met, and the assembling probability was satisfactory. However, the expansion of the tolerance interval caused a 35.6% reduction of the manufacturing cost. The other two listed example designed schemes can also be considered depending on the acceptable manufacturing cost considered.

Table 3. Design requirement of the object tolerance term and the nominal and designed value of the to-be-designed tolerance terms.

Objective Tolerance Term	e_f	**Assembly Functional Requirement**		$e_f \leq 0.020$	
To-be-designed tolerance terms	e_1	e_2	e_3	Assembling probability	Relative manufacturing cost
Nominal value (mm)	0.010	0.012	0.005	1	1
Designed scheme 1 (mm)	0.013	0.016	0.009	0.957	0.644
Designed scheme 2 (mm)	0.012	0.014	0.016	0.959	0.679
Designed scheme 3 (mm)	0.014	0.015	0.006	0.969	0.754

5. Conclusions

A tolerance design method based on Skin Model Shapes considering processing feature degradation is proposed in this paper. To include the geometric form deviation and degrading processing feature, the machined surface model was constructed in the form of 3D point clouds based on Skin Model Shapes. A uniform sampling was implemented on the grid nodes of the assembly surface. Using machine part samples in mass production, the point dataset was acquired by a high-precision coordinate measuring machine. Then, a continuous-time multi-dimensional Markov process was trained to model the feature degradation process; it was also used in further numerical experiments. To improve the reliability and rationality of the numerical experiments, the assembly force constraints and assembly entity constraints were applied to the assembly simulation. Then, the static and dynamic tolerance indices were analyzed and synthesized. The values of the to-be-designed tolerance terms were designed with the aim of conforming to the assembly tolerance requirement, guaranteeing the assembling probability and reducing the manufacturing cost as much as possible.

The tolerance design method in this paper was applied to an example assembly tolerance design problem regarding a five-axis machine tool rotary feed system. Data analysis indicated that the predictive machined surface model is more accurate than that employed in common Skin Model Shape methods. The designed tolerance scheme has a larger tolerance interval and lower manufacturing costs. That is, the generation of the feature degradation model comprises an in-depth profile and dynamic investigation for production systems based on sampling machining data, which improved the ability of self-configuration of the designed tolerance scheme. Also, the designing reliability and robustness was improved through the improved assembly simulation considering multiple assembling constraints. In addition, the collection and analysis of the manufacturing information and the process of virtual simulation is closely related to the deployment of Internet of Things (IoT) systems and Cyber-Physical Production Systems (CPPS), especially manufacturing equipment with sensors, automation, and information flow. As a result, the proposed method helps to promote the design capability and production flexibility, and improves competence in an increasingly competitive business environment. It provides a new way to design with digitality and intelligence to help fill in the gaps between virtual engineering processes and virtual engineering factories, which would contribute to completing the structure of Industry 4.0. However, evenly distributed sampling points on the assembly surface may cause model distortion when local geometric features are complicated. Boundary treatment in point cloud combination, advanced sampling methods, and dimensionality reduction in solution space also need attention in further research.

Author Contributions: Conceptualization, C.H. and L.Q.; methodology, C.H.; software, C.H.; validation, C.H., S.Z. and L.Q.; formal analysis, S.Z.; investigation, C.H.; resources, C.H.; data curation, C.H.; writing—original draft preparation, C.H.; writing—review and editing, C.H., S.Z., L.Q., Z.W. and X.L.; visualization, C.H.; supervision, L.Q., S.Z., Z.W. and X.L.; project administration, L.Q.; funding acquisition, L.Q.

Funding: The work is supported by the National Science Foundation of China under grant No. 51675478.

Conflicts of Interest: The authors declare no conflict of interest.

References

1. Erboz, G. How to define industry 4.0: The main pillars of industry 4.0. In Proceedings of the 7th International Conference on Management (ICoM 2017), Nitra, Slovakia, 1–2 June 2017.
2. Hofmann, E.; Rüsch, M. Industry 4.0 and the current status as well as future prospects on logistics. *Comput. Ind.* **2017**, *89*, 23–34. [CrossRef]
3. Lu, Y. Industry 4.0: A survey on technologies, applications and open research issues. *J. Ind. Inf. Integr.* **2017**, *6*, 1–10. [CrossRef]
4. Geetha, K. Tolerance allocation and scheduling for complex assembly. *Int. J. Appl. Eng. Res.* **2015**, *10*, 4000–4003.
5. Shoukr, D.S.L.; Gadallah, M.H.; Metwalli, S.M. The reduced tolerance allocation problem. In Proceedings of the ASME's International Mechanical Engineering Congress and Exposition (IMECE2016), Phoenix, AZ, USA, 11–17 November 2016.
6. Khodaygan, S. Meta-model based multi-objective optimisation method for computer-aided tolerance design of compliant assemblies. *Int. J. Comput. Integr. Manuf.* **2019**, *32*, 27–42. [CrossRef]
7. Delos, V.; Arroyave-Tobón, S.; Teissandier, D. Introducing a projection-based method to compare three approaches computing the accumulation of geometric variations. In Proceedings of the ASME 2018 International Design Engineering Technical Conferences and Computers and Information in Engineering Conference, Quebec City, QC, Canada, 26–29 August 2018.
8. Lin, E.E.; Zhang, H.C. Theoretical tolerance stackup analysis based on tolerance zone analysis. *Int. J. Adv. Manuf. Technol.* **2001**, *17*, 257–262. [CrossRef]
9. Wang, Y. Closed-loop analysis in semantic tolerance modeling. *J. Mech. Des.* **2008**, *130*, 061701. [CrossRef]
10. Geis, A.; Husung, S.; Oberänder, A.; Weber, C.; Adam, J. Use of vectorial tolerances for direct representation and analysis in CAD-systems. *Proc. CIRP* **2015**, *27*, 230–240. [CrossRef]
11. Heling, B.; Aschenbrenner, A.; Walter, M.S.J.; Wartzack, S. On connected tolerances in statistical tolerance-cost-optimization of assemblies with interrelated dimension chains. *Proc. CIRP* **2016**, *43*, 262–267. [CrossRef]
12. Ameta, G.; Davidson, J.K.; Shah, J.J. Tolerance-maps applied to a point-line cluster of features. *J. Mech. Des.* **2007**, *129*, 782–792. [CrossRef]

13. Bhide, S.; Ameta, G.; Davidson, J.K.; Shah, J.J. Tolerance-maps applied to the straightness and orientation of an axis. In *Models for Computer Aided Tolerancing in Design and Manufacturing*; Springer: Dordrecht, The Netherlands, 2007.
14. Chitale, A.N.; Davidson, J.K.; Shah, J.J. Statistical tolerance analysis with sensitivities established from tolerance-maps and deviation spaces. *J. Comput. Inf. Sci. Eng.* **2019**, *19*, 041002. [CrossRef]
15. Cheng, H.; Li, Y.; Zhang, K.F.; Su, J. Bin Efficient method of positioning error analysis for aeronautical thin-walled structures multi-state riveting. *Int. J. Adv. Manuf. Technol.* **2011**, *55*, 217–233. [CrossRef]
16. 1Wang, H.; Liu, J. Tolerance simulation of thin-walled c-section composite beam assembling with small displacement torsor model. *Proc. CIRP* **2016**, *43*, 274–279.
17. Li, H.; Zhu, H.; Zhou, X.; Li, P.; Yu, Z. A new computer-aided tolerance analysis and optimization framework for assembling processes using DP-SDT theory. *Int. J. Adv. Manuf. Technol.* **2016**, *86*, 1299–1310. [CrossRef]
18. Du, Q.; Zhai, X.; Wen, Q. Study of the ultimate error of the axis tolerance feature and its pose decoupling based on an area coordinate system. *Appl. Sci.* **2018**, *8*, 435. [CrossRef]
19. Yan, H.; Cao, Y.; Yang, J. Statistical tolerance analysis based on good point set and homogeneous transform matrix. *Proc. CIRP* **2016**, *43*, 178–183. [CrossRef]
20. Laperrière, L.; Elmaraghy, H.A. Tolerance analysis and synthesis using Jacobian transforms. *CIRP Ann.* **2000**, *49*, 359–362. [CrossRef]
21. Desrochers, A.; Ghie, W.; Laperrière, L. Application of a unified jacobian—Torsor model for tolerance analysis. *J. Comput. Inf. Sci. Eng.* **2003**, *3*, 2–14. [CrossRef]
22. Zeng, W.; Rao, Y.; Wang, P.; Yi, W. A solution of worst-case tolerance analysis for partial parallel chains based on the Unified Jacobian-Torsor model. *Precis. Eng.* **2017**, *47*, 276–291. [CrossRef]
23. Kopardekar, P.; Anand, S. Tolerance allocation using neural networks. *Int. J. Adv. Manuf. Technol.* **1995**, *10*, 269–276. [CrossRef]
24. Luo, C.; Franciosa, P.; Ceglarek, D.; Ni, Z.; Jia, F. A novel geometric tolerance modeling inspired by parametric space envelope. *IEEE Trans. Autom. Sci. Eng.* **2018**, *15*, 1386–1398. [CrossRef]
25. Samper, S.; Formosa, F. Form Defects tolerancing by natural modes analysis. *J. Comput. Inf. Sci. Eng.* **2007**, *7*, 44–51. [CrossRef]
26. Homri, L.; Goka, E.; Levasseur, G.; Dantan, J.Y. Tolerance analysis—Form defects modeling and simulation by modal decomposition and optimization. *CAD Comput. Aided Des.* **2017**, *91*, 46–59. [CrossRef]
27. Lin, E.E. Graph-Matrix-Based Automated Tolerance Analysis and Setup Planning in Computer-Aided Process Planning. Ph.D. Thesis, Texas Tech University, Lubbock, TX, USA, 2000.
28. Zhang, K.; Li, Y.; Tang, S. An integrated modeling method of unified tolerance representation for mechanical product. *Int. J. Adv. Manuf. Technol.* **2010**, *46*, 217–226. [CrossRef]
29. Schleich, B.; Wartzack, S.; Anwer, N.; Mathieu, L. Skin model shapes: Offering new potentials for modelling product shape variability. In Proceedings of the ASME 2015 International Design Engineering Technical Conferences and Computers and Information in Engineering Conference, Boston, MA, USA, 2–5 August 2015.
30. ISO 17450-1. *Geometrical Product Specifications (GPS): General Concepts: Part 1: Model for Geometrical Specification and Verification*; ISO: Geneva, Switzerland, 2011.
31. ISO 17450-2. *Geometrical Product Specifications (GPS)—General Concepts: Part 2: Basic Tenets, Specifications, Operators, Uncertainties and Ambiguities*; ISO: Geneva, Switzerland, 2012.
32. Ballu, A.; Mathieu, L. Univocal expression of functional and geometrical tolerances for design, manufacturing and inspection. In *Computer-Aided Tolerancing*; Springer: Dordrecht, The Netherlands, 2011.
33. Anwer, N.; Ballu, A.; Mathieu, L. The skin model, a comprehensive geometric model for engineering design. *CIRP Ann. Manuf. Technol.* **2013**, *62*, 143–146. [CrossRef]
34. Zhang, M.; Anwer, N.; Mathieu, L.; Zhao, H. A discrete geometry framework for geometrical product specifications. In Proceedings of the 21st CIRP Design Conference, Kaist, Korea, 1 January 2011.
35. Schleich, B.; Anwer, N.; Mathieu, L.; Wartzack, S. Skin Model Shapes: A new paradigm shift for geometric variations modelling in mechanical engineering. *CAD Comput. Aided Des.* **2014**, *50*, 1–15. [CrossRef]
36. Yacob, F.; Semere, D.; Nordgren, E. Octree-based generation and variation analysis of skin model shapes. *J. Manuf. Mater. Process.* **2018**, *2*, 52. [CrossRef]
37. Yan, X.; Ballu, A. Generation of consistent skin model shape based on FEA method. *Int. J. Adv. Manuf. Technol.* **2017**, *92*, 789–802. [CrossRef]

38. Schleich, B.; Anwer, N.; Mathieu, L.; Wartzack, S. Contact and mobility simulation for mechanical assemblies based on skin model shapes. *J. Comput. Inf. Sci. Eng.* **2015**, *15*, 021009. [CrossRef]
39. Dantan, J.-Y. Comparison of skin model representations and tooth contact analysis techniques for gear tolerance analysis. *J. Comput. Inf. Sci. Eng.* **2015**, *15*, 021010. [CrossRef]
40. Yan, X. Assembly Simulation and Evaluation Based on Generation of Virtual Workpiece with Form Defect. Ph.D. Thesis, Université de Bordeaux, Bordeaux, France, 2018.
41. Wang, J.; Sanchez, J.; Iturrioz, J.; Ayesta, I. Geometrical defect detection in the wire electrical discharge machining of fir-tree slots using deep learning techniques. *Appl. Sci.* **2018**, *9*, 90. [CrossRef]
42. Sun, J.; Rahman, M.; Wong, Y.S.; Hong, G.S. Multiclassification of tool wear with support vector machine by manufacturing loss consideration. *Int. J. Mach. Tools Manuf.* **2004**, *44*, 1179–1187. [CrossRef]
43. Dai, W.; Chi, Y.; Lu, Z.; Wang, M.; Zhao, Y. Research on reliability assessment of mechanical equipment based on the performance–feature model. *Appl. Sci.* **2018**, *8*, 1619. [CrossRef]
44. Ozcelik, B.; Bayramoglu, M. The statistical modeling of surface roughness in high-speed flat end milling. *Int. J. Mach. Tools Manuf.* **2006**, *46*, 1395–1402. [CrossRef]
45. Shu, M.H.; Hsu, B.M.; Kapur, K.C. Dynamic performance measures for tools with multi-state wear processes and their applications for tool design and selection. *Int. J. Prod. Res.* **2010**, *48*, 4725–4744. [CrossRef]
46. Hsu, B.M.; Shu, M.H.; Wu, L. Dynamic performance modelling and measuring for machine tools with continuous-state wear processes. *Int. J. Prod. Res.* **2013**, *51*, 4718–4731. [CrossRef]
47. Moghaddass, R.; Zuo, M.J. A parameter estimation method for a condition-monitored device under multi-state deterioration. *Reliab. Eng. Syst. Saf.* **2012**, *106*, 94–103. [CrossRef]
48. Cannarile, F.; Compare, M.; Baraldi, P.; Di Maio, F.; Zio, E. Homogeneous continuous-time, finite-state hidden semi-markov modeling for enhancing empirical classification system diagnostics of industrial components. *Machines* **2018**, *6*, 34. [CrossRef]

Article

Cloud-Based Collaborative 3D Modeling to Train Engineers for the Industry 4.0

José Luis Saorín [1], Jorge de la Torre-Cantero [1], Dámari Melián Díaz [1,*] and Vicente López-Chao [2,*]

1 Engineering Graphic Area, Universidad de La Laguna, Avenida Ánguel Guimerá Jorge s/n, 38204 La Laguna, Spain; jlsaorin@ull.edu.es (J.L.S.); jcantero@ull.edu.es (J.d.l.T.-C.)
2 Department of education, University of Almería, 04120 La Cañada, Spain
* Correspondence: dmeliand@ull.edu.es (D.M.D.); valchao@ual.es (V.L.-C.); Tel.: +34-950-015750 (V.L.-C.)

Received: 4 October 2019; Accepted: 25 October 2019; Published: 27 October 2019

Abstract: In the present study, Autodesk Fusion 360 software (which includes the A360 environment) is used to train engineering students for the demands of the industry 4.0. Fusion 360 is a tool that unifies product lifecycle management (PLM) applications and 3D-modeling software (PDLM—product design and life management). The main objective of the research is to deepen the students' perception of the use of a PDLM application and its dependence on three categorical variables: PLM previous knowledge, individual practices and collaborative engineering perception. Therefore, a collaborative graphic simulation of an engineering project is proposed in the engineering graphics subject at the University of La Laguna with 65 engineering undergraduate students. A scale to measure the perception of the use of PDLM is designed, applied and validated. Subsequently, descriptive analyses, contingency graphical analyses and non-parametric analysis of variance are performed. The results indicate a high overall reception of this type of experience and that it helps them understand how professionals work in collaborative environments. It is concluded that it is possible to respond to the demand of the industry needs in future engineers through training programs of collaborative 3D modeling environments.

Keywords: collaborative learning; engineering graphics; PLM; 3D modeling; engineering education

---

1. Introduction

At the end of the 20th century, companies indicated that engineers were very individualistic. This profile did not meet the new needs of the industry, such as task management between several work teams or design and manufacturing time. For this, communication between all project members was a major requirement [1]. This lack inevitably relates to the fact that in engineering education the lecture method was the main teaching approach [2]. Today, human collaboration and interoperability of software systems are essential requirements in industry 4.0 [3].

In the industry, product lifecycle management (PLM) applications have been used for decades to reduce design time and to manage collaborative work of the product engineering teams [4]. So, 3D modeling software companies for engineering such as Pro-Engineer, Autodesk or Solidworks offer collaborative work platforms (Autodesk Vault, PTC PLM Cloud or 3D Experience) that are mainly aimed at large companies and combine PLM tasks and 3D modeling, that we will abbreviate as PDLM (product design and life management). Nevertheless, the cost and use complexity of these collaborative platforms makes their implementation in educational engineering environments very difficult.

Meanwhile, collaborative pedagogies and project-based learning methods are becoming increasingly popular. Their combination with cloud-based software creates a simulation of the professional context. So, this approach fosters coordination and communication between team members while solving challenges. For this, several cloud technologies have been tested in education to improve collaborative learning (Google Docs, Skype, Dropbox, Facebook, Prezi, Skype, etc.) [5]. Collaborative writing tasks in distance engineering education have also been tested with web-based tool experiences and have obtained high student acceptability [6], deeper learning and high motivation [7]. Moreover, they promote a self-motivational environment and help students to comprehend the engineering profession [8]. Likewise, e-journals allow sharing laboratory resources among colleagues with certain limitations due to technical compatibility problems [9]. However, they focus on communication and writing [10]. They are not entirely appropriate for particular engineering activities such as design and 3D modeling.

In Spain, engineering graphics professors have begun to assume the shift towards multi-method active teaching [11] and the inclusion of ICT (Information and Communication Technologies) methodologies [12,13]. Furthermore, project-based learning stands as one of the most appropriate methods for the development of professional competencies of the engineer [14].

In this context, a collaborative experience has been carried out with civil engineering students at the University of La Laguna in the academic year 2018–2019. The novelty of this study relies on the use of categories to deepen the student perception of usability of a PDLM software through a categorical analysis of variance, which is essential to know how to design training experiences for industry 4.0. Autodesk Fusion 360 software is used to develop the graphic simulation of an engineering project. This cloud-based software incorporates PLM tools and a 3D modeling synchronous environment. This paper focuses on the explanation of the educational experience and the student's perception of the resolution of an engineering project under the approach of collaborative design as training for industry 4.0.

At the University of La Laguna, engineering graphics professors teach under a project-based learning approach. In groups, students design an engineering product made of five components, which they model in 3D solids. Then, they develop 2D drawings and produce dynamic animations and create some realistic infographics. In 2013, Autodesk Inventor and Dropbox were combined to achieve collaboration in the 3D modeling process [15,16]. This system was a simulation of a PDLM application in terms of functionality, which allowed several students to work simultaneously in the creation of the project graphic documentation. However, this system had the following limitation: students could not use simultaneously the same design file. A similar experience took place at the School of Engineering Arts et Metiers ParisTech by combining Catia V5 and Dropbox [17].

The recent arrival of Autodesk Fusion 360 solves some of the tasks reserved for the PLM programs that the engineering companies use. Fusion 360 combines 3D modeling with the advantages of cloud computing tools (Figure 1) through cloud-stored files in the Autodesk cloud (called A360). It naturally allows collaborative 3D modeling, which is appropriate for educational environments and fab labs [18].

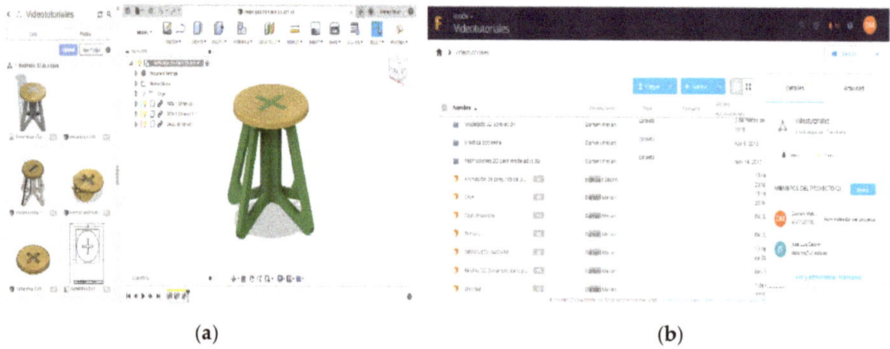

(a) (b)

Figure 1. (**a**) A 3D modeling environment and (**b**) file web environment (A360).

In 2017 Vila, Ugarte, Ríos and Abellán [19] proposed the use of PLM systems to encourage collaboration among engineering students. They compared Fusion 360 and 3D Experience in terms of data management, decision support, personal data management, project management and communication. They concluded that both environments allowed proper collaborative 3D-model management. Moreover, while 3D Experience allows better project planning and its data management is highly complex, Fusion 360 presented more limitations in the other categories. Nevertheless, Fusion 360 seemed to be easier to be implemented in educational environments in terms of economic cost (free for educational purposes and for start-ups generating less than $100,000/year in total revenue or wholly non-commercial hobbyist users [20]) and ease of learning.

Likewise, in the Youngstown State University, Brozina and Sharma used Fusion 360 in their first year of engineering design course [21]. They conducted a collaborative workshop using Cloud Computing to allow the exchange of notes and views using browsers, mobile devices and the software itself. They concluded that students better understand the collaboration process using a collaborative 3D modeling environment.

Rassovytska and Striuk analyzed the suitability of cloud tools for the professional activity of mechanical engineers [22]. The objective of this work was to identify mobile and cloud services for professional activity in mechanical engineering and how to implement them in higher education. More than 30 services in the cloud and mobile applications were evaluated. The analysis found that the use of the services in the Autodesk cloud and its integration with Google Cloud is appropriate for teaching and professional use in mechanical engineering. They applied a questionnaire to professionals and professors to figure out which competencies engineers should enhance. CAD software was pointed out as the most important, followed in the seventh position by collaboration. They analyzed On-Shape, a 3D CAD cloud-based application that works on the web browser, which has been developed from scratch to solve the current problems of engineers, taking into account aspects such as collaboration in 3D environments. They analyzed functionality, availability, access, integration and collaboration; Fusion 360 obtained the highest value.

At the University of Illinois, the professors decided to move from non-collaborative CAD software (Inventor, Rhinoceros and Alias) towards Autodesk Fusion 360 for the creation of digital prototyping in multidisciplinary teams [23]. Most students indicated that they used the Fusion 360 cloud options and also that the use of this application reduced the need for face-to-face meetings. Moreover, they indicated that they usually worked simultaneously in several design stages which improved their productivity.

Furthermore, at the University of Tianjin (China), the different environments offered by Fusion 360 were analyzed and it was concluded that this application is an ideal tool for collaborative design in engineering, since it integrates modeling, rendering, manufacturing analysis and data management [24].

Besides, some synchronous collaboration tools may not be successful in face-to-face teaching. Students seem to prefer direct interaction and when they need an alternative, they choose their asynchronous communication solutions such as SMS [9]. Furthermore, one of the educational challenges in the inclusion of PLM software is that students must know how to divide the work among the different stakeholders [16].

2. Materials and Methods

The research design corresponds to a quantitative quasi-experimental approach. A collaborative graphic simulation of an engineering project was proposed in the engineering graphics subject of the civil engineering degree at the University of La Laguna in the academic year 2018–2019. After the elaboration and presentation of each team's results, a questionnaire was designed and applied to measure student's perception of the resolution of an engineering project under the approach of integral collaborative design as training for industry 4.0. Groups were asked to agree on their answers, but each member delivered their own questionnaire sheet. This criterion was based on the need to receive a group perception of the collaborative practice, meanwhile we took into consideration whether the number of students in each group varied. Then, Cronbach's alpha and descriptive analysis were conducted (mean, median and standard deviation). Consequently, line graphs were generated to observe whether Usability of Product Design and Life Management platform (UPDLM) scale values varied concerning three grouping variables. Finally, non-parametric analyses were applied to check whether the previous differences were statistically meaningful.

2.1. Sample

The sample consisted of 65 civil engineering undergraduate students from the University of La Laguna. A proportion of 30% the students knew PLM applications for collaborative work in CAD environments. However, 95% of the them had not worked in a collaborative 3D modeling experience. They considered that group work in engineering environments is very important for their professional activity and 97% reflected that working with cloud-based data is fundamental for an engineer.

2.2. Software

Autodesk Fusion 360 was the selected parametric design software due to its collaborative cloud-based environment. Moreover, A360 (the Autodesk cloud) was used for data management.

2.3. Measurement Instrument

The usability of product design and life management platform (UPDLM) scale (Table 1) was designed to measure the perception of the participants about the use of the Fusion 360 software for the accomplishment of collaborative work. UPDLM scale is a 1–5 Likert scale style, in other words, students were indicated to score each item from 1 (low importance) to 5 (high importance).

Table 1. Usability of product design and life management platform (UPDLM) scale.

Factor	Items	Abbreviation
F1	Fusion 360 boosts communication and collaboration	V1
	Fusion 360 reduces the need for face-to-face group meetings.	V2
	Fusion 360 allows simultaneous 3D modeling on different components.	V3
	Fusion 360 improves the coordination of the group	V4
	Fusion 360 is intuitive in terms of geometry drawing	V5
F2	Fusion 360 is easier to collaborate with under assigned roles within the group	V6
	A360 Comment is useful for communication	V7
	A360 Calendar is useful for coordination	V8
	A360 is useful for data management	V9
	A360 Conversations foster communication	V10
F3	Fusion 360 helps to understand professional collaborative work environments	V11
	A360 public 3D viewer is useful	V12
	Fusion 360 is a worthy PDLM application	V13
	Fusion 360 should incorporate permissions (editing, visualization, etc.)	V14

This questionnaire is composed of 14 questions divided into three categories: collaboration and design features (Factor 1 or F1), PLM features (Factor 2 or F2) and collaborative engineering (Factor 3 or F3). The design of the measurement instrument was based on the theoretical ground that has been referenced in the introduction section by combining collaborative-based learning and PLM theories with the needs of the industry 4.0.

In addition, students rated three grouping variables from 1 to 5: PLM knowledge, individual practices utility and collaborative engineering perception.

2.4. Procedure

The educational experience was to perform the graphic simulation of an engineering project. The graphic documentation includes the 3D models of each component of the object, their assembly, the 2D engineering drawings, an animation of the assembly–disassembly process, the creation of infographics and a video-presentation (Figure 2). Finally, each group made a presentation to report their work.

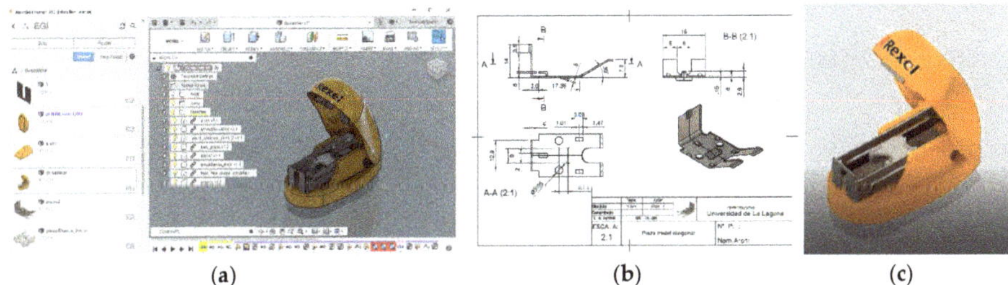

Figure 2. Example views of group 3: (**a**) Fusion 360 environment; (**b**) 2D-drawings; (**c**) infographic.

For this, each group assigned different roles to the participants: project manager, 3D modeler, 2D drawings assistant, infographic assistant, video editing manager and supervisor of additional materials (models, augmented reality, etc.). The objective of roles assignment was to distribute the responsibilities of the project. However, all the members participated in all the tasks. After, each student group had to propose an object to work with, which must fulfil the following characteristics. The object must be composed of at least five different components. The pieces must be simple to model, but students must use at least three different modeling operations (extrusion, revolution, sweep or loft). The pieces must also need editing operations. This activity was distributed in ten sessions (2 h each), in addition to the time they needed outside the classroom:

- Session 1: Students must form groups of five members and install Fusion 360.
- Session 2: Groups must assign roles and distribute tasks to each member. Moreover, they have to create a project in A360 and share it among the team members, so that everyone has access to the files of the modeling process. Besides, comment, calendar and data management are explained by the professor.
- Session 3: The object must be proposed to the professor who must accept its viability (some examples are watches, PlayStation controls are Wi-Fi routers). Then, students begin the sketch phase and 3D modeling with Fusion 360.
- Sessions 4–9: The group simultaneously generates all the graphic documentation, so that one of the members can be modeling a piece, while other is doing the 2D drawing of the same piece and a third is doing the assembly of the set. Moreover, project management in terms of model and file changes in A360 is taught (Figure 3). At this stage, groups are encouraged to access the A360 utilities and each student is asked to make at least one comment with the members of their group.
- Session 10: At the end of this experience, each group makes a presentation of the work done. Some groups present additional materials such as 3D printed models, online 3D repositories and models in augmented or virtual reality (Figure 4). After that, each group fills the UPDLM scale.

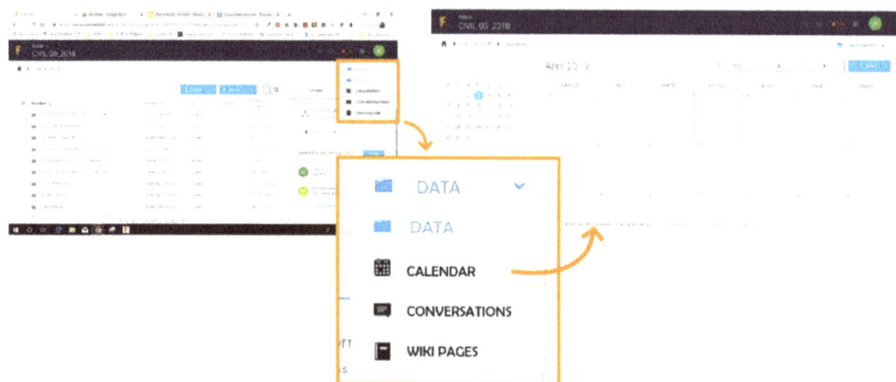

Figure 3. A360 interface with data management options, calendar and conversations.

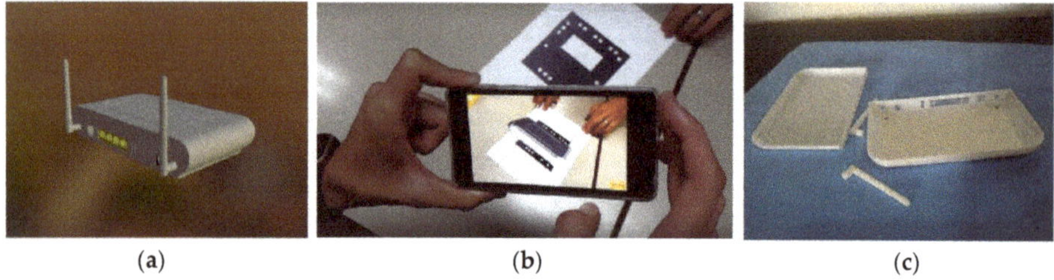

Figure 4. (**a**) 3D models in Sketchfab; (**b**) Augmented reality; (**c**) 3D printed model.

3. Results

Cronbach's alpha statistic was calculated to assess the reliability of the scale and a high value was obtained ($\alpha = 0.838$). Likewise, the reliability of F1 ($\alpha = 0.829$), F2 ($\alpha = 0.782$) and F3 ($\alpha = 0.714$) have been calculated. Consequently, descriptive analyses of the mean, median and standard deviation were carried out (Table 2).

Table 2. Mean and standard deviation results.

Factor	Item	Mean	Median	Std. Dev.	Factor Reliability (Cronbach alpha)
F1	V1	3.85	4.00	0.96	0.829
	V2	4.46	5.00	0.64	
	V3	4.62	5.00	0.49	
	V4	3.46	4.00	1.01	
	V5	4.69	5.00	0.61	
F2	V6	3.46	4.00	1.22	0.782
	V7	2.62	2.00	1.34	
	V8	2.15	2.00	1.41	
	V9	2.00	2.00	1.04	
	V10	2.00	2.00	1.05	
F3	V11	4.54	5.00	0.50	0.714
	V12	3.92	4.00	1.15	
	V13	3.69	4.00	1.14	
	V14	4.5	5.00	0.75	

Regarding F1 (collaboration and design features), the students scored all the items close to two values: 3.60 (coordination, collaboration and communication features) and 4.50 (geometry and design usability and its viability to reduce face-to-face meetings). In terms of PLM features, students scored values close to 2.00 for all the PLM tools. However, they assigned a 3.46 to the effectiveness of the roles to collaborate. Finally, students highly scored the third factor (collaborative engineering); they scored values close to 3.8 (for the 3D public viewer and the app as a PDLM) and 4.5 (for utility for understanding a collaborative engineering environment and the need to add permissions).

Moreover, Figure 5 represents the distribution of UPDLM and grouping variables. The range of values varies among one and five points. Students scored F1 and F3 variables with high values (mostly from 3 to 5 points) and F2 with low values (mostly from 1 to 3 points). In a few cases, groups provided data points that deviate significantly from the mean values in a rather consistent manner (See the Table S1 in Supplementary Materials). For example, students normally rated the variable 8 with a low value (1, 2 or 3). However, two groups perceived this variable with a value of 5. Both groups that rated V8 with 5 points, also gave a high value (4 points) to the grouping variable of PLM previous knowledge. Meanwhile, the students that rated V8 with a low value indicated their PLM previous knowledge with a similar value.

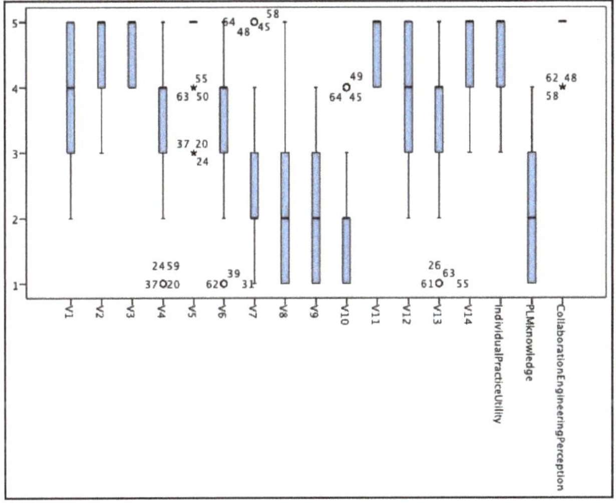

Figure 5. Box plot of UPDLM and grouping variables.

Subsequently, UPDLM mean values were represented in line graphs (see Figure 6) according to the score that the students attributed to the grouping variables: individual practices utility, PLM knowledge, and collaborative engineering perception. A color code has been used to easily differentiate the membership factor (F1 = green, F2 = blue and F3 = red).

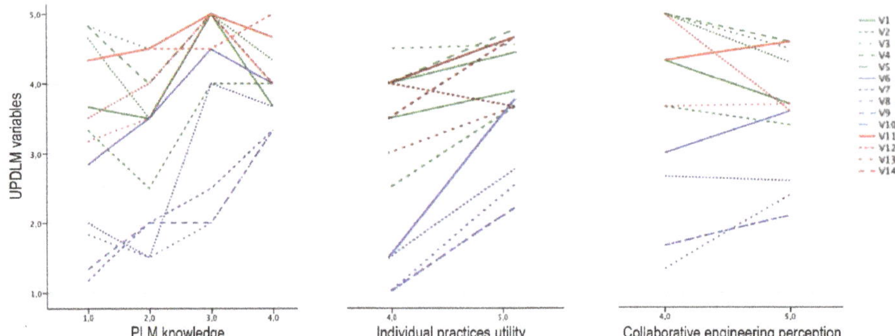

Figure 6. UPDLM scale line graphs. Grouping variables: product lifecycle management (PLM) knowledge, individual practices utility and collaborative engineering perception.

On the one hand, the grouping variable PLM knowledge exhibits the following trends. The variables belonging to F1 oscillate their scores by an average of 1.2 points, F2 variables by 2.04 points and F3 variables by 1.10 points. Also, there is an inverse trend from the 3-point valuation of the grouping variable for most variables (except for three variables of F2 and one of the F1). On the other hand, in the individual practices utility group, the results indicate a direct relationship in all the variables except one (V12). In this case, the scores range from 0.50, 1.46 and 0.57 average points for F1, F2 and F3, respectively. Finally, in the collaborative engineering perception grouping, the results of the scale show an inverse relationship for all the F1 variable and for two variables of the F3. In F2, V7 remains and the others increase by an average of 0.7 points.

Afterwards, the Kolmogorov–Smirnov test was applied to assess the normality of the variables PLM knowledge, individual practices utility and collaborative engineering perception. None of them followed a normal distribution ($p < 0.001$), so non-parametric analyses were applied.

Finally, the non-parametric variance analysis test for the PLM knowledge cluster shows significant differences in all variables (Table 3). The individual practices utility group shows significant differences concerning all the variables of F2 and F3 and for V2 (face-to-face meetings), V3 (simultaneous 3D modeling) and V5 (ease of geometry drawing) of F1. The collaborative engineering perception group shows significant differences in four of the five variables belonging to F1, as well as in the V8 (A360 Calendar is useful for coordination) and V12 (A360 public 3D viewer is useful).

Table 3. Kruskal–Wallis test.

Factor	Variable	Grouping Variable					
		PLM Knowledge		Individual Practices Utility		Collaborative Engineering Perception	
		X^2	Sig.	X^2	Sig.	X^2	Sig.
F1	V1	18.35	<0.001	0.63	0.729	5.02	0.025
	V2	30.00	<0.001	10.44	0.005	15.60	<0.001
	V3	36.27	<0.001	7.38	0.025	12.00	0.001
	V4	12.07	0.007	4.75	0.093	0.18	0.669
	V5	10.91	0.012	9.62	0.008	5.67	0.017
F2	V6	18.57	<0.001	24.44	<0.001	1.45	0.229
	V7	31.20	<0.001	9.37	0.009	0.16	0.688
	V8	18.16	<0.001	14.92	0.001	6.18	0.013
	V9	45.28	<0.001	11.67	0.003	1.09	0.296
	V10	32.76	<0.001	15.06	0.001	0.69	0.407
F3	V11	14.48	0.002	14.48	0.001	3.25	0.071
	V12	13.47	0.004	12.23	0.002	18.93	<0.001
	V13	26.58	<0.001	10.02	0.007	0.043	0.836
	V14	8.33	0.04	27.12	<0.001	0.516	0.473

4. Discussion and Conclusions

In most Spanish universities, engineering education includes learning to model in 3D but does not include the possibility of working collaboratively in 3D modeling. In fact, professors have recently begun to include ICT teaching methods [12,13]. This research aims to address the new reality of industry 4.0 through a training experience of engineering graphics students in a cloud-based collaborative 3D modeling platform. For this, we proposed an engineering training through Fusion 360 and A360 environment, which merge the functionalities of a traditional PLM and simultaneous 3D modeling, and which we have named as PDLM: product design and life management. The didactic proposal consisted of the collaborative graphic simulation of an engineering project.

Students highly valued the implementation of a PDLM concerning the factors of collaboration and design features as well as collaborative engineering. Specifically, students considered that simultaneous 3D modeling is indispensable and that its application reduces the need for face-to-face meetings to serve the current connected industry, which supports previous research [23]. Likewise, this training helped students to understand how specialists work in collaborative engineering environments, obtaining similar results to other authors [21].

Furthermore, the variables of the PLM features factor obtained much lower results, except for the role-assignment as a teamwork facilitator, which agrees with the literature as a challenge to overcome since students need to know how to divide work in PLM environments [17]. These low results in PLM features are consistent with previous research that indicates that students prefer to use their asynchronous communication solutions such as SMS [9].

Moreover, these results match with the graphic contingency analysis outcomes. They indicated that the better the PLM knowledge and individual practices are valued by students, the better students will embrace PDLM software. Besides, the analysis of variance by categories of dependent variables brings new information to the literature. The results evidenced that the student perception of the implementation of a PDLM statistically varies concerning their value towards the individual practices and their previous knowledge of PLM software. So, both factors are essential in the implementation and training in these engineering environments, which are a must for industry 4.0. However, students' preconception of collaborative engineering work hardly shows significant variations, and when they occur, the mean variations are not high.

To conclude, this research has applied training for future engineers adapted to the needs of industry 4.0 through a cloud-based collaborative 3D modeling platform. This has demonstrated a good reception with statistically significant changes with dependence on PLM knowledge and individual practices. Likewise, the analysis of relationships carried out shows that different pieces of training can serve to further improve some perceptions of the students and consequently their future application in professional practice.

Additionally, it is necessary to emphasize the importance of this type of study in engineering. Literature and companies have evidenced a relationship between the profiles of engineers and their training in the university. Therefore, other authors are encouraged to address the needs of industry 4.0 from the university, not only through proposals but also through applied research that generates knowledge on how to address their training.

Finally, and despite the difficulties of implementing PLM programs in education (due to their cost and difficulty), the results of this research provide some suggestions for making possible PDLM training in the first year of engineering studies by using Autodesk Fusion360. Learning must be continuous and addressed as early as possible. At first, individual practices should focus on learning the basic tools of the chosen environment (both modeling and management), PLM knowledge and applicability. Later, the teacher will be vital to monitor and detect weaknesses in time. Likewise, it is essential to organize groups and roles to balance responsibilities and simulate from the beginning of the training the experience of a collaborative project.

Supplementary Materials: The following are available online at http://www.mdpi.com/2076-3417/9/21/4559/s1, Table S1: UPDLM ULL database.

Author Contributions: Conceptualization, J.L.S.; methodology, software investigation and resources, J.d.l.T.-C. and D.M.D.; formal analysis, V.L.-C.; original draft preparation of the manuscripts J.d.l.T.-C. and J.L.S.; writing—review and editing V.L.-C. and D.M.D.

Funding: This research received no external funding.

Conflicts of Interest: The authors declare no conflict of interest.

References

1. Koehn, E.E. Assessment of communications and collaborative learning in civil engineering education. *J. Prof. Issues Eng. Educ. Pract.* **2001**, *127*, 160–165. [CrossRef]
2. McCuen, R.H. Constructivist learning model for ethics education. *J. Prof. Issues Eng. Educ. Pract.* **1995**, *120*, 273–278. [CrossRef]
3. Rajala, S.A. Beyond 2020: Preparing engineers for the future. *Proc. IEEE* **2012**, *100*, 1376–1383. [CrossRef]
4. Garetti, M.; Terzi, S.; Bertacci, N.; Brianza, M. Organisational change and knowledge management in PLM implementation. *Int. J. Prod. Lifecycle Manag.* **2005**, *1*, 43–51. [CrossRef]
5. Jung, Y.W.; Lim, Y.K.; Kim, M.S. Possibilities and limitations of online document tools for design collaboration: The case of Google Docs. In Proceedings of the 2017 ACM Conference on Computer Supported Cooperative Work and Social Computing, Portland, OR, USA, 25 February–1 March 2017; pp. 1096–1108.
6. Hadjileontiadou, S.J.; Sakonidis, H.N.; Balafoutas, G.J. Lin2k: A novel web-based collaborative tool-application to engineering education. *J. Eng. Educ.* **2003**, *92*, 313–324. [CrossRef]
7. Moreno, L.; Gonzalez, C.; Castilla, I.; Gonzalez, E.; Sigut, J. Applying a constructivist and collaborative methodological approach in engineering education. *Comput. Educ.* **2007**, *49*, 891–915. [CrossRef]
8. do Carmo, B.B.T.; Pontes, R.L.J. Collaborative learning concept implementation through web. 2.0 tools: The case of industrial engineering fundamentals' discipline. *Int. J. Eng. Educ.* **2013**, *29*, 205–214.
9. Gillet, D.; Ngoc, A.V.N.; Rekik, Y. Collaborative web-based experimentation in flexible engineering education. *IEEE Trans. Educ.* **2005**, *48*, 696–704. [CrossRef]
10. Al-Samarraie, H.; Saeed, N.A. Systematic review of cloud computing tools for collaborative learning: Opportunities and challenges to the blended-learning environment. *Comput. Educ.* **2018**, *124*, 77–91. [CrossRef]

11. López-Pena, V.; López-Chao, V.A.; López-Chao, A. Analysis of teaching methods in graphic design in the Galician University System (GUS) in Spain. *Anthropologist* **2016**, *25*, 214–219. [CrossRef]
12. López-Chao, V.; López-Pena, V.; Ramiro-Aparicio, D. Beyond graphic expression: A diagnosis about the use of ICT teaching methods in engineering and architecture degrees. In Proceedings of the 12th international technology, education and development conference (INTED), Valencia, Spain, 5–7 March 2016; pp. 9290–9295.
13. Ramiro-Aparicio, D.; López-Pena, V.; López-Chao, V. Comparative study of the acquisition of ICT competence in graphic expression between the UDC and the US. In Proceedings of the 10th International Conference of Education, Research and Innovation, Seville, Spain, 16–18 November 2017; pp. 5150–5154.
14. Arri, Z.A.; Mujika, M.G.; Albisua, M.J.B.; Mendez, E.S. El desarrollo de habilidades profesionales en los estudios de ingeniería en la Universidad del País Vasco: ¿Aprendizaje basado en problemas o en proyectos? *DYNA* **2019**, *94*, 22–25. [CrossRef]
15. Saorin, J.L.; de La Torre, J.; Martín, N.; Carbonell, C. Education working group management using digital tablets. *Procedia Soc. Behav. Sci.* **2013**, *93*, 1569–1573. [CrossRef]
16. Segonds, F.; Maranzana, N.; Veron, P.; Aoussat, A. Collaborative reverse engineering design experiment using PLM solutions. *Int. J. Eng. Educ.* **2011**, *27*, 1037–1045.
17. Maranzana, N.; Segonds, F.; Buisine, S. Collaborative design tools in engineering education: Insight to choose the appropriate PLM software. *Int. J. Mech. Eng. Educ.* **2018**. [CrossRef]
18. Saorín, J.L.; Lopez-Chao, V.; de la Torre-Cantero, J.; Díaz-Alemán, M.D. Computer aided design to produce high-detail models through low cost digital fabrication for the conservation of aerospace heritage. *Appl. Sci.* **2019**, *9*, 2338. [CrossRef]
19. Vila, C.; Ugarte, D.; Ríos, J.; Abellán, J.V. Project-based collaborative engineering learning to develop Industry 4.0 skills within a PLM framework. *Procedia Manuf.* **2017**, *13*, 1269–1276. [CrossRef]
20. Autodesk Fusion 360: Fusion 360 for Hobbyist and Makers. Available online: https://www.autodesk.com/campaigns/fusion-360-for-hobbyists (accessed on 2 September 2019).
21. Brozina, C.; Sharma, A. Workshop: Implementing cloud collaboration using fusion 360 into a first-year engineering design course. In Proceedings of the FYEE Conference, Daytona Beach, FL, USA, 6–8 August 2017.
22. Rassovytska, M.; Striuk, A. Mechanical engineers training in using cloud and mobile services in professional activity. *CEUR Workshop Proc.* **2017**, *1844*, 348–359.
23. Leake, J.M.; Weightman, D.; Batmunkh, B. Digital prototyping by multidisciplinary teams. In Proceedings of the ASEE Annual Conference Exposition, Columbus, OH, USA, 25–28 June 2017.
24. Song, P.P.; Qi, Y.M.; Cai, D.C. Research and application of autodesk fusion360 in industrial design. In *IOP Conference Series: Materials Science and Engineering*; IOP Publishing: Bristol, UK, 2018; p. 012037.

Article

A Digital Twin for Automated Root-Cause Search of Production Alarms Based on KPIs Aggregated from IoT

Alexios Papacharalampopoulos, Christos Giannoulis, Panos Stavropoulos and Dimitris Mourtzis *

Laboratory for Manufacturing Systems and Automation, Department of Mechanical Engineering and Aeronautics, University of Patras, 26504 Patras, Greece; apapacharal@lms.mech.upatras.gr (A.P.); cgian@lms.mech.upatras.gr (C.G.); pstavr@lms.mech.upatras.gr (P.S.)

* Correspondence: mourtzis@lms.mech.upatras.gr; Tel.: +30-2610-997-314

Received: 25 February 2020; Accepted: 27 March 2020; Published: 31 March 2020

Abstract: A dashboard application is proposed and developed to act as a Digital Twin that would indicate the Measured Value to be held accountable for any future failures. The current study describes a method for the exploitation of historical data that are related to production performance and aggregated from IoT, to eliciting the future behavior of the production, while indicating the measured values that are responsible for negative production performance, without training. The dashboard is implemented in the Java programming language, while information is stored into a Database that is aggregated by an Online Analytical Processing (OLAP) server. This achieves easy Key Performance Indicators (KPIs) visualization through the dashboard. Finally, indicative cases of a simulated transfer line are presented and numerical examples are given for validation and demonstration purposes. The need for human intervention is pointed out.

Keywords: digital twin; decision support system; factor analysis; KPI; quantitative analysis; root-cause analysis

---

1. Introduction

Todays' manufacturing era focuses on monitoring the process on shop-floor by utilizing various sensorial systems that are based on data collection [1–3]. The automated systems directly collect an enormous amount of performance data from the shop-floor (Figure 1), and are stored into a repository, in a raw or accumulated form [4,5]. However, automated decision making in the era of Industry 4.0 is yet to be discovered. The current work is a step towards holistic alarms management (independently of context [6]), and, in particular, automated root-cause analysis.

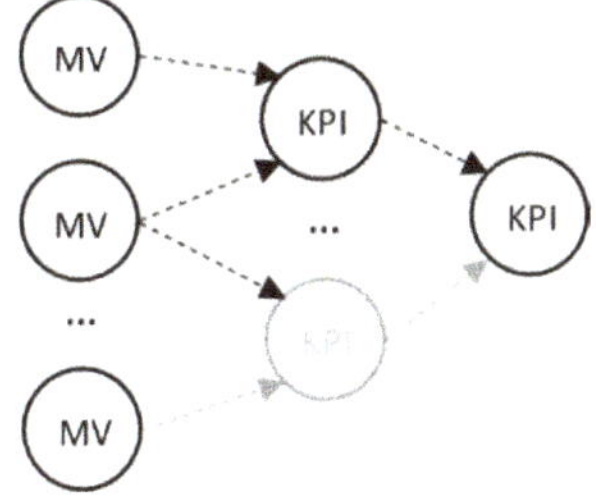

Figure 1. Key Performance Indications (KPIs) aggregation from Measured Values (MV).

1.1. KPIs and Digital Twin Platforms in Manufacturing

A lot of KPIs have been defined in literature (Manufacturing, Environmental, Design, and Customer) [7], and, despite the fact that methods have been developed for the precise acquisition of measured values [8], the interconnection of the values (Figure 1) can lead to mistaken decision making when one searches for the root cause of an alarm occurring in a manufacturing environment.

In addition, regarding manufacturing-oriented dashboard systems, they aid in the visualization of complex accumulations, trends, and directions of Key Performance Indications (KPIs) [9]. However, despite the situation awareness [10] that they offer, they are not able to support the right decision made at management level by elaborating automatically the performance metrics and achieving profitable production [11]. In the meantime, multiple criteria methods are widely used for decision support or the optimization of production, plant, or machine levels [1,12], based (mainly) on the four basic manufacturing attributes i.e.,: Cost, Time, Quality, and Flexibility levels [1]. This set of course can be extended. The "performance" that is collected from the shop-floor has been considered as measured values (MV) and their aggregation or accumulation into higher level Key Performance Indicators, as initially introduced by [13]. The goal of all these is the achievement of digital manufacturing [14].

On the other hand, the term digital twin can be considered as an "umbrella", and it can be implemented with various technologies beneath, such as physics [15], machine learning [16], and data/control models [17]. A digital twin deals with giving some sort of feedback back to the system and it varies from process level [15] to system level [18], and it can even handle design aspects [19].

Regarding commercial dashboard solutions, they enable either manual or automated input of measured values for the production of KPIs [20–22]. However, their functionality is limited to the reporting features of the current KPI values, typically visualized with a Graphical User Interface (GUI), usually cluttered, with gauges, charts, or tabular percentages. Additionally, there is need to incorporate various techniques from machine learning or Artificial Intelligence in general [23] and signal processing techniques [24]. The typical functionality that is found in dashboards serves as a visual display of the most important information for one or more objectives, consolidated and arranged on a single screen, so as for the information to be monitored at a glance [9]. This is extended in order to analyze the aggregated KPIs and explore potential failures in the future by utilizing monitored production performances. Finding, however, the root cause of a problem occurring, utilizing the real time data in an efficient and fast way, is still being pursued.

1.2. Similar Methods and Constrains on Applicability

In general, categories in root-cause finding are hard to be defined, but if one borrows the terminology from Intrusion Detection Alarms [25], they can claim that there are three major categories: anomaly detection, correlation, and clustering. Specific examples of all three in decision making are given below, in the next paragraph. The alternative classifications of the Decision Support Systems that are given below permit this categorization; the first classification [26] regards: (i) Industry specific packages, (ii) Statistical or numerical algorithms, (iii) Workflow Applications, (iv) Enterprise systems, (v) Intelligence, (vi) Design, (vii) Choice, and (viii) Review. Using another criterion, the second classification [27] concerns: (a) communications driven, (b) data driven, (c) document driven, (d) knowledge driven, and (e) model driven.

There are several general purpose methods that are relevant and are mentioned here to achieve root-cause finding. Root Cause Analysis is a quite good set of techniques, however, it remains on the descriptive empirical strategy level [28,29]. Moreover, scorecards are quite descriptive and empirical, while the Analytical Hierarchical Process (AHP) requires criteria definition [30]. Finally, Factor Analysis, requires a specific kind of manipulation/modelling due to its stochastic character [31].

Regarding specific applications in manufacturing-related decision making, usually finding that the root-cause has to be addressed through identifying a defect in the production. Defects can refer to either product unwanted characteristics, or resources' unwanted behaviour. Methods that have been previously used—regardless of the application—are Case Based Reasoning [32], pattern recognition [33,34], Analysis of Variance (ANOVA) [35], neural networks [36], Hypothesis testing [37], Time Series [38], and many others. However, none of these methods is quick enough to give the results from a deterministic point of view and without using previous measurements for training. Additionally, they are quite focused in application terms, which means that they cannot be used without re-calibration to a different set of KPIs. On the other hand, traditional Statistical Process Control (SPC) does not offer solutions without context, meaning the combination of the application and method [31].

1.3. Research Gaps & Novelty

In this work, an analysis method on KPIs has been implemented in the developed dashboard to identify, automatically and without prior knowledge, for a given KPI, the variables that are responsible for an undesired production performance. This mechanism is triggered by predicting a threshold exceedance in management level KPIs. The prediction utilizes (linear) regression, which is applied on the historical trend of each variable for the estimation of the performance values for the upcoming working period in order for the weaknesses of the production to be elicited beforehand. The dashboard in which the current methodology has been framed acts as an abstractive Digital Twin for production managers and it supports automated decision making.

In the following sections, the analysis approach is presented, followed by its implementation details within dashboard. The results from case studies, the points of importance, and the future research trends have been discussed.

2. Materials and Methods

2.1. The Description of the Flow and the Calculations

The current method consists of two stages. Initially, the formulas of each KPI are analyzed and their dependent variables are collected. Thus, the relationships between the Measured Values (MV), Intermediate Values (IV), Performance Indicators (PI), and Key Performance Indicators (KPIs) are known. This stage, in reality, constitutes the modelling of the production and it runs only once, while the data are aggregated form IoT technologies. The exact methodology can be found as IoT-Production in previous literature [39]. Regarding the second stage, Figure 2 presents its flow of the analysis for a single KPI. This stage runs continuously. In the beginning, the trend for each measure of the current period that is examined is acquired from the OLAP. For reasons of flexibility and modularity, they are not directly acquired from the data acquisition system. The measure trend is supplied as input to the tool, whereas a linear regression is applied to the estimation of the data points until the end of the next period, namely, horizon (as defined in detail of Figure 3). Once an estimated value is over the measure of the supposed goal (A1/A2), it is marked as 'out-of-goal' and the user is notified by the user interface. The user can then choose to repeat the process for lower level KPIs (turning it into investigated metric), after the automated suggestion of the analysis tool as to who is accounted for this result (B). The analysis method utilizes differentials; this mathematical approach that utilizes differentials is explained hereafter. The decisions that have to be taken (C) are beyond the scope of the current work. Nevertheless, it is noted that their generation often is straightforward, even though the existence of a knowledge base, to this end, would be extremely useful.

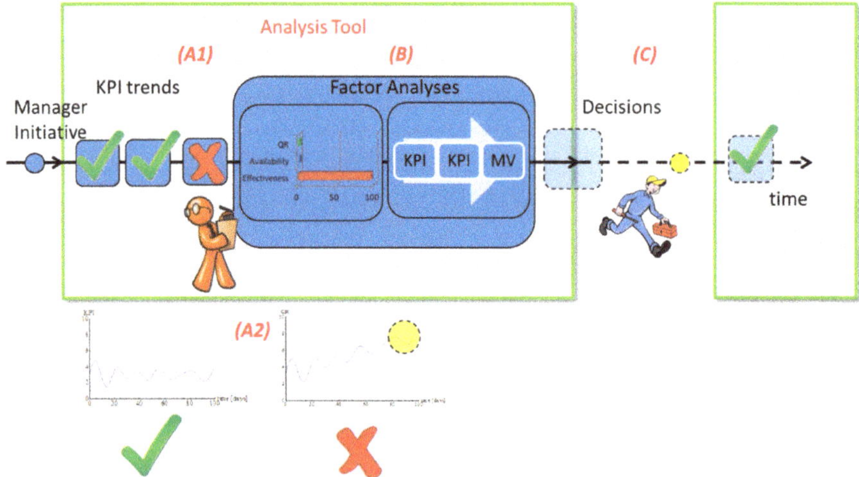

Figure 2. Schematic of the way the tool is run in real production.

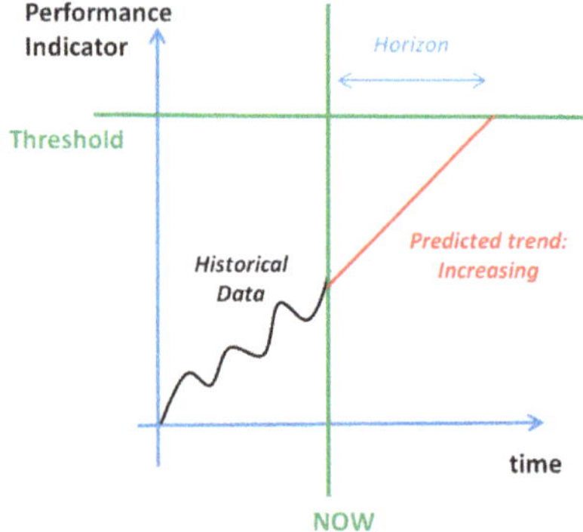

Figure 3. Alarm that is created by prediction of KPI exceeding threshold.

Figure 4 summarizes this procedure while using a flowchart. Additionally, for easy comprehension, the algorithmic description follows, along with some notes for each step.

1. All of the KPIs are checked for alarms (exceedance of specific value in the future) through regression. This KPI can be named an "investigated metric". KPIs historical values are retrieved with OLAP Queries. In continuation, (Linear) Extrapolation is used to predict the tendency and trigger alarms. To this end, thresholds and time horizons, as per Figure 3, have to be defined prior to running the tool, and their definition has been made via the production characteristics; for instance, in the case of monetary metrics, the desired profit is the criterion. In other cases, such as quality, tolerances are the basis for this definition.
2. If an investigated metric is found to create an alarm, all of its constituent PIs are evaluated in terms of contribution. This step is the Key Concept to the current algorithm and it is quite easy to run, as the relationship between the metrics is already known. Although, the exact relationship is required; in case only the constituents are known, the method cannot be applied. Additionally, the method used here is computationally light, as the relations are pure algebraic. At the time of the creation of the tool, the hierarchy of the indicators has been documented in a database-like structure, where information about the path is logged, including the operations.

In the context of a specific example, it is mentioned that Figure 5 is indicative of the graphical illustration of such information. The derivatives formulas are also pre-installed, as they are used during this step.

The partial differential is the tool that is applied within the analysis function used to estimate the cause of the variation. It is a powerful tool that deterministically quantifies the effect of the variation of one KPI to the variation of another KPI. More specifically, if KPI $A(t)$ is a function of PIs $A_1(t)$, $A_2(t)$, and $A_3(t)$, the percentage that each $A_n(t)$ contributes in the variation of $A(t)$ in time is the quantity

$$\delta_{A_n}^{A*} = \varepsilon\left\{\frac{\partial A}{\partial A_n}\Delta A_n\right\} \quad (1)$$

Given the corresponding notation, the operators of absolute value $|\cdot|$, mean valuc $\varepsilon\{\cdot\}$, partial derivative $\partial/\partial X$, and difference Δ are used for the computation of the mean partial differential. This helps in pointing out the direction that the production manager should focus on.

1. The constituent PI(s) that are found to cause this variation are turned into "investigated metric(s)" and #2 is run again. Its constituents KPIs are checked. Unless the investigated PIs are Measured Values, the loop continues. The operator can stop this loop at will at any level. However, in most cases, it is the Measured Value, which gives the maximum of information regarding the actions that have to be taken, as PIs of higher level depend on a variety of factors (the final case study is an appropriate case).
2. The MV(s) that has come up during the procedure is considered to be the root cause of the alarm. This result has been given without much effort, as in case other methods would have used, more information would be needed; SPC would require expert operators, AHP would require voting among experts, machine learning, ANOVA, and Hypothesis Testing would require exposure to similar situations and Time Series do not guarantee the optimal use of models.
3. Directives are given to the operators for actions through a knowledge base if it is not straight-forward.

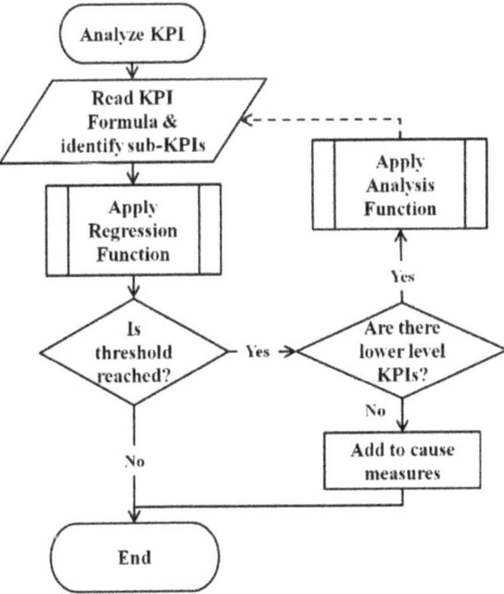

Figure 4. Flow chart of the algorithmic procedure.

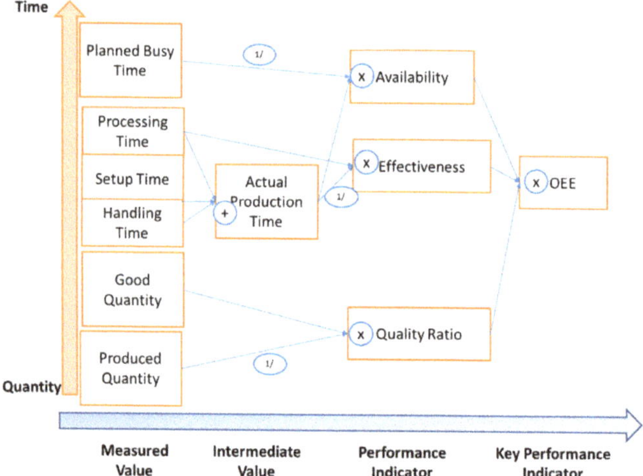

Figure 5. Relationship between Measured Values and KPI. Operations include summation, products and inversion.

2.2. Implementation within a Dashboard: Services and Hardware Framework

The dashboard is implemented in Java in an Object-Oriented (OO) paradigm as a Web Application that runs on a typical Java Servlet Container (Apache Tomcat); thus, it is accessible through a typical Internet browser. The Digital Twin is implemented in a Service-Oriented Architecture (SOA) that follows the N-Tier Architecture with multiple layers per tier. An integration layer handles the asynchronous communication of the browser with the server by enabling the user to begin the analysis at any time and without leaving the current screen. The browser sends HTTP requests to the server, whereas the JavaScript Object Notation (JSON) objects, which contain the results, are returned to the client.

Figure 6 illustrates the system's architecture comprising two individual processes. The Monitor Process records every equipment performance activity in the database, while the Actor directly interacts with the Analysis Process through the dashboard pages and requests an analysis for certain KPIs. Numbers 1 and 6 are indicative of communication between human (engineer/operator) and the dashboard, 2 and 5 denote a query request or result, and 3 and 4 are indicative of the OLAP/RDBMS communication.

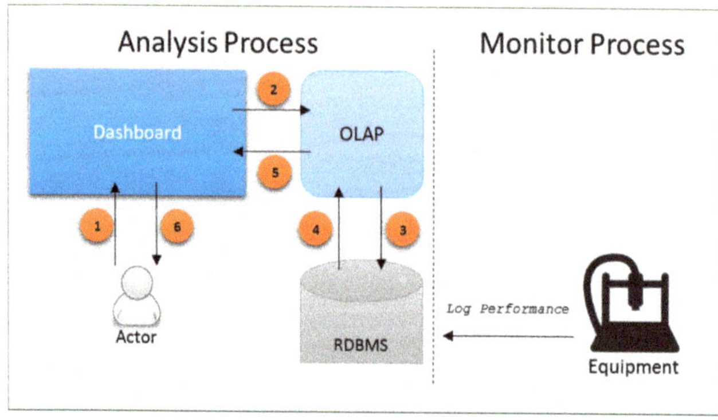

Figure 6. Application architecture.

In particular, the Actor interacts with the AnalyticAction of the dashboard, as depicted in Figure 7, which, in turn, cooperates with the AnalysisService, employing the Facade Design Pattern [40]. This architecture allows the usability of a software library and provides a context-specific interface launching a set of services without the user being exposed to the complexity of the algorithms beneath. This performed through a variety of entities. Firstly, the AnalysisService directly uses the OLAPQueryService to acquire the values of the KPIs requested, holds them in the KPIManager along with their formula expression stored into the database—this is due to an implementation limitation of the current OLAP servers providing the calculation function for a KPI. Next, the PredictionTool applies the regression function to the data, which are consecutively fed to the FactorAnalysisTool for the analysis of the influence factor on each dependent variable of the KPI. The PredictionTool and the FactorAnalysisTool both correspond the implementations of the extrapolation tool and the tool estimating differentials that are described in the previous section.

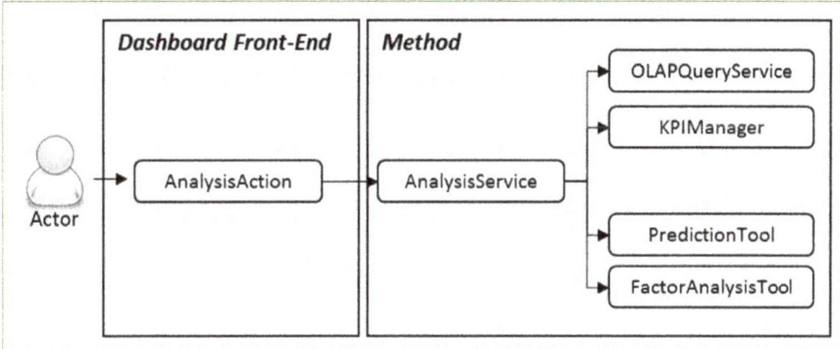

Figure 7. Unified Model Language (UML) of objects implemented in dashboard and their workflow.

The dashboard also uses the RDBMS to store the information that is required for the visualization of the customized KPI views in various forms. Each KPI is defined in a KPIDefinition instance along with its formula. The Measured Values are retained in the same structure but without a formula. Formulas are lexical names that correspond to definition names. For instance, the Availability KPI is defined by the lower case name 'availability', and it is referred to by the OEE KPI as #{availability} variable, whereas the interpreter replaces it at runtime with the actual value. The exact UML of the Informational Model used is given in Figure 8. Black rhombus indicates composition, white rhombus indicates aggregation, and pure arrow indicates association.

Historical data of the production's execution are stored into the relational database management system/RDBMS (Figure 6) as records regarding the depiction of the data acquisition method. A relational table holds the machine's cycle output, whereas each record represents a single machine cycle with attributes that relate to the corresponding entities in the production system. The smallest set of these attributes can be:

- The Part machined at that moment.
- The Process performed.
- The Equipment used in that cycle.
- The Operator of the equipment.
- The Work Order of the part machined.
- For each machine cycle what can be recorded is:
- The actual setup time spent for the preparation of the equipment.
- The actual processing time spent for the part's machining.
- The pre-processing time prior to machining being started.
- The post-processing time required for the part's unloading.

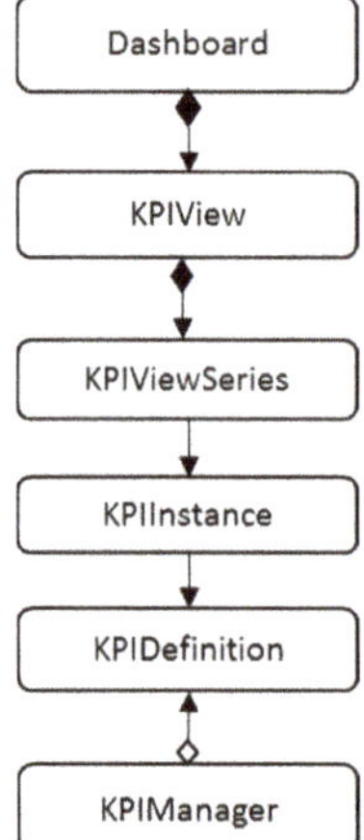

Figure 8. UML of the Informational Model.

These are attributes and records that mostly refer to productivity. Additional attributes, even of different character (i.e., sustainability related, such as energy consumption), can be added. A graph is then formed with involved KPIs, where each KPI is represented by a vertex v and their dependency by an edge e, similar notation to precedence diagrams. The analysis method exploits the data and follows their dependencies until the root measure has been identified (among Measured Values). Finally, that measure is presented to the user as an indication of the type of measure that will negatively affect the performance of production in the near future.

The production performance is loaded from the OLAP cube [41], where the measures are formed with the help of the aggregation functions. For instance, measure 'Good Quantity' is aggregated by the count function of the 'Good Part' attribute.

3. Results & Discussion

The functionality of the methodology that is presented in the above sections is proved herein utilizing adequately complex cases. Noise is also present in the first three cases, where dummy KPIs have been used, to simulate the short-time variations of the KPIs. These first case studies have been established in order to check the numerical success of the algorithm, where the fourth case study originates from a real industrial problem and its scope is to test the applicability of the digital twin and the validity of the main algorithm in problems set in environments of higher Technology Readiness Level.

3.1. Case I

In this case, the relationship between the KPI $B(t)$ and the PIs $A_1(t)$, $A_2(t)$, $A_3(t)$ is $B(t) = A_1(t)*A_2(t) + A_3(t)$. The evolution of the PIs trends are as follows:

1. $A_1(t)$ increases in time;
2. $A_2(t)$ is constant in time; and,
3. $A_3(t)$ is constant in time.

It is evident from the contribution diagram of Figure 9 that there is success in predicting that the contribution of the first factor should be high. The same happens if $A_1(t)$ decreases in time. Despite the noise existence, the algorithm worked to a satisfactory extent, which resulted in characterizing $A_1(t)$ as the root-cause of the variation.

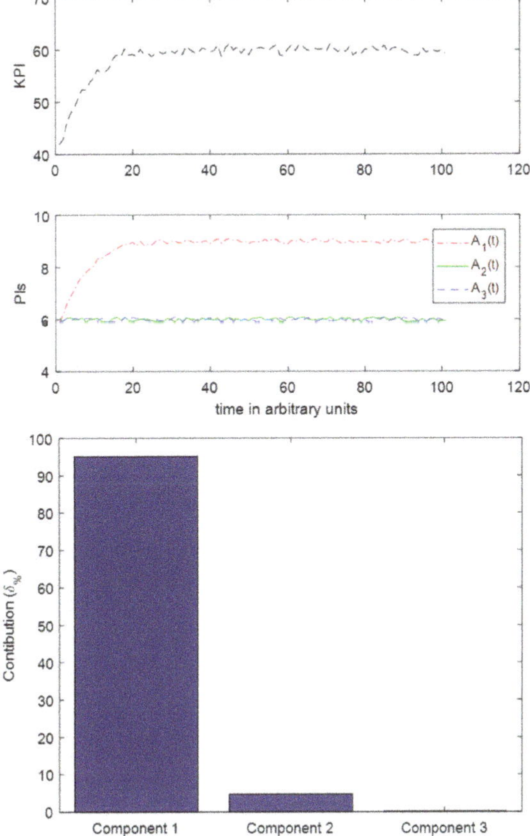

Figure 9. Case I. Top: Evolution of KPI, Middle: Evolution of PIs, Bottom, Contribution of each PI.

3.2. Case II

In this case, the relationship between the KPI $B(t)$ and the PIs $A_1(t)$, $A_2(t)$, $A_3(t)$ is $B(t) = A_1(t)*A_2(t) + A_3(t)$. The evolution of the PIs trends is, as follows:

1. $A_1(t)$ is constant in time;
2. $A_2(t)$ is constant in time; and,
3. $A_3(t)$ increases in time.

It is evident from the contribution diagram of Figure 10 that there is success in predicting that the contribution of the third factor should be high. The same happens if $A_3(t)$ decreases in time. It seems that the algorithm runs successfully, regardless of the operations that are performed among the KPIs.

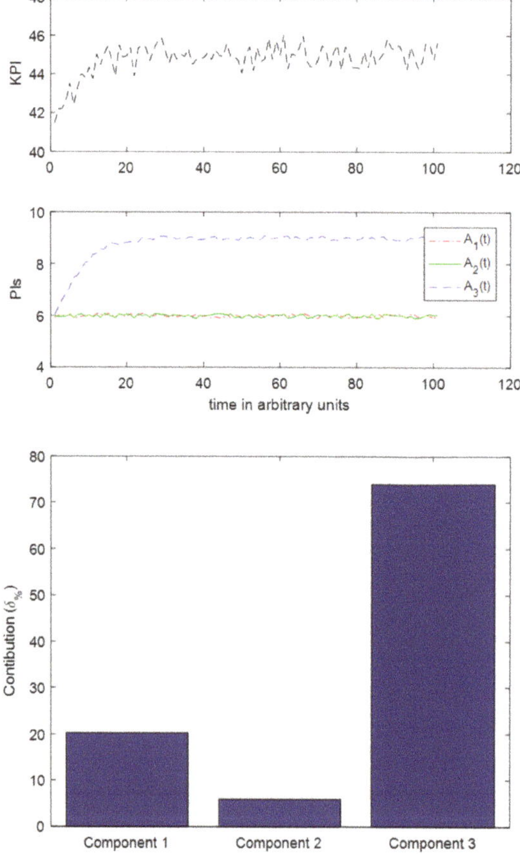

Figure 10. Case II. Top: Evolution of KPI, Middle: Evolution of PIs, Bottom, Contribution of each PI.

3.3. Case III

In this case, the relationship between the KPI $B(t)$ and the PIs $A_1(t)$, $A_2(t)$, and $A_3(t)$ is $B(t) = A_2(t)/A_1(t) + A_3(t)$. The evolution of the PIs trends is, as follows:

1. $A_1(t)$ decreases in time;
2. $A_2(t)$ increases in time; and,
3. $A_3(t)$ is constant in time.

It is evident from the contribution diagram in Figure 11 that there is success in predicting that the contribution of the first signal should be higher than that of the second one. It is apparent that both of the first two PIs are causes of the variation, however, the denominator change is the one that affects more the result. Accordingly, the algorithm has pointed towards the correct correction.

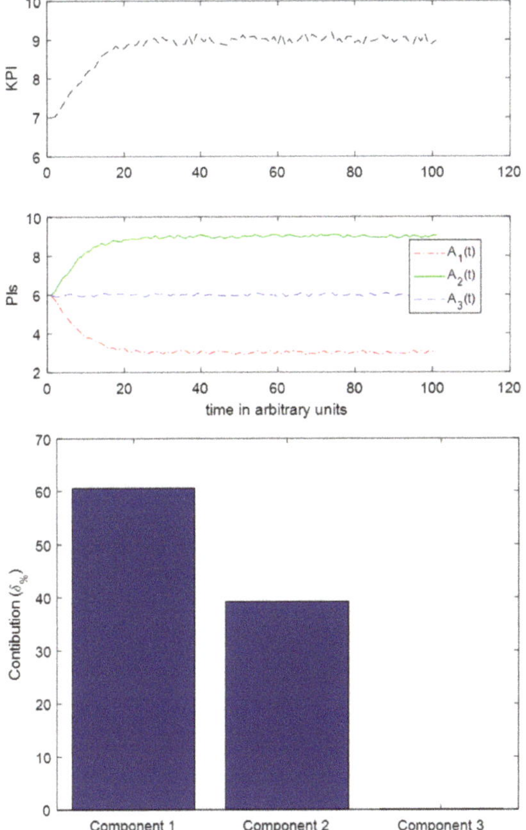

Figure 11. Case III. Top: Evolution of KPI, Middle: Evolution of PIs, Bottom, Contribution of each PI.

3.4. Case IV

This case regards a true production system and it is used to test the validity of the algorithm in terms of the physical relationship between the PIs. This is done through forcing the Measured Values to create an KPI-level alarm and check whether the algorithm will find the root cause and indicate towards the correct Measured Value. Thus, a production system of thirty (30) identical machines that are arranged in eight work-centers has been simulated by a commercial software [42] (Figure 12). The system simulates a real-life machine-shop producing two variants of Cylinder Head parts: (a) Petrol and (b) Diesel. Different processing times and setup times have been configured. Additionally, negative exponential distribution for Mean-Time-Between-Failures (MTBF), and a real distribution on Mean-Time-To-Repair (MTTR) considered the equipment breakdown on machines. Customer demand has been set by a normal distribution. Finally, the performances have been recorded and manually fed in the informational model.

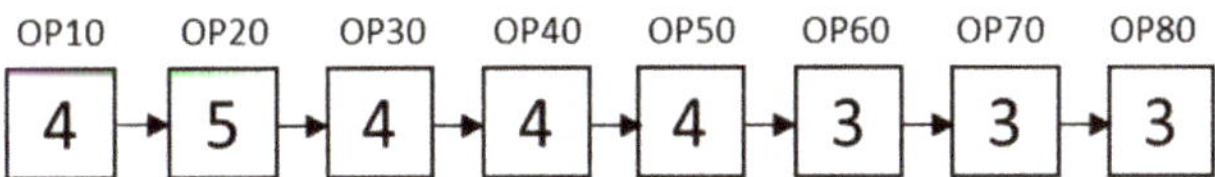

Figure 12. Workcenters arrangement of the production system. The number indicates the amount of machines for each workcenter.

The KPIs that are utilized in this case study (top to bottom) are mentioned below and their relationship is also given in Figure 5, beginning with Overall Equipment Efficiency (OEE).

- OEE = Availability × Effectiveness × Quality Rate
- Availability = Actual Production Time/Planned Busy Time
- Effectiveness = Processing Time/Actual Production Time
- Quality Ratio = Good Quantity/Produced Quantity
- Actual Production Time = Processing Time + Setup Time + Handling Time

A simulation has been conducted with the machine breakdowns deliberately increased with time. Subsequently, the algorithm of finding the root-cause is run once. In Figure 13, it is clearly shown (yellow shaded rectangles and curved arrows) that the algorithm dictates the major role of Effectiveness. This definitely helps in pointing out the direction that the production manager should focus on.

As a matter of fact, if one repeats the procedure of running the algorithm, but this time utilizing Effectiveness instead of OEE (68% contribution), and then the Measured Value that comes up is Processing Time. This means that the engineer should focus on why the machines do not work that much, which is indeed the root cause of the OEE drop (Figure 13).

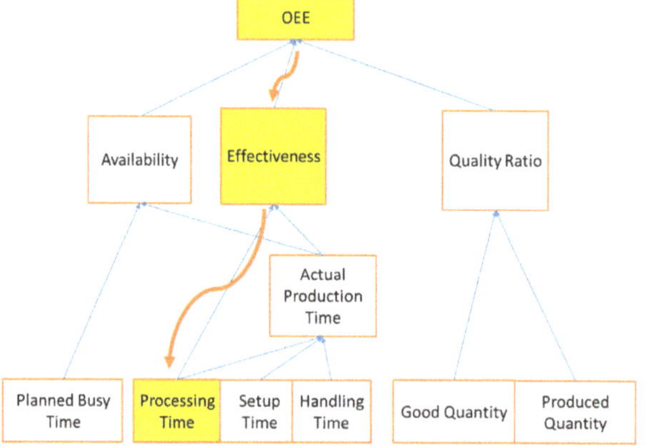

Figure 13. Tracking the cause of the OEE alarm back to Processing Time.

It seems that the algorithm works successfully without setting any rules or training. On the contrary, in the literature, works coming up often have to use one or the other, so a direct comparison cannot be made. For instance, in another alarm root cause detection system [43], the authors, even though they are describing the impact on case studies in brief, explicitly point out the need for rules in an internal expert system. Additionally, the use of KPIs at process level [44] is complementary to the current work. A framework that aims to detect the quality of the process, as per classifiers utilizing depth of cut and spindle rate could be integrated; corresponding indicators can be regarded as processed Measured Values. Additionally, the issue of validity of the indicators that are aggregated from IoT is also an issue that has not been raised herein, as the data have been considered to be valid. In addition to this, architectures that are based on complexity handling [45] and algorithms, such as Blockchain [46], can guarantee this goal, while, as indicated by corresponding results in literature [47], the introduction of a manipulation strategy, such as KPI-ML, will boost the performance of any algorithm, including the current. In other pieces of literature [48], the Multiple Alarms Matrix is utilized and hierarchical clustering is applied. In their results, they seem to be presenting the usability of the method towards the correlation of the alarms and extraction of the unique ones. This has not pointed out herein, as the goal has been to search for immediate causal relationships. Inferring Bayesian

Networks are utilized in a different work [49]; this approach can also be useful towards a causal model [50]. However, Abele et al., in their own discussion, state the use of probabilities and simulations, while expert knowledge is needed, whereas, in the current work, knowledge use is solely used in the actions generation. Furthermore, the enhancement with Max–Min Hill Climbing algorithm seems to be useful, however it is characterized as time consuming [51]. Case Based Reasoning seems to be similar to the present work, given the results in the literature [52]; however, as per the authors, it is best applied when records of previously successful solutions exist.

Moreover, the environment plays a significant role in the efficiency of such aggregation. More specifically, regarding manufacturing, large networks that share decision making, alternative bandwidth reservation [53] can be adopted for reasons of performance. Connection to control layer, as performed in other works [54], could be potentially addressed in an extension of the current study. Along the same lines, KPIs related to CPPS issues [55] could potentially extend the applicability of the current framework. Finally, regarding innovation and technology readiness, the upscale of technologies that are integrated in shop-floor is also relevant, in terms of decision making for designing systems and process planning integrating Industry 4.0 Key Enabling Technologies. The Technology Readiness Level of the various technologies could also be described in terms of KPIs, as per the literature [56].

4. Conclusions

This paper presents an approach for identifying the measured values that may negatively affect the production performance by eliciting historical data with the aid of Key Performance Indicators. Differentials have been adopted to this effect, being implemented in the developed dashboard and applied to a data set generated by simulation. To achieve this, performances have been stored into an RDBMS system and they have been aggregated through an OLAP server. The key advantage of the method is the quick identification of the root measured value, which will lead the production performance to undesirable directions from the ones expected by the managers, but in a non-supervised way. Therefore, it provides a technique for the development of functional dashboards beyond the visualization of accumulated indicators. Additionally, this paper proposes that dashboard applications, despite being complex and perhaps becoming cluttered up, could be used for automated decision support, as in the case of production performances. The results have shown that managers using the dashboard may be alerted about the weaknesses of their production performance before it is actually done at the shop-floor.

It seems that a digital twin for real time management of production alarms is feasible, and the algebraic notion of differentials is a powerful tool towards this direction. It covers several aspects of a digital twin; it uses the data themselves instead of elaborated models, it is deterministic towards variations (by default), it is immune to noise (at least in the described examples), and, of course, it can be used in loops of automated decision making. However, the use of Human-In-the-Loop Optimization techniques might be inevitable, as the actions derivation, as well as triggering of the tool, may have to include the knowledge that operators have accumulated.

Regarding extensions of the work, knowledge-based libraries for action identification have to be implemented as a continuation to the current approach. Additionally, it seems logical that, besides the mean value, moments of higher order, such as variance, skewness, and kurtosis, could be elaborated to assist decision making. However, the interpretation of these values exceeds the purposes of this study. Another extension is that of the differential order. Derivatives of any order, as well as integrators, could be used to enrich the information that is given to the user. In addition, it has to be mentioned that the definition of the relations between the KPIs is of great importance both to the calculation and to the interpretation of the differentials.

Author Contributions: Conceptualization, P.S. and D.M.; Formal analysis, P.S.; Methodology, A.P.; Software, A.P. and C.G.; Supervision, D.M.; Writing—original draft, A.P. and C.G.; Writing–review & editing, D.M. All authors have read and agreed to the published version of the manuscript.

Funding: This research received no external funding.

Conflicts of Interest: The authors declare no conflict of interest.

References

1. Chryssolouris, G. *Manufacturing Systems: Theory and Practice*; Springer: New York, NY, USA, 2006.
2. Larreina, J.; Gontarz, A.; Giannoulis, C.; Nguyen, V.K.; Stavropoulos, P.; Sinceri, B. Smart manufacturing execution system (SMES): The possibilities of evaluating the sustainability of a production process. In Proceedings of the 11th Global Conference on Sustainable Manufacturing, Berlin, Germany, 23–25 September 2013; pp. 517–522. [CrossRef]
3. Ding, K.; Chan, F.T.; Zhang, X.; Zhou, G.; Zhang, F. Defining a digital twin-based cyber-physical production system for autonomous manufacturing in smart shop floors. *Int. J. Prod. Res.* **2019**, *57*, 6315–6334. [CrossRef]
4. Stavropoulos, P.; Chantzis, D.; Doukas, C.; Papacharalampopoulos, A.; Chryssolouris, G. Monitoring and control of manufacturing processes: A review. *Procedia CIRP* **2013**, *8*, 421–425. [CrossRef]
5. Zendoia, J.; Woy, U.; Ridgway, N.; Pajula, T.; Unamuno, G.; Olaizola, A.; Fysikopoulos, A.; Krain, R. A specific method for the life cycle inventory of machine tools and its demonstration with two manufacturing case studies. *J. Clean. Prod.* **2014**, *78*, 139–151. [CrossRef]
6. Al-Kharaz, M.; Ananou, B.; Ouladsine, M.; Combal, M.; Pinaton, J. Evaluation of alarm system performance and management in semiconductor manufacturing. In Proceedings of the 6th International Conference on Control, Decision and Information Technologies (CoDIT), Paris, France, 23–26 April 2019; pp. 1155–1160.
7. Mourtzis, D.; Fotia, S.; Vlachou, E. PSS design evaluation via kpis and lean design assistance supported by context sensitivity tools. *Procedia CIRP* **2016**, *56*, 496–501. [CrossRef]
8. Peng, K.; Zhang, K.; Dong, J.; Yang, X. A new data-driven process monitoring scheme for key performance indictors with application to hot strip mill process. *J. Frankl. Inst.* **2014**, *351*, 4555–4569. [CrossRef]
9. Few, S. *Information Dashboard Design: The Effective Visual Communication of Data*; O'Reilly Media, Inc.: Sebastopol, CA, USA, 2006; Volume 3, Edv 161245. [CrossRef]
10. Ghimire, S.; Luis-Ferreira, F.; Nodehi, T.; Jardim-Goncalves, R. IoT based situational awareness framework for real-time project management. *Int. J. Comput. Integr. Manuf.* **2016**, *30*, 74–83. [CrossRef]
11. Epstein, M.J.; Roy, M.J. Making the business case for sustainability: Linking social and environmental actions to financial performance. *J. Corp. Citizsh.* **2003**, 79–96.
12. Apostolos, F.; Alexios, P.; Georgios, P.; Panagiotis, S.; George, C. Energy efficiency of manufacturing processes: A critical review. *Procedia CIRP* **2013**, *7*, 628–633. [CrossRef]
13. Kaplan, R.S.; Norton, D.P. *The Strategy-Focused Organization: How Balanced Scorecard Companies Thrive in the New Business Environment*; Harvard Business School Press: Boston, MA, USA, 2001. [CrossRef]
14. Mourtzis, D.; Papakostas, N.; Mavrikios, D.; Makris, S.; Alexopoulos, K. The role of simulation in digital manufacturing: Applications and outlook. *Int. J. Comput. Integr. Manuf.* **2015**, *28*, 3–24. [CrossRef]
15. Papacharalampopoulos, A.; Stavropoulos, P.; Petrides, D.; Motsi, K. Towards a digital twin for manufacturing processes: Applicability on laser welding. In Proceedings of the 13th CIRP ICME Conference, Gulf of Naples, Italy, 17–19 July 2019.
16. Athanasopoulou, L.; Papacharalampopoulos, A.; Stavropoulos, P. Context awareness system in the use phase of a smart mobility platform: A vision system utilizing small number of training examples. In Proceedings of the 13th CIRP Conference on Intelligent Computation in Manufacturing Engineering, Gulf of Naples, Italy, 17–19 July 2019.
17. Papacharalampopoulos, A.; Stavropoulos, P. Towards a digital twin for thermal processes: Control-centric approach. *Procedia CIRP* **2019**, *86*, 110–115. [CrossRef]
18. Gallego-García, S.; Reschke, J.; García-García, M. Design and simulation of a capacity management model using a digital twin approach based on the viable system model: Case study of an automotive plant. *Appl. Sci.* **2019**, *9*, 5567. [CrossRef]
19. Madni, A.M.; Madni, C.C.; Lucero, S.D. Leveraging digital twin technology in model-based systems engineering. *Systems* **2019**, *7*, 7. [CrossRef]
20. Klipfolio Site 2019. Available online: http://www.klipfolio.com/features#monitor (accessed on 10 February 2020).
21. KPI Monitoring 2014. Available online: http://www.360scheduling.com/solutions/kpi-monitoring/ (accessed on 23 October 2016).
22. SimpleKPI 2014. Available online: http://www.simplekpi.com/ (accessed on 23 October 2016).

23. Monostori, L. AI and machine learning techniques for managing complexity, changes and uncertainties in manufacturing. In Proceedings of the 15th Triennial World Congress, Barcelona, Spain, 21–26 July 2002; pp. 119–130. [CrossRef]
24. Teti, R. Advanced IT methods of signal processing and decision making for zero defect manufacturing in machining. *Procedia CIRP* **2015**, *28*, 3–15. [CrossRef]
25. Julisch, K. Clustering intrusion detection alarms to support root cause analysis. *ACM Trans. Inf. Syst. Secur.* **2003**, *6*, 443–471. [CrossRef]
26. Sheshasaayee, A.; Jose, R. A theoretical framework for the maintainability model of aspect oriented systems. *Procedia Comput. Sci.* **2015**, *62*, 505–512. [CrossRef]
27. Ada, Ş.; Ghaffarzadeh, M. Decision making based on management information system and decision support system. *Eur. Res.* **2015**, *93*, 260–269. [CrossRef]
28. Energy, U.S.D. *Doe Guideline—Root Cause Analysis*; US Department of Energy: Washington, DC, USA, 1992; DOE-NE-STD-1004-92.
29. Nelms, C.R. The problem with root cause analysis. In Proceedings of the 2007 IEEE 8th Human Factors and Power Plants and HPRCT 13th Annual Meeting, Monterey, CA, USA, 26–31 August 2007; pp. 253–258. [CrossRef]
30. Kurien, G.P. Performance measurement systems for green supply chains using modified balanced score card and analytical hierarchical process. *Sci. Res. Essays* **2012**, *7*, 3149–3161. [CrossRef]
31. Apley, D.W.; Shi, J. A factor-analysis method for diagnosing variability in mulitvariate manufacturing processes. *Technometrics* **2001**, *43*, 84–95. [CrossRef]
32. Mourtzis, D.; Vlachou, E.; Milas, N.; Dimitrakopoulos, G. Energy consumption estimation for machining processes based on real-time shop floor monitoring via wireless sensor networks. *Procedia CIRP* **2016**, *57*, 637–642. [CrossRef]
33. Masood, I.; Hassan, A. Pattern recognition for bivariate process mean shifts using feature-based artificial neural network. *Int. J. Adv. Manuf. Technol.* **2013**, *66*, 1201–1218. [CrossRef]
34. Stavropoulos, P.; Papacharalampopoulos, A.; Vasiliadis, E.; Chryssolouris, G. Tool wear predictability estimation in milling based on multi-sensorial data. *Int. J. Adv. Manuf. Technol.* **2016**, *82*, 509–521. [CrossRef]
35. Jin, J.; Guo, H. ANOVA method for variance component decomposition and diagnosis in batch manufacturing processes. *Int. J. Flex. Manuf. Syst.* **2003**, *15*, 167–186. [CrossRef]
36. Jeng, J.Y.; Mau, T.F.; Leu, S.M. Prediction of laser butt joint welding parameters using back propagation and learning vector quantization networks. *J. Mater. Process. Technol.* **2000**, *99*, 207–218. [CrossRef]
37. Koufteros, X.A. Testing a model of pull production: A paradigm for manufacturing research using structural equation modeling. *J. Oper. Manag.* **1999**, *17*, 467–488. [CrossRef]
38. Asakura, T.; Ochiai, K. Quality control in manufacturing plants using a factor analysis engine. *Nec Tech. J.* **2016**, *11*, 58–62.
39. Mourtzis, D.; Vlachou, K.; Zogopoulos, V. An IoT-based platform for automated customized shopping in distributed environments. *Procedia CIRP* **2018**, *72*, 892–897.
40. Freeman, E.; Freeman, E. *Head First Design Patterns*; O'Reilly Media, Inc.: Sebastopol, CA, USA, 2013. [CrossRef]
41. OLAP Server Site 2019. Available online: http://www.iccube.com/ (accessed on 10 February 2020).
42. Lanner Witness Site 2020. Available online: https://www.lanner.com/en-us/technology/witness-simulation-software.html (accessed on 10 February 2020).
43. Rollo, M.; Novák, P.; Kubalík, J.; Pěchouček, M. Alarm root cause detection system. In Proceedings of the International Conference on Information Technology for Balanced Automation Systems, Vienna, Austria, 27–29 September 2004; pp. 109–116.
44. Palacios, J.A.; Olvera, D.; Urbikain, G.; Elías-Zúñiga, A.; Martínez-Romero, O.; de Lacalle, L.L.; Rodríguez, C.; Martínez-Alfaro, H. Combination of simulated annealing and pseudo spectral methods for the optimum removal rate in turning operations of nickel-based alloys. *Adv. Eng. Softw.* **2018**, *115*, 391–397. [CrossRef]
45. Lee, J.; Noh, S.D.; Kim, H.J.; Kang, Y.S. Implementation of cyber-physical production systems for quality prediction and operation control in metal casting. *Sensors* **2018**, *18*, 1428. [CrossRef]
46. Fernández-Caramés, T.M.; Blanco-Novoa, O.; Froiz-Míguez, I.; Fraga-Lamas, P. Towards an autonomous industry 4.0 warehouse: A uav and blockchain-based system for inventory and traceability applications in big data-driven supply chain management. *Sensors* **2019**, *19*, 2394.

47. Brandl, D.L.; Brandl, D. KPI Exchanges in smart manufacturing using KPI-ML. *IFAC-PapersOnLine* **2018**, *51*, 31–35. [CrossRef]
48. Chen, Y.; Lee, J. Autonomous mining for alarm correlation patterns based on time-shift similarity clustering in manufacturing system. In Proceedings of the IEEE Conference on Prognostics and Health Management, Denver, CO, USA, 20–23 June 2011; pp. 1–8.
49. Abele, L.; Anic, M.; Gutmann, T.; Folmer, J.; Kleinsteuber, M.; Vogel-Heuser, B. Combining knowledge modeling and machine learning for alarm root cause analysis. *IFAC Proc. Vol.* **2013**, *46*, 1843–1848. [CrossRef]
50. Büttner, S.; Wunderlich, P.; Heinz, M.; Niggemann, O.; Röcker, C. Managing complexity: Towards intelligent error-handling assistance trough interactive alarm flood reduction. In Proceedings of the International Cross-Domain Conference for Machine Learning and Knowledge Extraction, Reggio, Italy, 29 August–1 September 2017; pp. 69–82.
51. Wunderlich, P.; Niggemann, O. Structure learning methods for Bayesian networks to reduce alarm floods by identifying the root cause. In Proceedings of the 22nd IEEE International ETFA Conference, Limassol, Cyprus, 12–15 September 2017; pp. 1–8.
52. Amani, N.; Fathi, M.; Dehghan, M. A case-based reasoning method for alarm filtering and correlation in telecommunication networks. In Proceedings of the Canadian Conference on Electrical and Computer Engineering, Saskatoon, SK, Canada, 1–4 May 2005; pp. 2182–2186.
53. Zuo, L.; Zhu, M.M.; Wu, C.Q.; Hou, A. Intelligent bandwidth reservation for big data transfer in high-performance networks. In Proceedings of the IEEE International Conference on Communications, Kansas City, MO, USA, 20–24 May 2018; pp. 1–6.
54. Contuzzi, N.; Massaro, A.; Manfredonia, I.; Galiano, A.; Xhahysa, B. A decision making process model based on a multilevel control platform suitable for industry 4.0. In Proceedings of the 2019 II Workshop on Metrology for Industry 4.0 and IoT, Naples, Italy, 4–6 June 2019; pp. 127–131.
55. Lüder, A.; Schmidt, N.; Hell, K.; Röpke, H.; Zawisza, J. Description means for information artifacts throughout the life cycle of CPPS. In *Multi-Disciplinary Engineering for Cyber-Physical Production Systems*; Springer: Berlin/Heidelberg, Germany, 2017; pp. 169–183.
56. Sastoque Pinilla, L.; Llorente Rodríguez, R.; Toledo Gandarias, N.; López de Lacalle, L.N.; Ramezani Farokhad, M. TRLs 5–7 advanced manufacturing centres, practical model to boost technology transfer in manufacturing. *Sustainability* **2019**, *11*, 4890. [CrossRef]

Article

Statistical Process Control with Intelligence Based on the Deep Learning Model

Tao Zan, Zhihao Liu *, Zifeng Su, Min Wang, Xiangsheng Gao and Deyin Chen

Beijing Key Laboratory of Advanced Manufacturing Technology, Beijing University of Technology, Beijing 100124, China; zantao@bjut.edu.cn (T.Z.); s201801048@emails.bjut.edu.cn (Z.S.); wangm@bjut.edu.cn (M.W.); gaoxsh@bjut.edu.cn (X.G.); chendeyin@emails.bjut.edu.cn (D.C.)

* Correspondence: liuzhihao@emails.bjut.edu.cn

Received: 5 December 2019; Accepted: 30 December 2019; Published: 31 December 2019

Abstract: Statistical process control (SPC) is an important tool of enterprise quality management. It can scientifically distinguish the abnormal fluctuations of product quality. Therefore, intelligent and efficient SPC is of great significance to the manufacturing industry, especially in the context of industry 4.0. The intelligence of SPC is embodied in the realization of histogram pattern recognition (HPR) and control chart pattern recognition (CCPR). In view of the lack of HPR research and the complexity and low efficiency of the manual feature of control chart pattern, an intelligent SPC method based on feature learning is proposed. This method uses multilayer bidirectional long short-term memory network (Bi-LSTM) to learn the best features from the raw data, and it is universal to HPR and CCPR. Firstly, the training and test data sets are generated by Monte Carlo simulation algorithm. There are seven histogram patterns (HPs) and nine control chart patterns (CCPs). Then, the network structure parameters and training parameters are optimized to obtain the best training effect. Finally, the proposed method is compared with traditional methods and other deep learning methods. The results show that the quality of extracted features by multilayer Bi-LSTM is the highest. It has obvious advantages over other methods in recognition accuracy, despite the HPR or CCPR. In addition, the abnormal patterns of data in actual production can be effectively identified.

Keywords: statistical process control; pattern recognition; long short-term memory; feature learning; control chart; histogram

---

1. Introduction

Since 1920s, statistical process control (SPC) theory has played an important role in product quality improvement and quality supervision [1]. SPC mainly uses a statistical analysis method to monitor the production process, and scientifically distinguishes the random fluctuation and abnormal fluctuation of product quality in the production process [2,3]. Thus, the abnormal trend of production process is expected, so that production managers can take timely measures to eliminate abnormalities and restore the stability of the process, so as to achieve the purpose of improving and controlling the quality. Intelligent SPC data analysis is realized, which can reduce production time and cost and improve product quality. It will likely become an integral part of industry 4.0 technology. Among the many theories of SPC, control charts and histograms are the most important and practical visual indicators [4]. If the manufacturing process is affected not only by random factors, but also by other specific factors, there will be specific abnormal patterns in the control chart and histogram [5]. For example, as far as the control chart is concerned, the cyclic patterns might be related to the periodic variation in the power supply. Trend patterns may be related to factors that change slowly, while changes in raw materials, workers and machine tools may cause shift patterns.

In the early application of control charts, it is necessary to manually determine whether or not there is any abnor–mality in the control charts and what kind of abnormality occurs. It is easy to detect abnormalities beyond the control limit, but difficult to identify abnormal patterns within the control limit, which is easily affected by the experience level of quality control personnel [6]. Then, many non-natural pattern detection methods based on supplementary rules were proposed by scholars [7–9]. The Nelson criterion is one of these methods [10]. However, control charts contain a lot of information about the production process. Supplementary rules cannot describe the specific modes of a process in detail. In addition, a large number of rules are used, which is not conducive to the real-time monitoring of production process, and will cause many false alarms [11,12]. In order to realize automated control chart pattern recognition (CCPR), scholars have designed a series of expert systems so that quality management personnel can take timely remedial measures for uncontrolled manufacturing process [13–16]. They are mainly based on statistical tests and heuristic algorithms [17]. Because of the shortcomings of these judgment rules, early expert systems cannot get satisfactory results [18]. This stimulates interest in developing more accurate CCPR algorithms.

Because machine learning algorithms have strong pattern recognition capabilities, their applications in the CCPR field have received more and more attention and achieved some success. It mainly includes an artificial neural network (ANN) and a support vector machine (SVM). The multilayer perceptron (MLP) are the most widely used in ANN-based methods [19–21]. In [22], an effective CCPR system based on MLP is studied. The effects of different layers of MLP and different training algorithms on the results are also compared. It is found that the resilient back-propagation (RBP) algorithm has the best convergence speed and the Levenberg–Marquardt (LM) algorithm has the best optimization effect. In [23], a new learning algorithm based on bees algorithm is adopted, and an optimized radial basis function neural network (RBFNN) is trained, which shows good performance in CCPR tasks. In addition, the probability neural network (PNN) [24], spiking neural network (SNN) [25] and learning vector quantization (LVQ) [26–28] are also widely used. They all belong to supervised learning algorithms. At the same time, unsupervised algorithms such as self-organizing mapping (SOM) [29] and adaptive resonance theory (ART) are also used in CCPR. Because of its shallow structure and limited learning ability, ANN has some shortcomings, such as difficulty in convergence and can easily fall into local minimum, which limits its further development. Subsequently, SVM and its variants have been proposed by scholars and used to solve CCPR problems. For example, weighted SVM [30], multi-kernel SVM [31] and fuzzy SVM [1]. They generally show better recognition accuracy than traditional ANN.

Histogram is a graphical representation of the distribution of quality data. Its distribution degree is a common tool for judging and predicting the quality of production process. Similar to the control chart pattern (CCP), the abnormality of the histogram pattern (HP) corresponds to specific factors. For example, skewed patterns are caused by poor processing habits of workers, and flat-top patterns indicate that production processes are affected by slow-changing factors. As far as we know, only one paper has studied histogram pattern recognition (HPR) in SCI and EI database in recent years [4]. In [4], an HPR method based on a fuzzy ART neural network is proposed. The advantage of this method is that it can cluster HPs adaptively and generate new classes for unknown patterns. At the same time, there is a big risk to classify the known patterns into new categories, which is not conducive to practical application. Moreover, the simulation data used in this paper are not random enough. The quality data of real processes cannot be fully simulated. The lack of research on HPR is a major drawback of SPC intelligent research. This is because histograms and control charts complement each other, as shown in Figure 1. According to the theory of a control chart, the quality data is normal. However, its histogram pattern is an island pattern, which means that the production process is abnormal. Control charts are suitable for identifying short-term anomalies, while histograms can reflect the long-term distribution of quality data. Therefore, the role of a histogram cannot be ignored, and enterprises are also in urgent need of accurate automatic HPR.

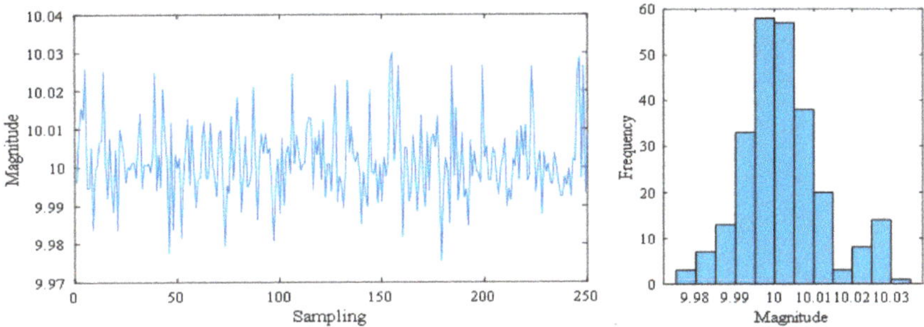

Figure 1. Quality data and its histogram.

It is well known that in the field of pattern recognition, the input form of data has a great influence on classification, which is called the input representation problem. One form is to take the raw data as input [24,32,33], such as quality data in the control chart and the frequency of each interval in the histogram. Another form is to take the features extracted from the raw data as input, such as wavelet features [11], shape features [19,34,35] and statistical features [23,34]. The latter is called feature engineering, that is, experts design favorable features for pattern recognition problems based on experience. Most scholars' research shows that if the dimension of input data is too large, the classifier size of traditional machine learning algorithm will be too large, which is unfavorable. Its accuracy and efficiency are often lower than the feature-based method, because the dimension of feature set is usually very small [11,29]. There is no doubt that the key to improve classification accuracy is to select the most advantageous feature set. To achieve this goal, a genetic algorithm (GA) [1] and local linear embedding (LLE) [36] are used to optimize features from high-dimensional feature sets. However, they only select good features from the known expert feature set, and do not improve the quality of feature set from the root. The discarded raw data still has potential values, so more advanced methods, such as feature learning, still need to be used. Feature learning refers to adaptively extract the most advantageous features from the raw data by using the deep neural network (DNN), without relying on any expert experience. Since the feature set is obtained by learning and minimizing the loss function, it can be considered that this feature set is the best choice for this classification task [37]. The most representative deep learning algorithms include the deep belief network (DBN) [38,39] based on the restricted Boltzmann machine (RBM), the convolutional neural network (CNN) [40–42] based on the convolution layer and pooling layer and the recurrent neural network (RNN) [43,44] based on the recursive layer. They have made remarkable achievements in machine vision, natural language processing and fault diagnosis [45].

On the basis of relevant research in recent years [18,46,47], this paper proposes to use multilayer bidirectional long short-term memory networks (Bi-LSTM) to learn the features of histogram and control chart, and finally realize HPR and CCPR. The Bi-LSTM is an improved method of RNN. Its special gate structure enables it to capture both short-term dependencies and long-term dependencies. This paper is an extension of previous studies [18]. Different from previous studies, multilayer Bi-LSTM is used to replace the former one-dimensional CNN (1D-CNN), because it is specially used to process one-dimensional data such as time series, and has a stronger ability to process the relationship between before and after the sequence. It has also been compared with the traditional method and many typical deep learning methods. In addition, besides automatic CCPR, accurate HPR is also realized. Until now, HPR has been ignored by scholars. In this way, the automation and intelligence level of SPC has been improved more comprehensively. Furthermore, the six CCPs were studied before they were expanded to nine, making the CCPR more comprehensive and refined. In order to detect anomalies in production as quickly as possible, the length of the data we use for pattern recognition is 25. This is because longer data may mean that more defective products have been produced in the factory when abnormalities are identified [18].

The rest of this paper is organized as follows. Section 2 explains the basic structure of LSTM and the mathematical representation of HPs and CCPs. Section 3 introduces the details and process of the proposed method. Section 4 carries on the experiment and completes the related discussion. Finally, Section 5 summarizes the paper and gives the prospect of future work.

2. Methodology

2.1. Simulation Method of Histogram Patterns

Seven typical HPs of the production process are shown in Figure 2, including normal (NOR) patterns, bimodal (BIM) patterns, left and right island (LI and RI) patterns, left and right skew (LS and RS) patterns and flat top (FT) patterns.

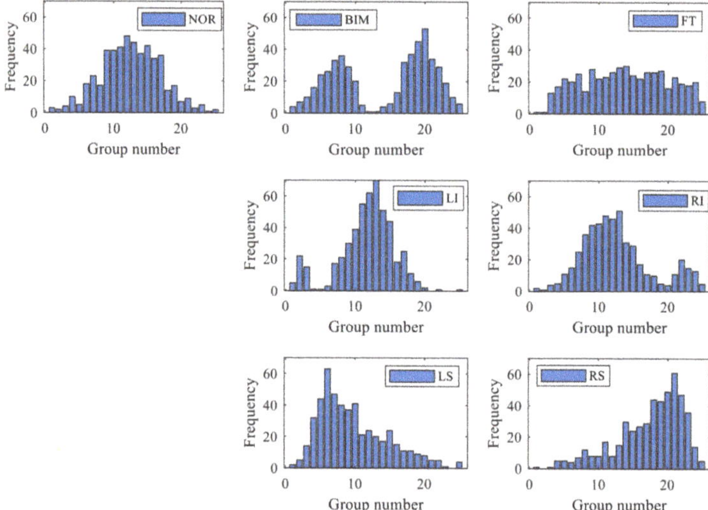

Figure 2. Seven typical histogram patterns.

The Monte Carlo simulation algorithm is recognized as a HP simulation method in the SPC field [48]. In this paper, the method is used to simulate the quality data of each HP.

If $y(t)$ is the value of quality data measured at t, then the quality data in the NOR pattern is normal distribution:

$$y(t) \sim N\left(\mu, \sigma^2\right) \tag{1}$$

where μ is the mean of quality data under controlled conditions, σ is the standard deviation.

The quality data of BIM pattern is a combination of two normal distributions, which can be simulated by the following formula [48]:

$$y(t) \sim aN\left(\mu - b_1, \left(\frac{3\sigma - b_1}{3}\right)^2\right) + (1-a)N\left(\mu + b_2, \left(\frac{3\sigma - b_2}{3}\right)^2\right) \tag{2}$$

where, a and $(1-a)$ are the proportion of two normal distributions, and b_1 and b_2 are the distance between the center of two normal distributions and μ.

The LI and RI pattern can be seen as a small normal distribution next to the normal distribution., which can be simulated by the following formula:

$$y(t) \sim aN\left(\mu, \sigma^2\right) + (1-a)N\left(\mu \pm b, \left(\frac{3\sigma - b}{3}\right)^2\right) \tag{3}$$

where, a and $(1-a)$ are the proportion of two normal distributions, and b is the distance between the center of small normal distributions and μ; "+" represents RI pattern, and "−" represents LI pattern.

The quality data of LS and RS pattern can be combined by several normal distributions [48], which can be simulated by the following formula [48]:

$$y(t) \sim \sum_{i=0}^{m-1} \frac{1}{m} N\left(\mu \pm 3\sigma\left(\left(\frac{2}{3}\right)^i - 1\right),\ \left(\left(\frac{2}{3}\right)^i \sigma\right)^2\right) \tag{4}$$

where, m is the number of normal distributions; "+" represents LS pattern, and "−" represents RS pattern.

FT pattern quality data can be formed by mixing uniform distribution with normal distribution, which can be simulated by the following formula:

$$y(t) \sim aN\left(\mu,\ \sigma^2\right) + (1-a)U(\mu - 3\sigma,\ \mu + 3\sigma) \tag{5}$$

where, a and $(1-a)$ are the proportion of normal distribution and uniform distribution respectively.

2.2. Simulation Method of Control Chart Patterns

Nine typical CCPs of the production process are shown in Figure 3, including normal (NOR) patterns, cycle (CYC) patterns, systematic (SYS) patterns, upward and downward trend (UT and DT) patterns, stratification (STA) patterns, upward and downward shift (US and DS) patterns and mixture (MIX) patterns.

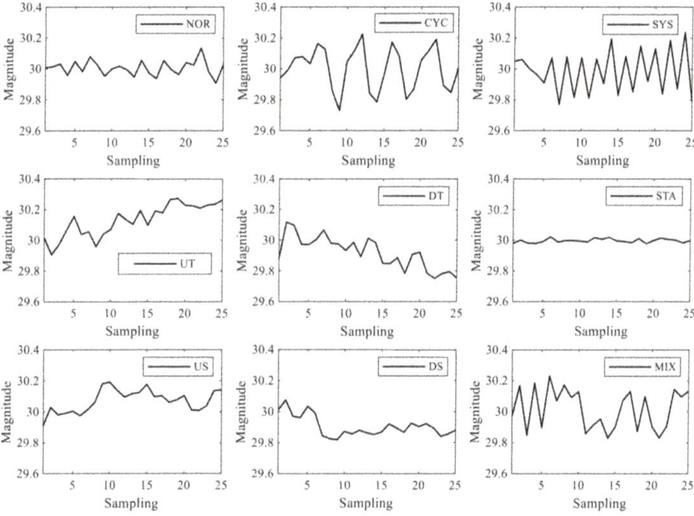

Figure 3. Nine typical CCPs.

Equation (6) is a general formula for simulating various CCP, which includes the process mean and two noise components [5]: $x(t)$ is random noise and $d(t)$ is a special fluctuations from specific factors in manufacturing process.

$$y(t) = \mu + x(t) + d(t) \tag{6}$$

where $y(t)$ is the quality data at time t. μ is the mean value of the product. Random noise $x(t)$ obeys normal distribution, $x(t) \sim N\left(0,\ \sigma^2\right)$.

A detailed description of the simulation methods of nine typical CCPs can be found in Section 4.6 or [49]. We will not repeat the description here.

2.3. *Long Short-Term Memory Network Model*

In recent years, deep learning has made remarkable achievements in various fields. The deep learning algorithm has advantages that the traditional shallow machine learning algorithm does not have, such as complex data preprocessing and feature engineering are no longer needed. The raw data can be used as the input of the model. The deep-seated of neural network and the special design of network structure make it can learn the potential deeper knowledge from the raw data and be competent for more complex tasks. The features it learns are the most suitable for this classification task and fares better than the features designed by human experts.

LSTM is a typical deep learning model and a variant of RNN. The basic idea is still to take the previous output of the network as the next input, as shown in Equation (7). Where $\boldsymbol{x}_t$ is the tth input of the network, $\boldsymbol{h}$ is the output of the network, and H is the nonlinear transformation function.

$$\boldsymbol{h}_t = H(\boldsymbol{x}_t, \boldsymbol{h}_{t-1}) \tag{7}$$

Compared with traditional RNN, LSTM can capture long-distance dependency. It is different from the standard RNN in structure. LSTM adds some gate structures for each cell, which allows information to pass selectively. It can be understood as a mechanism of feature learning selection and update [45]. The cell structure of LSTM is shown in Figure 4.

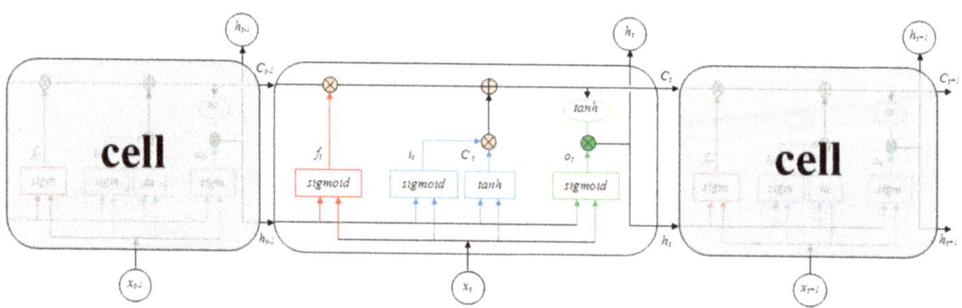

Figure 4. Typical LSTM cell structure.

The forget gate (the red part in Figure 4) is used to decide what information will be discarded from the cell state $\boldsymbol{C}_{t-1}$. A forget gate is mathematically represented by [45]:

$$\boldsymbol{f}_t = sigmoid\left(\boldsymbol{W}_f \cdot [\boldsymbol{x}_t, \boldsymbol{h}_{t-1}] + \boldsymbol{b}_f\right) \tag{8}$$

The input gate (the blue part in Figure 4) is used to determine what new information $\boldsymbol{C}'_t$ will be stored in the cell state. A input gate is mathematically represented by [45]:

$$\boldsymbol{i}_t = sigmoid(\boldsymbol{W}_i \cdot [\boldsymbol{x}_t, \boldsymbol{h}_{t-1}] + \boldsymbol{b}_i) \tag{9}$$

$$\boldsymbol{C}'_t = tanh(\boldsymbol{W}_c \cdot [\boldsymbol{x}_t, \boldsymbol{h}_{t-1}] + \boldsymbol{b}_c) \tag{10}$$

Then the cell state is updated, and the decisions made in the previous steps are implemented to get a new cell state $\boldsymbol{C}_t$. The mathematical representation of the state update is as follows:

$$\boldsymbol{C}_t = \boldsymbol{f}_t * \boldsymbol{C}_{t-1} + \boldsymbol{i}_t * \boldsymbol{C}'_t \tag{11}$$

The output gate (the green part in Figure 4) determines the final output h_t according to the updated cell state. An output gate is mathematically represented by [45]:

$$o_t = sigmoid(W_o \cdot [x_t, h_{t-1}] + b_o) \tag{12}$$

$$h_t = o_t * tanh(C_t) \tag{13}$$

where W_f, W_i, W_c and W_o represent the weight of forget gate, input gate, current cell and output gate respectively. b_f, b_i, b_c and b_o represent the bias.

The final output h_T of the last cell of LSTM is the feature extracted adaptively from the raw data (where T refers to the last cell). The final recognition can be achieved by inputting this feature into the traditional ANN. The last layer is the output layer, whose activation function is Softmax, and the number of neurons is N, representing the number of types that want to identify patterns. Finally, the stochastic gradient descent (SGD) algorithm is used to optimize the weight (W_f, W_i, W_c and W_o) and bias (b_f, b_i, b_c and b_o) in LSTM. Commonly used SGD methods include SGD with momentum (SGDM), root mean square propagation (RMSProp), and adaptive moment estimation (Adam) [50].

In this paper, Bi-LSTM is used to complete CCPR and HPR, which is slightly different from standard LSTM, that is, Bi-LSTM consists of forward LSTM and backward LSTM. More details of the Bi-LSTM structure are introduced in the following sections.

3. Proposed Method

This section describes the details of the proposed method. The input of the LSTM is the unprocessed data, such as quality data in the control chart and the frequency of each interval in the histogram, and the output is the category of the pattern. Feature extraction, and feature selection are all completed by LSTM through learning. Compared with unidirectional LSTM, Bi-LSTM can learn the front and back relation of sequence better.

Figure 5 is a structural diagram of the proposed method. It has a bidirectional recursive layer as well as a multilayer structure. When the input of the model passes through a multilayer structure, the information transmitted by each layer will be represented in different dimension spaces. Therefore, data is gradually learned by increasing the number of layers of the network. The connection between input and output is improved to better describe the characteristics of the system [51]. In other words, bidirectional and multilayer recursive structure can raise the learning space and flexibility of the model [45]. Each small square in Figure 5 represents a LSTM cell.

The network consists of multilayer Bi-LSTM and a Softmax classifier. The former is used to extract features from the raw data, and the latter is used to classify various patterns. The input of Softmax layer is the combination of the last forward LSTM output $h_T^{\rightarrow}$ and the last backward LSTM output $h_T^{\leftarrow}$, as shown in Equation (14). The initial learning rate was 0.05, and each LSTM cell has 10 neurons. The next section discusses and optimizes other structural and training parameters of multilayer Bi-LSTM, such as optimization algorithm, training batch size and network layer number.

$$h_T = \left[h_T^{\rightarrow}, h_T^{\leftarrow}\right] \tag{14}$$

Because of the end-to-end ability of deep learning method, the specific implementation process of the method proposed in this paper is as follows. Firstly, the Monte Carlo algorithm is used to simulate training set and test set of HPR and CCPR respectively. The one-hot encoding method is used to label the samples. Then, training sets are used to train two Bi-LSTMs, which are used for HPR and CCPR respectively. Finally, the performance of the optimized two Bi-LSTMs are verified by the test set and real production data.

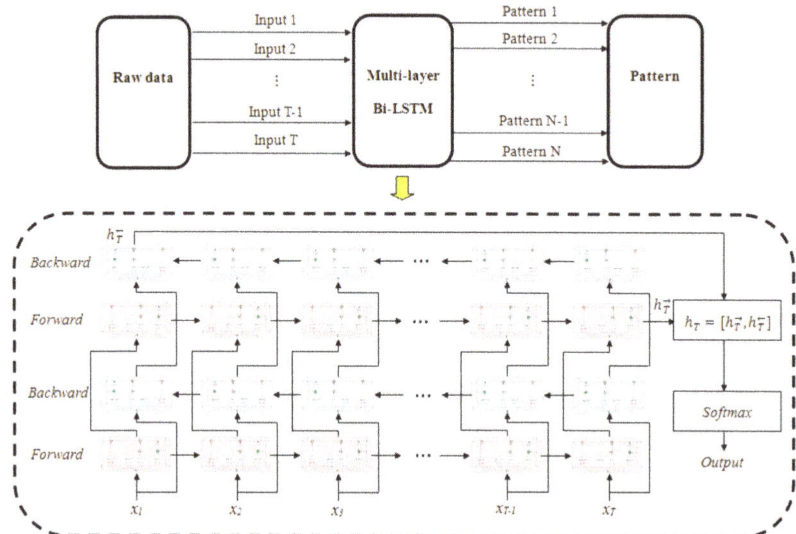

Figure 5. The structure of the multilayer Bi-LSTM.

4. Experiment and Discussion

In this section, some simulation data experiments and production data experiments were used to verify superiority of the multilayer Bi-LSTM. It was implemented by the MATLAB environment, and the experiment was carried out on a 3.10 GHz CPU with 4 GB RAM. The correct recognition rate (CRR) was used as the evaluation standard of model performance. At the same time, the discussion related to the experiment results was completed.

4.1. Simulation Parameters of HPs

In order to make the simulation data closer to the complex production data, the Monte Carlo simulation parameters of the seven HPs were randomly selected within a certain range, using the uniform distribution. The range of parameters is shown in Table 1. For example, parameter a of all BIM pattern samples is uniform distribution in the range $[0.4, 0.6]$.

Table 1. The simulation parameters of seven typical HPs.

Patterns	Parameters	Parameter Value/Range
NOR	Mean μ, standard deviation σ	$\mu = 10, \sigma = 0.01$
BIM	Proportion a, offset distance b_1 and b_2	$a \in [0.4, 0.6]$, $b_1 \in [\sigma, 2\sigma]$, $b_2 \in [\sigma, 2\sigma]$
LI	Proportion a, offset distance b	$a \in [0.8, 0.9]$, $b \in [1.5\sigma, 2.5\sigma]$
RI	Proportion a, offset distance b	$a \in [0.8, 0.9]$, $b \in [1.5\sigma, 2.5\sigma]$
LS	The number of normal distributions m	$m \in \{3, 4, 5\}$
RS	The number of normal distributions m	$m \in \{3, 4, 5\}$
FT	Proportion a	$a \in [0.6, 0.8]$

It is worth noting that the simulation results are the quality data of each HP, not the histogram. After that, interval statistics is needed to get histogram. The purpose of this simulation method is to be more in line with the actual situation. The quality data length of each sample is 500, and the number of groups of histogram is 25. The data set consists of 14,000 samples (2000 for each HP), which are randomly divided into two parts, of which 80% samples were used to train Bi-LSTM, and the rest was used for testing.

4.2. Performance Comparison of Optimization Algorithm

The SGD algorithm is a common optimization algorithm, which is often used to optimize the weight and bias of neural network in the training stage. However, the performance of its sub algorithm and variant is different. For this reason, in order to obtain the best training results, we compared several popular optimization algorithms, namely SGDM, RMSProp, and Adam [50]. Different optimization algorithms are used to train the Bi-LSTM. In this experiment, the training data set generated in Section 4.1 is used to complete the training. The initial learning rate is 0.05, and the batch size is 100. They were trained for 20 epochs and collected the corresponding training losses. The results are shown in Figure 6.

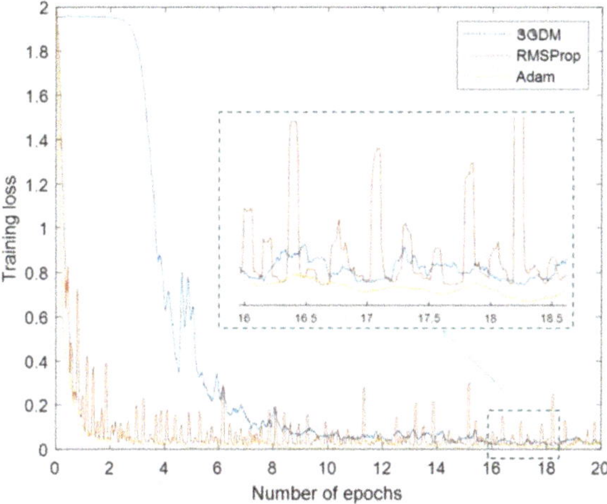

Figure 6. The training loss under different SGD algorithms.

It can be seen from Figure 6 that the SGDM algorithm has a slow convergence speed, and has a large training loss. The convergence speed of the RMSProp algorithm is fast, but the training process fluctuates greatly, and the training loss is not ideal. The Adam algorithm has the fastest convergence speed, the smallest loss, and the process is stable. Therefore, this optimization algorithm with the best performance is applied in the following experiments.

SGDM algorithm uses a single learning rate in the whole training process. Other optimization algorithms use different learning rates to improve the network training, and can automatically adapt to the loss function being optimized. This is how the RMSProp algorithm works. Adam updates with parameters similar to RMSProp, but adds a momentum term to that [50]. Therefore, the neural network can obtain fast and stable training effect.

4.3. The Influence of Batch Size on Training Process

The batch size determines the number of samples for network learning in each iteration., it is an important part of network training parameters. According to the past experience, this parameter has great influence on the training result and training time. Therefore, in the experiment, different batch sizes were compared and the results are shown in Figure 7.

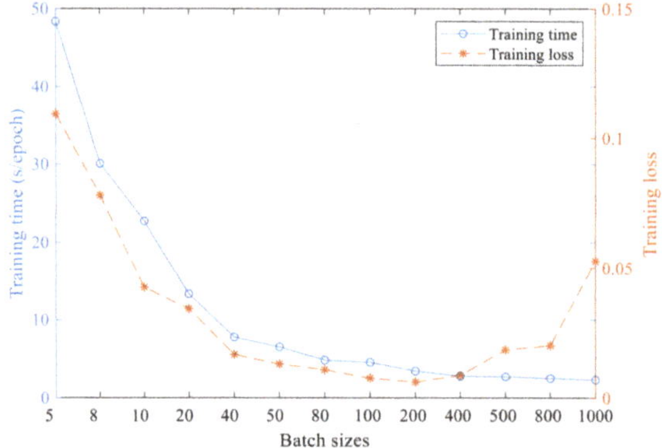

Figure 7. The training time and loss under different batch sizes.

It can be seen from Figure 7 with the increase of batch size, the training time of each epoch decreases continuously, and tends to be stable when the batch size exceeds 200. However, with the increase of the batch size, the training loss first decreases and then increases, and it is the best when the batch size is 200. Therefore, when the batch size is 200, the recognition accuracy of Bi-LSTM can be guaranteed and the training time can be greatly reduced.

4.4. The Influence of the Number of Network Layers

In the field of deep learning, this view is recognized by scholars. With the deepening of the network, the performance of it will be improved. This is because multilayer structure can improve the capacity and flexibility of the model [45]. However, some scholars a point out that the increase of network layers will increase the risk of over-fitting [51]. Therefore, this experiment compares the influence of the number of different network layers on training loss and testing accuracy, as shown in Figure 8. In this experiment, the network structure of Figure 5 is used, that is, the output of the first forward layer of the network is the input of the second forward layer, and the output of the first backward layer is the input of the second backward layer. According to this stacking rule, a multilayer network is formed. The data and training parameters used are the same as the above experiments. The training loss and testing accuracy of different networks are collected.

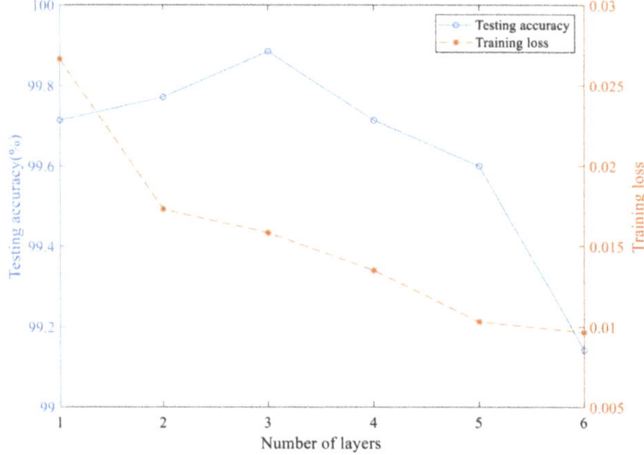

Figure 8. The testing accuracy and training loss under different number of layers.

It can be seen from Figure 8, with the increase of the number of layers of the Bi-LSTM network, the training loss of the training set data continues to decrease, which shows that the network fits these training data better and better. On the contrary, the recognition rate of the trained multilayer Bi-LSTM network to the data of the test set increases first and then decreases. This is a typical over fitting phenomenon, that is, with the increase of the number of layers, the network continues to deepen, the fitting degree of the training set data is too high, and the generalization ability of the test set data is reduced. According to the above experimental results, it is finally determined that the number of layers of multilayer Bi-LSTM is 3.

4.5. Comparison of HPR Results of Several Methods

Through the above comparison experiments, the structural parameters and training parameters finally determined for multilayer Bi-LSTM are shown in Table 2.

Table 2. The structural parameters and training parameters of multilayer Bi-LSTM.

Initial Learning Rate	Optimization Algorithm	Batch Size	Number of Layers
0.05	Adam	200	3

In order to further verify the effectiveness of the proposed method in HPR tasks, we compare multilayer Bi-LSTM with traditional machine learning methods (MLP) and several deep learning methods (DBN and 1D-CNN). These methods are described in detail as follows:

(1) MLP: There are three layers of MLP used in the experiment. The first layer of MLP is the input layer with 25 neurons, which are used to receive the frequency of 25 intervals of histogram. The second layer is the hidden layer with 30 neurons. The last layer is the output layer with seven neurons, corresponding to seven typical HPs. The activation function of all layers of MLP is Sigmoid function.

(2) DBN: The input of DBN is the frequency of 25 intervals of histogram. The DBN in this experiment is composed of three layers of RBM. Each RBM layer is pre-trained unsupervised. After the layer by layer pre-training is completed, the supervised global optimization is carried out. Each RBM layer contains 60 neurons, and the activation function of all layers is the Sigmoid function.

(3) 1D-CNN: The 1D-CNN used in this experiment is the structure recommended in [18]. The input of 1D-CNN is the frequency of 25 intervals of histogram. It consists of two convolution layers, two pooling layers and a full connection layer, and they are all one-dimensional. The number of feature maps of the two convolution layers is 6 and 12 respectively, and the convolution kernel size is 2*1 and 9*1, respectively. The activation function of the convolution layer was the rectified linear units (ReLU) function, and the output layer was the Softmax function.

In order to observe the results of pattern recognition more intuitively, confusion matrix is used. The values of the elements on the diagonal of the confusion matrix respectively represent the proportion of the correct recognition of various patterns. Other values represent the misjudgment of the classifier. The total CRR is equal to the average of all elements on the diagonal [1]. The confusion matrix of various identification methods is shown in Figure 9, and the total CRR and the time consumption of each epoch in the training process are shown in Table 3.

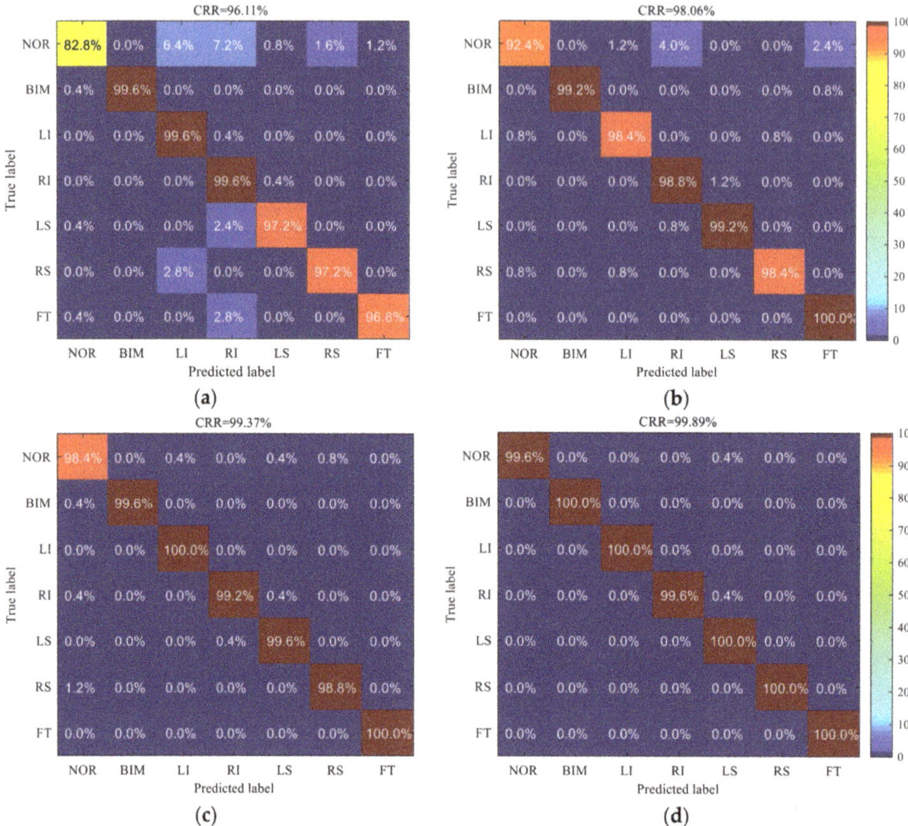

Figure 9. The HPR confusion matrix for (a) the MLP and frequency, (b) the DBN and frequency, (c) the 1D-CNN and frequency, and (d) the multilayer Bi-LSTM and frequency.

Table 3. Comparison of different HPR methods.

MLP & Frequency		DBN & Frequency		1D-CNN & Frequency		Multilayer Bi-LSTM & Frequency	
CRR (%)	Time (s/epoch)	CRR (%)	Time (s/epoch)	CRR (%)	Time (s/epoch)	CRR (%)	Time (s/epoch)
96.11	4.16	98.06	0.74	99.37	1.37	99.89	4.67

Figure 9 shows that there are different levels of confusion in the recognition results of several HPR methods. However, the performance of three recognition methods based on deep learning is better than that of traditional machine learning methods. In the recognition results of several methods, there is a trend of confusion from the NOR pattern to the island pattern. This problem is the most serious in MLP method, followed by DBN method. This may be due to the similarity between the NOR pattern and the island pattern. This is an unacceptable result, which means that the HPR system based on these methods will have a very serious Type I error (false alarm), which will bring unnecessary trouble to the quality control work of the enterprise. On the contrary, the method based on 1-DCNN effectively improves the recognition results of nor pattern, and proves the effectiveness of processing one-dimensional data and the ability to capture the details of data. Due to the multilayer and bidirectional structure, the training time of multilayer Bi-LSTM is a little longer, but it gets a very satisfactory recognition result. In 2800 test samples, only 4 samples are misclassified, and the CRR is as high as 99.89%, achieving accurate HPR. Compared with other methods, it has obvious advantages.

4.6. Simulation Parameters of CCPs

The simulation parameters of nine common CCPs are shown in Table 4.

Table 4. The simulation parameters of nine typical CCPs.

Pattern	Parameters	Mathematical Representation	Parameter Value/Range
NOR	Mean μ, standard deviation σ	$y(t) = \mu + x(t)$	$\mu = 30$, $\sigma = 0.05$
UT	Gradient d	$y(t) = \mu + x(t) + d \times t$	$d \in [0.15\sigma,\ 0.3\sigma]$
DT	Gradient d	$y(t) = \mu + x(t) - d \times t$	$d \in [0.15\sigma,\ 0.3\sigma]$
US	Shift magnitude s	$y(t) = \mu + x(t) + v \times s$	$s \in [1.5\sigma,\ 3\sigma]$
DS	Shift magnitude s	$y(t) = \mu + x(t) - v \times s$	$s \in [1.5\sigma,\ 3\sigma]$
CYC	Amplitude a, Period ω	$y(t) = \mu + x(t) + v \times a \times \sin(2\pi t/\omega)$	$a \in [1.5\sigma,\ 4\sigma], \omega \in \{4,\ 5,\ 6,\ 7,\ 8\}$
SYS	Magnitude g	$y(t) = \mu + x(t) + v \times g \times (-1)^t$	$g \in [1\sigma,\ 3\sigma]$
STA	Standard deviation σ'	$y(t) = \mu + x'(t)$	$\mu = 30$, $\sigma' \in [0.2\sigma,\ 0.4\sigma]$
MIX	Magnitude m	$y(t) = \mu + x(t) + v \times m \times (-1)^p$	$m \in [1.5\sigma, 2.5\sigma]$

For the MIX pattern, p is a binary integer, which is randomly selected from 0 or 1 at each sampling time t in the sample. The value of v is determined by where the mutation occurs. It is equal to 0 before the mutation occurs and 1 after the mutation occurs. The starting position obeys uniform distribution in the range [4,9]. The quality data length of each CCP is 25. The data set consists of 18,000 samples (2000 for each CCP), which are randomly divided into two parts, of which 80% samples were used to train multilayer Bi-LSTM, and the rest was used for testing.

4.7. Comparison of CCPR Results of Several Methods

In order to further verify the effectiveness of the proposed method in CCPR tasks, we compare multilayer Bi-LSTM with traditional machine learning methods (MLP) and several deep learning methods (DBN and 1D-CNN). Among them, MLP is widely used in CCPR field [19,21,22], which is the reason why we compare with it in this paper. The network structure of these methods is exactly the same as was used in Section 4.5. The difference is the input of the network.

The input of MLP becomes CCP feature set, which includes statistical features (mean, standard deviation, skewness and kurtosis) and shape features (S, NC1, NC2, APML and APSL). These features are designed by experts in this field, and have been proved to be very effective in many years of application [18,19,33–35]. Therefore, the first input layer of MLP has nine neurons, which are used to receive the above nine features respectively. However, the input of the three methods based on deep learning is still the raw data, that is, the quality data on the control chart. This is because they can adaptively extract the best features from the raw data. The results are shown in Figure 10 and Table 5.

Figure 10 shows that there are different levels of confusion in the recognition results of several CCPR methods. The most serious confusion is the DBN method. There are several of Type I errors and Type II errors (missed disturbances) in its identification results, which cannot be applied to the quality control of actual production. Although it is a deep learning method, which can extract features from the raw data through RBM pre-training, its learning ability is limited and cannot retrieve satisfactory recognition results. On the contrary, an expert feature set helps MLP get a better result, which effectively reduces the occurrence of two type errors. The validity of recognition method based on the expert feature set is proved. However, the disadvantage is that there is a little confusion between UT and US, DT and DS. More serious confusion occurs between CYC and MIX, which indicates that the existing expert feature set is not sensitive to distinguish the two CCPs. More acute features are yet to be explored. In addition, 1D-CNN got a better classification result, which reduced the confusion between CYC and MIX, and proves the effectiveness of processing one-dimensional data and the ability to capture the details of raw data. The recognition result of multilayer Bi-LSTM is the most satisfactory and the confusion rate is the lowest, which shows that multilayer Bi-LSTM has a strong ability of self-adaptive extraction of the best features from the raw data. The CRR reaches 99.26%, which can achieve accurate CCPR. Compared with other methods, it has obvious advantages.

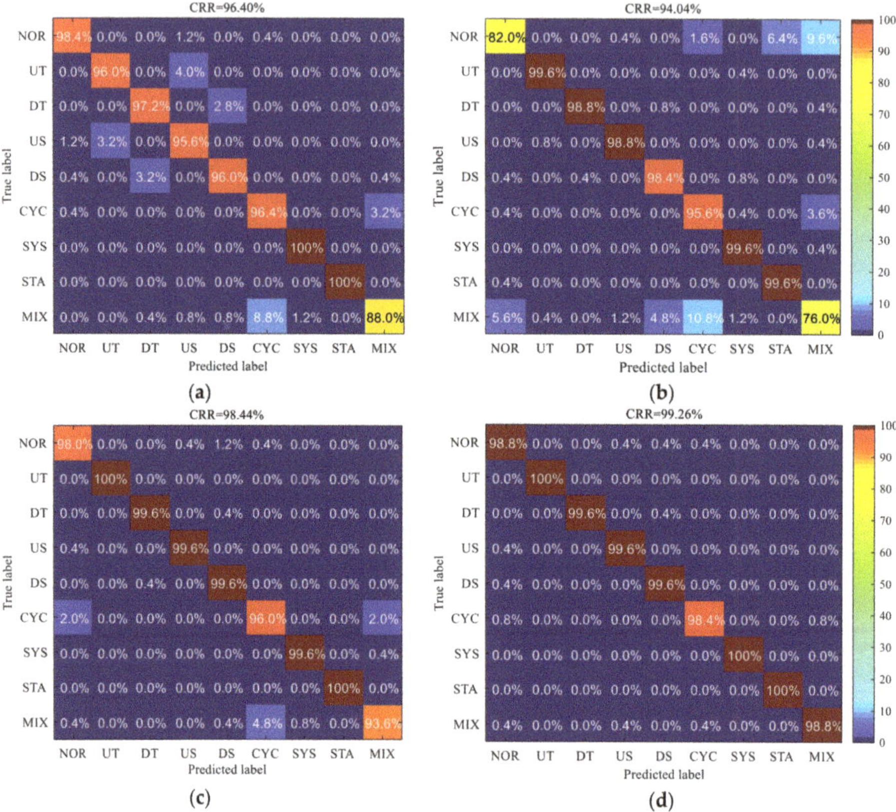

Figure 10. The CCPR confusion matrix for (**a**) the MLP and feature set, (**b**) the DBN and quality data, (**c**) the 1D-CNN and quality data, and (**d**) the multilayer Bi-LSTM and quality data.

Table 5. Comparison of different CCPR methods.

MLP & Feature Set		**DBN & Quality Data**		**1D-CNN & Quality Data**		**Multilayer Bi-LSTM & Quality Data**	
CRR (%)	Time (s/epoch)	CRR (%)	Time (s/epoch)	CRR (%)	Time (s/epoch)	CRR (%)	Time (s/epoch)
96.40	2.03	94.04	0.91	98.44	1.83	99.26	6.32

In order to further verify the feature learning ability of multilayer Bi-LSTM to the CCPs, the features extracted from the raw data are visually compared with expert features and the features extracted by 1D-CNN, and the results are shown in Figure 11. In this paper, the t-distributed stochastic neighbor embedding (t-SNE) [52] algorithm is used to reduce the dimensions of feature sets, so that they can be drawn in two-dimensional space for data visualization.

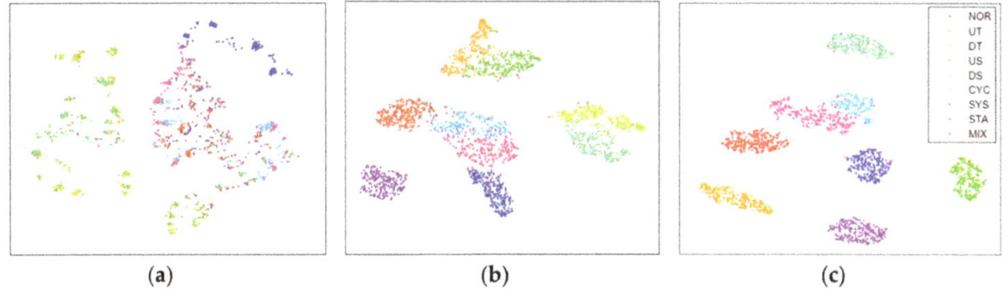

Figure 11. The feature visualization results for (**a**) the expert features, (**b**) the features extracted by 1-CNN, and (**c**) the features extracted by multilayer Bi-LSTM.

As we know, in the field of pattern recognition, as far as pattern classification is concerned, feature sets should present clustering distribution in feature space. The smaller the distance within the same class, the larger the distance between different classes, indicating that the higher the quality of this feature set, the more conducive to the classification of the classifier. As shown in Figure 11a, the expert feature set has the trend of clustering distribution in the feature space, but the serious cross between each pattern indicates that the quality of existing expert features is not good. This is the root cause of the confusion in Figure 10a. In Figure 11b, the features extracted by 1D-CNN show clustering distribution in the feature space, and the clustering phenomenon increases significantly, but the distance between different patterns is relatively close, and the classifier has the risk of misclassification. In contrast, the features extracted by multilayer Bi-LSTM are more closely distributed in the same pattern, and are better separated from other patterns in the feature space. This shows that it has learned more excellent features from the raw data. In addition, since the features are automatically extracted by the network, the accuracy of pattern recognition is improved, and the workload of quality control personnel is greatly reduced.

4.8. Application in Real Production Data

In order to verify the engineering value of the proposed method, it is applied to the quality control of real production. The diameter of the controlled object is the key to judge whether its quality is good or not, and its standard size is $\varphi 750^{+0.235}_{+0.158}$ mm. The width of the detection window is still 25. The multilayer Bi-LSTM is used to monitor the data of its production process, and some recognition results are listed in Figure 12.

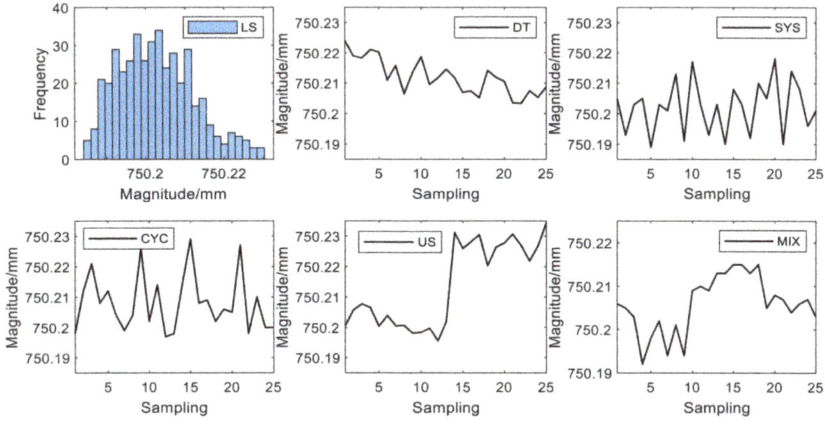

Figure 12. Abnormal HPs and CCPs found by the multilayer Bi-LSTM from production data.

As shown in Figure 12, the proposed method can identify a variety of abnormal patterns from the quality data of the production environment. At the same time, the product quality distribution in this stage is skewed, which is also effectively recognized by the multilayer Bi-LSTM. It can be found that the key dimensions of this batch of products are generally smaller. According to the results of communication with factory engineers, the reason for this histogram pattern is likely to be the result of conscious processing by workers. It is worth noting that the multilayer Bi-LSTM trained with the above simulation data set is used in the quality control of real production data. So before the real data is input into the network, linear transformation is carried out to scale the data to the same range of simulation data. This transformation is very simple and will not affect the speed and automation of recognition. At the same time, it shows that the network trained by the simulation data can identify the real data well, and does not need to train the network repeatedly according to the different product specifications. The proposed method can be easily integrated into the existing industry 4.0 system to provide intelligent SPC data analysis scheme for manufacturing enterprises and improve production efficiency and product quality. In addition, the SYS pattern in Figure 12 was recognized as CYC pattern by 1D-CNN in the previous study [18], which shows that the recognition results of the previous method are not precise enough. In this paper, the CCP is increased to nine, which can achieve more refined CCPR.

5. Conclusions

In order to realize intelligent SPC, this paper proposes a pattern recognition method based on multilayer Bi-LSTM for quality control, which uses feature learning to realize end-to-end HPR and CCPR. After experimental study, the following conclusions can be drawn. First of all, the convergence speed, recognition accuracy and over fitting degree of the network are related to the optimization algorithm, batch size and network layer number, and these parameters should be properly selected. Secondly, after learning, the network can extract the optimal feature set adaptively from the raw data, which has higher quality than the traditional manual expert feature set. Finally, with the raw data as input, the recognition rate of multilayer Bi-LSTM is 99.89% for HPs and 99.26% for CCPs. The recognition accuracy of this model is significantly better than that of traditional methods and other deep learning methods.

To sum up, the proposed multilayer Bi-LSTM method reduces the trouble of manual feature extraction, is competent for HPR and CCPR with high accuracy, and can effectively improve the level of intelligence and automation of SPC. It will likely become an integral part of industry 4.0 technology.

There is still a problem in this study. The length of the control chart and the number of histogram groups input into the network must be fixed, for example, the length of the data in this study is 25. However, in practical application, it may be necessary to adjust the data length to adapt to the production process. The current method must retrain the network, which will cause inconvenience. In the future, we will solve this problem, so that the network can adapt to different data lengths.

Author Contributions: Conceptualization, T.Z. and Z.L.; software, T.Z. and Z.L.; validation, M.W., X.G.; resources, Z.S. and D.C.; writing—original draft preparation, Z.L.; writing—review and editing, T.Z., M.W. and X.G. All authors have read and agreed to the published version of the manuscript.

Funding: This research is supported by China Scholarship Council, grant number 201806545032, and National Natural Science Foundation of China, grant number 51575014, 51875008, and 51975020.

Acknowledgments: The production data used in this study are from Tongyu Heavy Industry Co., Ltd.

Conflicts of Interest: The authors declare no conflict of interest.

References

1. Zhou, X.; Jiang, P.; Wang, X. Recognition of control chart patterns using fuzzy SVM with a hybrid kernel function. *J. Intell. Manuf.* **2018**, *29*, 51–67. [CrossRef]

2. Haghtalab, S.; Xanthopoulos, P.; Madani, K. A robust unsupervised consensus control chart pattern recognition framework. *Expert Syst. Appl.* **2015**, *42*, 6767–6776. [CrossRef]
3. Gutierrez, H.D.L.T.; Pham, D.T. Estimation and generation of training patterns for control chart pattern recognition. *Comput. Ind. Eng.* **2016**, *95*, 72–82. [CrossRef]
4. Wang, M.; Zan, T.; Fei, R.Y. Statistical process control with intelligence using fuzzy art neural networks. *Front. Mech. Eng.* **2010**, *5*, 149–156. [CrossRef]
5. Zan, T.; Wang, M.; Fei, R.Y. Pattern recognition for control charts using AR spectrum and fuzzy ARTMAP neural network. *Adv. Mater. Res.* **2010**, *97–101*, 3696–3702. [CrossRef]
6. Yang, W.A.; Zhou, W.; Liao, W.; Guo, Y. Identification and quantification of concurrent control chart patterns using extreme-point symmetric mode decomposition and extreme learning machines. *Neurocomputing* **2015**, *147*, 260–270. [CrossRef]
7. Roberts, S.W. Control chart tests based on geometric moving averages. *Technometrics* **1959**, *1*, 239–244. [CrossRef]
8. Nelson, L.S. Interpreting Shewhart X-bar control charts. *J. Qual. Technol.* **1985**, *17*, 114–116. [CrossRef]
9. Ducan, A.J. *Quality Control and Industrial Statistics*, 5th ed.; Richard D. Irwin: Homewood, IL, USA, 1986.
10. Nelson, L.S. The Shewhart control chart: Test for special causes. *J. Qual. Technol.* **1984**, *16*, 237–239. [CrossRef]
11. Ranaee, V.; Ebrahimzadeh, A. Control chart pattern recognition using a novel hybrid intelligent method. *Appl. Soft Comput.* **2011**, *11*, 2676–2686. [CrossRef]
12. Cheng, C.S. A neural network approach for the analysis of control chart patterns. *Int. J. Prod. Res.* **1997**, *35*, 667–697. [CrossRef]
13. Al-Ghanim, A.M.; Ludeman, L.C. Automated unnatural pattern recognition on control charts using correlation analysis techniques. *Comput. Ind. Eng.* **1997**, *32*, 679–690. [CrossRef]
14. Swift, J.A.; Mize, J.H. Out-of-control pattern recognition and analysis for quality control charts using LISP-based systems. *Comput. Ind. Eng.* **1995**, *28*, 81–91. [CrossRef]
15. Cheng, C.; Hubele, N.F. Design of a knowledge based expert system for statistical process control. *Comput. Ind. Eng.* **1992**, *22*, 501–517. [CrossRef]
16. He, S.; He, Z.; Wang, G.A. Online monitoring and fault identification of mean shifts in bivariate processes using decision tree learning techniques. *J. Intell. Manuf.* **2013**, *24*, 25–34. [CrossRef]
17. Kuo, T.; Mital, A. Quality control expert systems: A review of pertinent literature. *J. Intell. Manuf.* **1993**, *4*, 245–257. [CrossRef]
18. Zan, T.; Liu, Z.; Wang, H.; Wang, M.; Gao, X. Control chart pattern recognition using the convolutional neural network. *J. Intell. Manuf.* **2019**. [CrossRef]
19. Pham, D.T.; Wani, M.A. Feature-based control chart pattern recognition. *Int. J. Prod. Res.* **1997**, *35*, 1875–1890. [CrossRef]
20. Guh, R.S.; Tannock, J. A neural network approach to characterize pattern parameters in process control charts. *J. Intell. Manuf.* **1999**, *10*, 449–462. [CrossRef]
21. Al-Assaf, Y. Recognition of control chart patterns using multiresolution wavelets analysis and neural networks. *Comput. Ind. Eng.* **2004**, *47*, 17–29. [CrossRef]
22. Ranaee, V.; Ebrahimzadeh, A. Control chart pattern recognition using neural networks and efficient features: A comparative study. *Pattern Anal. Appl.* **2013**, *16*, 321–332. [CrossRef]
23. Addeh, A.; Khormali, A.; Golilarz, N.A. Control chart pattern recognition using RBF neural network with new training algorithm and practical features. *Isa Trans* **2018**, *79*, 202–216. [CrossRef] [PubMed]
24. Cheng, Z.; Ma, Y.Z. A Research about Pattern Recognition of Control Chart Using Probability Neural Network. In Proceedings of the Isecs International Colloquium on Computing, Communication, Control, & Management, Guangzhou, China, 3–4 August 2008; IEEE: Piscatway, NJ, USA.
25. Awadalla, M.H.A.; Sadek, M.A. Spiking neural network-based control chart pattern recognition. *Alex. Eng. J.* **2012**, *51*, 27–35. [CrossRef]
26. Gauri, S.K. Control chart pattern recognition using feature-based learning vector quantization. *Int. J. Adv. Manuf. Technol.* **2010**, *48*, 1061–1073. [CrossRef]
27. Guh, R.S. Real-time recognition of control chart patterns in autocorrelated processes using a learning vector quantization network-based approach. *Int. J. Prod. Res.* **2008**, *46*, 3959–3991. [CrossRef]
28. Yang, W.A.; Zhou, W. Autoregressive coefficient-invariant control chart pattern recognition in autocorrelated manufacturing processes using neural network ensemble. *J. Intell. Manuf.* **2015**, *26*, 1161–1180. [CrossRef]

29. Hachicha, W.; Ghorbel, A. A survey of control-chart pattern-recognition literature (1991–2010) based on a new conceptual classification scheme. *Comput. Ind. Eng.* **2012**, *63*, 204–222. [CrossRef]
30. Xanthopoulos, P.; Razzaghi, T. A weighted support vector machine method for control chart pattern recognition. *Comput. Ind. Eng.* **2014**, *70*, 134–149. [CrossRef]
31. Hu, S.; Zhao, L. A Support Vector Machine Based Multi-kernel Method for Change Point Estimation on Control Chart. In Proceedings of the 2015 IEEE International Conference on Systems, Man, and Cybernetics (SMC), Kowloon, China, 9–12 October 2015.
32. Pham, D.T.; Oztemel, E. Control chart pattern recognition using learning vector quantization networks. *Int. J. Prod. Res.* **1994**, *32*, 721–729. [CrossRef]
33. Hassan, A.; Baksh, M.; Shaharoun, A.M.; Jamaluddin, H. Improved SPC chart pattern recognition using statistical features. *Int. J. Prod. Res.* **2003**, *41*, 1587–1603. [CrossRef]
34. Pelegrina, G.D.; Duarte, L.T.; Jutten, C. Blind source separation and feature extraction in concurrent control charts pattern recognition: Novel analyses and a comparison of different methods. *Comput. Ind. Eng.* **2016**, *92*, 105–114. [CrossRef]
35. Gauri, S.K.; Chakraborty, S. Recognition of control chart patterns using improved selection of features. *Comput. Ind. Eng.* **2009**, *56*, 1577–1588. [CrossRef]
36. Zhao, C.; Wang, C.; Hua, L.; Liu, X.; Zhang, Y.; Hu, H. Recognition of control chart pattern using improved supervised locally linear embedding and support vector machine. *Procedia Eng.* **2017**, *174*, 281–288. [CrossRef]
37. Janssens, O.; Slavkovikj, V.; Vervisch, B.; Stockman, K.; Loccufier, M.; Verstockt, S.; Van De Walle, R.; Van Hoecke, S. Convolutional Neural Network Based Fault Detection for Rotating Machinery. *J. Sound Vib.* **2016**, *377*, 331–345. [CrossRef]
38. Chen, Z.; Li, W. Multisensor feature fusion for bearing fault diagnosis using sparse autoencoder and deep belief network. *IEEE Trans. Instrum. Meas.* **2017**, *66*, 1693–1702. [CrossRef]
39. Gao, Z.; Ma, C.; Song, D.; Liu, Y. Deep quantum inspired neural network with application to aircraft fuel system fault diagnosis. *Neurocomputing* **2017**, *238*, 13–23. [CrossRef]
40. Fatemeh, A.; Antoine, T.; Marc, T. Tool condition monitoring using spectral subtraction and convolutional neural networks in milling process. *Int. J. Adv. Manuf. Technol.* **2018**, *98*, 3217–3227.
41. Zan, T.; Wang, H.; Wang, M.; Liu, Z.; Gao, X. Application of Multi-Dimension Input Convolutional Neural Network in Fault Diagnosis of Rolling Bearings. *Appl. Sci.* **2019**, *9*, 2690. [CrossRef]
42. Liu, E.; Chen, K.; Xiang, Z.; Zhang, J. Conductive particle detection via deep learning for ACF bonding in TFT-LCD manufacturing. *J. Intell. Manuf.* **2019**. [CrossRef]
43. Zhao, R.; Yan, R.; Wang, J.; Mao, K. Learning to monitor machine health with convolutional bi-directional lstm networks. *Sensors* **2017**, *17*, 273. [CrossRef]
44. Zhao, R.; Wang, D.; Yan, R.; Mao, K.; Wang, J. Machine health monitoring using local feature-based gated recurrent unit networks. *IEEE Trans. Ind. Electron.* **2017**, *65*, 1539–1548. [CrossRef]
45. Rui, Z.; Ruqiang, Y.; Zhenghua, C.; Kezhi, M.; Peng, W.; Gao, R.X. Deep learning and its applications to machine health monitoring. *Mech. Syst. Signal. Process.* **2019**, *115*, 213–237.
46. Miao, Z.; Yang, M. Control Chart Pattern Recognition Based on Convolution Neural Network. In *Smart Innovations in Communication and Computational Sciences. Advances in Intelligent Systems and Computing*; Panigrahi, B., Trivedi, M., Mishra, K., Tiwari, S., Singh, P., Eds.; Springer: Singapore, 2019; Volume 670.
47. Wu, C.; Zhu, B.; Wan, Y.; Zhao, S. Quality Control Chart Pattern Recognition Based on Bidirectional LSTM. *Comput. Eng. Softw.* **2019**, *40*, 89–95. (In Chinese)
48. Simonoff, J.S.; Udina, F. Measuring the stability of histogram appearance when the anchor position is changed. *Comput. Stat. Data. Anal.* **1997**, *23*, 335–353. [CrossRef]
49. Bag, M.; Gauri, S.K.; Chakraborty, S. An expert system for control chart pattern recognition. *Int. J. Adv. Manuf. Technol.* **2012**, *62*, 291–301. [CrossRef]
50. Kingma, D.; Ba, J. Adam: A method for stochastic optimization. *arXiv* **2014**, arXiv:1412.6980.

51. Zhang, J.; Wang, P.; Yan, R.; Gao, R.X. Deep Learning for Improved System Remaining Life Prediction. *Procedia CIRP* **2018**, *72*, 1033–1038. [CrossRef]
52. Van der Maaten, L.; Hinton, G. Visualizing data using t-SNE. *J. Mach. Learn. Res.* **2008**, *9*, 2579–2605.

Article

Methodology of Employing Exoskeleton Technology in Manufacturing by Considering Time-Related and Ergonomics Influences

Christian Dahmen [1,2,*] and Carmen Constantinescu [2]

1 BMW AG, Knorrstraße 147, 80807 München, Germany
2 Fraunhofer IAO, Nobelstraße 12, 70569 Stuttgart, Germany; Carmen.Constantinescu@iao.fraunhofer.de
* Correspondence: Christian.Dahmen@bmw.de

Received: 29 January 2020; Accepted: 20 February 2020; Published: 27 February 2020

Abstract: This article presents a holistic methodology for planning, optimization and integration of exoskeletons for human-centered workplaces, with a focus on the automotive industry. Parts of current and future challenges in this industry (i.e., need of flexible manufacturing but as well having demographic change) are the motivation for this article. This challenges should be transformed in positive effectiveness by integrating of exoskeletons regarding this article. Already published research work from authors are combined in a form of summary, to get all relevant knowledge, and especially results, in a coherent and final context. This article gives interested newcomers, as well as experienced users, planners and researchers, in exoskeleton technology an overview and guideline of all relevant parts: from absolute basics beginning until operative usage. After fixing the motivation with resulting three relevant research questions, an introduction to the exoskeleton technology and to the current challenges in planning and optimizing the ergonomics and efficiency in manufacturing are given. A first preselection method (called ExoMatch) is presented to find the most suitable exoskeleton for workplacesm by filtering and matching all the important analyzed attributes and characteristics under consideration to all relevant aspects from environments. The next section treats results regarding an analysis of influencing factors by integrating exoskeletons in manufacturing. In particular, ergonomic-related and production-process-related (especially time-management) influences identified and researched in already published works are discussed. The next important step is to present a roadmap as a guideline for integration exoskeleton. This article gives relevant knowledge, methodologies and guidelines for optimized integrating exoskeleton for human-centered workplaces, under consideration of ergonomics- and process-related influences, in a coherent context, as a result and summary from several already published research work.

Keywords: exoskeletons; planning methods; manufacturing; ergonomics; time management

1. Introduction

The objectives of this work are derived from the effectiveness of integrating disruptive innovations like exoskeleton technology into the manufacturing workplace and into the strategic planning of manufacturing factories. The research questions, hypotheses and approaches are specific consequences of these objectives. The current challenges for manufacturing at all scales, from network to site of production, production area, production system, workplace and production process, are derived from the final goal, the improvement of effectiveness. Of course there are many more impacts from all different areas, like psychophysical and financial. This goal can be achieved by answering the research questions stated by the authors and depicted in Figure 1, with the focus on disruptive innovation and promising exoskeleton technology. However, the next steps of this research will deepen this focus and will approach it in synergy with other technologies, which cannot be further extensively discussed here.

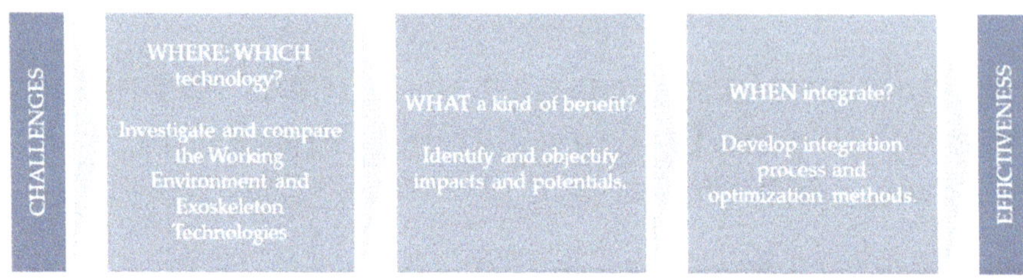

Figure 1. Objective: research questions, solution approach and hypotheses.

The first section of the paper presents an overview of the exoskeleton technology and the challenges of planning and optimizing the human-centred workplaces in manufacturing, based on the already published work [1]. Excerpts from this publication regarding the classification of workplaces and exoskeletons types answer the first research question on the way to improve effectiveness in manufacturing. In order to answer the second question, all the impacts and identified improvement potentials must be recognized first. This is accomplished with excerpts from the publication [2]. The current state-of-the art does not reveal ergonomics evaluation methods of workplaces with integrated exoskeleton technology. Excerpts from the publication [3] are used to investigate previously common procedures for their influence. With regard to the effects of time, in the context of this paper, the publication [4] of the results of wearing/unwearing times is presented.

The last section, Section 6, presents a short overview of a roadmap to integrate and optimize exoskeletons step by step.

2. Problem Statement and Motivation

Exoskeletons are worn externally and support body movement like a power suit does. The concept was first mentioned in 1966 [5]. Further uses and developments were found in the military sector, to enhance soldiers' strengths. Medical exoskeletons were developed to support disabled or handicapped people during their rehabilitation process and to support everyday life, for example, of patients with a disability. The motivation to use exoskeletons in the industry to assist and help workers to achieve daily tasks is based on these two different approaches. Exoskeletons are predominantly used preemptively in production to enhance the actual ergonomic work situation of the workers. Studies prove that such a preemptive investment has long-term financial benefits for the company in the coefficient of 1.445 [6]. In a future vision, exoskeletons will be able to assist disabled workers and therefore give them the ability to reintegrate. As a result, the reduction of lost workdays can be expected. At this time, in Germany, 26% of all lost workdays are caused by muscle–skeletal disorders (MSD) [7]. Furthermore, studies prove that, in Europe, charges of 240 billion euros (which is about 2% of GDP) are caused by MSD [8]. Equally positive side effects can be expected in the manufacturing area, like increased productivity and efficiency. For instance, increased productivity in specific sub-processes can be monitored because working with an exoskeleton is more intuitive and therefore faster than working with expensive mobile lifting assistance [9]. These systems lead to bad ergonomic situations, although the workplace is rated positive, and are therefore seldom used. Such results and ergonomic evaluations are conducted with analysis of simulations [10], where exoskeletons are worn virtually [11] in manufacturing facilities. Likewise, an increase in quality [12] and optimization was traceable in some areas [13,14]. Compared to other static technical assistant systems [15], bodyworn exoskeletons are intuitively operable. This increases the flexibility of the processes in manufacturing and is very important for consistent competitiveness [16] and mass customization, which will consist of manual processes or the processes that are difficult to be automated [17]. In summy, all facts mentioned above sustain the employment of exoskeletons for industry.

3. Methodology for the Integration of Exoskeleton Technology in Manufacturing Workplaces

3.1. Challenges and Specific Requirements for Integration of Exoskeletons in Industry

The selection of the suitable exoskeletons is done currently on a subjective base, depending on the specific workplace and worker. It should be scientificly founded, objectified and process-driven, but flexible, in the future. The following methods for preselecting suitable exoskeletons depending on the workplace are new in the scientific state. On this base, this paper describes objective methods (ExoScore and ExoMatch) which consider, filter and match important attributes from exoskeletons to workplaces in order to find the most suitable. Likewise, adjustments of workplaces and exoskeletons are needed to optimize the processes. The paper defines *Smart Exoskeletons* as devices which can be customized in variable combinations for supporting the worker's body. The required modules can be arms and the trunk [13]. We enhance our definition toward technical adaptivity of the *Smart Adaptive Exoskeletons*, which are easily adjustable for the specific workplace (working over the head, assembly, etc.),worker's tasks (e.g., tool holding, supporting force, etc.) and worker's status (performance, exhausting, fatigue, etc.).

The exoskeleton technology, as shown in Figure 2, is not sufficiently mature, since very few exoskeletons are available as certified products on the market. Technical definitions and certifications, or rather a declaration of conformity, are still non-final and not clearly solved challenges for most world's markets [18,19]. Due to the recency of this innovative technology, the product is not mature enough, because current development focuses on technical details and not human or organizational challenges. This leads to inadequacy of documented experiences, including the absence of long-term studies and structured literature for considerations in practice. The abovementioned situations represent reasons for the low deployment of exoskeletons in industry. Appropriate workplaces are in most cases characterized by simple and one-sided sequences of motions, and they can be enhanced with passive exoskeletons to support the current work.

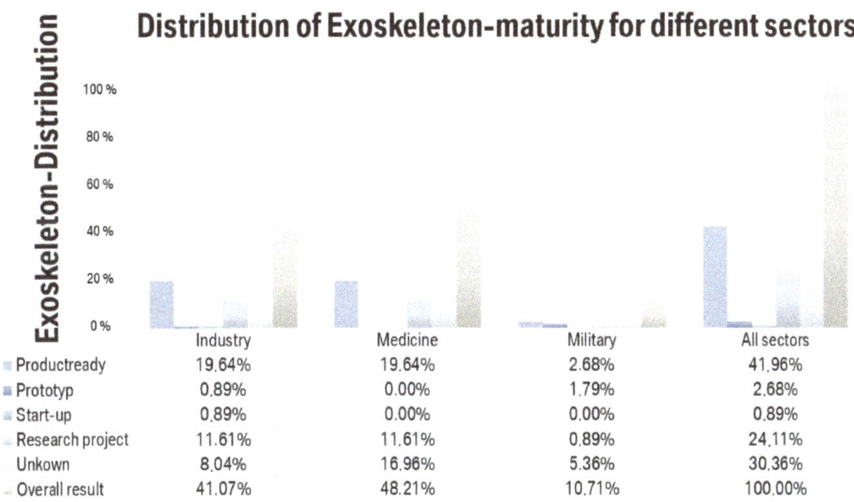

Figure 2. Overview and statistic from state of exoskeleton maturity on each sector © [1].

3.2. Classification of Smart Adaptive Exoskeletons for Their Deployment in Manufacturing

This section classifies the exoskeletons and the workplace's attributes. With the aim to develop a method supporting the selection of the most suitable device for a specific workplace, a database system for exoskeletons management, further on called *ExoData*, and workplaces management, named *UseCaseData*, was developed. The classified databased attributes are the base of the preassessment procedure, called *ExoScore*, and preselection method, named ExoMatch. The filtering of suitable exoskeletons for specific workplaces operates with the following categories, characterized by *attributes*, which are stated below [5]:

- General: number, priority, state of experience, etc.;
- Technical features: active/passive, power support, weight, size, material, storage, etc.;
- Compatibility: scratching, hygiene, etc.;
- Legal: conformity, occupational safety, instructions, etc.;
- Specific pro and cons: cheap, light, robust, etc.;
- Costs: acquirement, maintenance, service, training, etc.

Each of these detailed attributes has to be assessed with a weighted score-system, which will be developed. This category has a specific meaning for the subsequent specific workplace assessment and filtering. The exoskeleton Database (*ExoData*) has been populated with almost 140 commercial and prototype exoskeletons. The next step consists of sorting the *ExoData* by exoskeleton attributes for specific workplace requirements and identifying the favoured device. In addition, the procedure *ExoScore* enables the rough assessment of exoskeletons based on their functionalities. Further on, specific attributes were added, advanced with weighted factors (c_1, c_2, c_i). These factors are defined depending on different planning objectives. Below, Calculation (1) is partially highlighted:

$$SCORE_{Exo}(c_1, c_2, c_i) = Maturity \cdot c_1 + Costs \cdot c_2 + \ldots \quad (1)$$

The identification of suitable exoskeletons for one specific workplace requires categorized attributes as follows:

- Project state: number, priority, processing state, etc.;
- Workplace: description, contact, location, additional tasks, required tools, etc.;
- Ergonomics: load, weight (time and attitude), etc.;
- Environment: organization, division, access ways, escape exits, stairs, etc.

This section also presents how to find the most suitable of all existing exoskeletons (from *ExoData*) for any workplace, with a method called ExoMatch, as shown in Figures 3 and 4. The aim is to replace the previous selection with a transparent method that additionally increases the quality and decreases the working effort. In comparison to ExoScore, ExoMatch filters all available exoskeleton in *ExoData*, depending on the specific workplace conditions. The simple overview of the *ExoMatch* method is that *UseCaseData* and *ExoData* are related to each other. In Figure 3, these dependencies are solved with matching rules. These matching tables are the core of this method and are implemented as workplace conditions, and, based on these, it is possible to formulate exoskeleton requirements. The matching tables (Figure 4), where workplace conditions (*orange)* were translated to exoskeleton requirements (green), are structured in different filter clusters. For individual planning and complex influences, the threshold values (blue) can be adjusted or built on new tables. In Figure 4, two selected conceptual cluster tables are shown. If the ergonomic index for stress in shoulder exceeds the threshold value, there is a need for exoskeletons to support this body structure. Other examples and matching tables formulate requirements for exoskeletons with respect to the production environment. This concept makes it possible to realize general or situation-specific formulations of workplace conditions in which exoskeletons are explicitly not recommended. Additionally, the workplace planner has the chance to manually set filters which are normally not assessed in the matching tables.

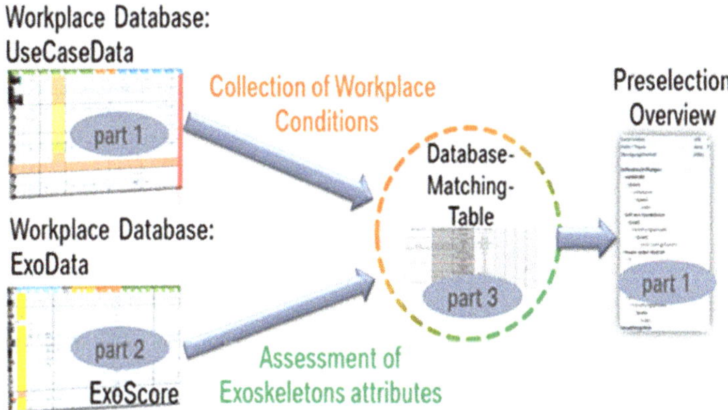

Figure 3. ExoMatch: method for preselecting most suitable exoskeletons depending on workplace © [1].

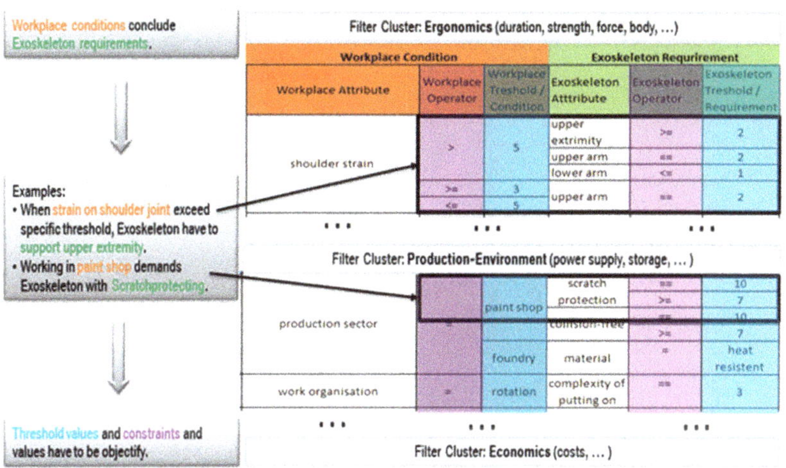

Figure 4. Database matching tables (DBMT): implement workplace-specific filter. Workplace conditions conclude exoskeleton requirements © [1].

3.3. Holistic Impacts by Integrating Exoskeleton Technology in Manufacturing

All relevant impacts of integrating exoskeletons have to be identified, described and structured. The results discussed were listed already in the form of a table [2]. In a first step, qualitative methods are needed to decide the relevance of different impacts. After this decision, adequate quantitative methods can be used [20]. The effects of exoskeletons have two significant perspectives. First, the ergonomic workplace improvement, and second, the impacts on the production system. The main focus, the ergonomic worker support, should not have under any circumstances an adverse effect on the production system and vice versa. However, the disadvantages have to be especially considered. The best case would be that all of these impacts and corresponding conversion functions can be combined in one unambiguous unit to estimate the holistic efficiency. However, because of the complexity and the lack of conversion methods, an assessment of clearly structured impact groups are necessary. Furthermore, new technologies, like exoskeletons, are establishing new possibilities. Different quantitative measurement methods are neccessary to investigate these impacts. Costs can be calculated and compared with other technical systems. Impacts on productivity and quality are researched in the following studies [21,22]. Time measurements are defined in [22].

4. Ergonomic Assessment of Human-Centred Workplaces with Integrated Exoskeleton Technology

4.1. Chain of Reasoning of Ergonomic Assessment

The chain of reasoning and process flow as motivation of provable and applicable assessment methods of workplaces with integrated exoskeleton technology is depicted in Figure 5. In the first step, it has to be discussed whether exoskeletons have an influence on the human body (ergonomic assessment), as well as on the workplace (production system), while performing a certain task. The hypothesis is that, when no effects can be expected, further steps are not necessary. Concluding that exoskeletons have a relevant impact, the next step is to demonstrate this impact. If neither effectiveness nor effects can be demonstrated, new methods or another type of exoskeleton has to be chosen, and the abovementioned loop has to be performed. The focus is to investigate the practical applicability of the study for the holistic integration of exoskeletons. If the applied methods within the performed studies cannot address the applicability, as well as the question of ergonomic benefits, there is currently no standardized way for the widespread integration of exoskeletons in a production system.

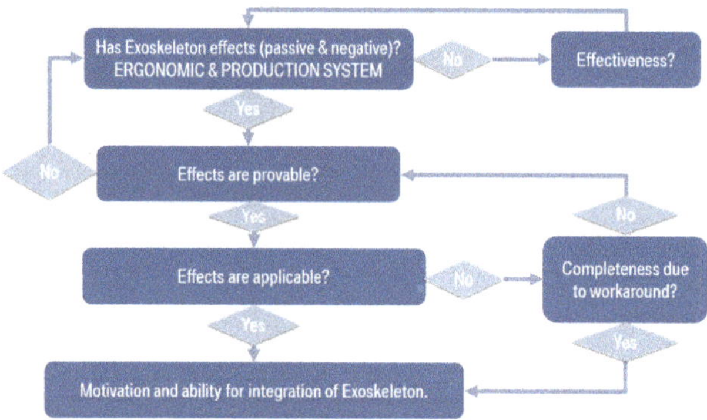

Figure 5. Chain of reasoning to employ exoskeletons in manufacturing workplaces [3].

4.2. Review of Studies Regarding Exoskeleton Ergonomics

Looking at the literature, one can find numerous articles regarding the effects of an exoskeleton device on particular muscle groups, using electromyography (EMG), while performing certain tasks. However, the majority of these articles focuses on the impact on the muscle groups which are intended to be supported by the assistant devices [23]. The potential negative aspects, such as the biomechanical load shift to other joints or muscle groups are rarely investigated, even though studies showed that the use of exoskeletons can cause significant postural changes or result in kinematic strains [24–26]. Additionally, a few investigations carried out the estimation of the effects of the integration of an exoskeleton device on the ergonomic risk assessment [27]. This might be an obstacle for the large-scale implementation of exoskeletons in industrial production systems, and it supports our assumption that the existing methods are somehow limited. Rashedi proved reduced muscle activity and discomfort up to 56% by using EMG and a subjective evaluation questionnaire (RPD—The rate of perceived discomfort). The measurement of an increased discomfort due to the weight of the device is performed in [28]. More test subjects and a specialized analysis are needed to validate this.

4.3. Assessment Requirements and Selected Methods Based on Manufacturing Needs

Existing studies, identified by the authors in previous papers (summarized in [2] with link to a lot of specific studies in detail), focus on the varying influences on specific parts of the human body, but they do not aim to present the results in a simplified "assessment-score" suitable for industrial demands, i.e., figures and colours that show the impact on the individual workplace. There is a critical discrepancy between the need for accuracy and the evaluation of complex work systems. Based on these contradictory requirements, there are many assessment methods designed to handle this challenge. Some large companies have even developed their own systems to assess their workplaces (i.e. APSA and SERA) [29,30].

Hence, the next step is to evaluate the exoskeletons on different workplace assessments with the currently existing methods from [27]. The research (detailed procedure and results in [3] based on ergonomic state-of-the-art: ergonomic scripts/books, ergonomic and industry training, internet and research) turned up 36 scientific assessment methods, that were categorized, each based on one of the following: forms for monitoring tasks, questionnaires, norms and threshold tables. Input parameters for these 36 assesment methods are, for example, shift-time, work-load, movement (duration, angle and velocity), temprature, pause, space, etc. for workplace charateristics and for for support-device/exoskeleton charateristics: weight, restriction of movements, support-leverl, heating, etc. The 36 differnet methods calculate for each input parameter with different weighting, combinations and tresholds specific scores as result (with and without exoskeleton). Characteristics are based on a description with relevant advantages and disadvantages, the scientific background, availability, field of application, considered body region and required input. All of these parameters were based on the industry requirements. The 36 methods were selected through a score system that selects and prioritizes the method with the highest amount of factors that the exoskeleton would have an impact on (i.e., force-and-exhaustion analysis, forced posture, heat, etc.), as well as the industrial applicability. After the score filter was applied with both conditions (exoskeleton impact and industry requirements), only five methods remained (shown in Table 1). For the sake of performing a holistic approach, negative aspects are included as well. The five methods selected are common industrial tools used to assess the workplace from an ergonomics perspective.

Table 1. Results for different assessments sheets (EAWS = Ergonomic Assessment Work Sheet, TB = total body, UP = upper body, KIM = key indicator method, REBA = Rapid Entire Body Assessment, RULA = Rapid Upper Limb Assessment) [3].

Method	Without Exoskeleton	With Exoskeleton
EAWS TB	61: High risk—Design measures are necessary.	66: High—Design measures
EAWS UB	78: High risk—Design measures are necessary	44.5: Potential risk—Design measures.
KIM MA	72.5: very high strain: measures are necessary.	67.5: High strain: design measures
REBA	4: Medium risk. Further, investigate.	3: Low risk. Change may be needed.
RULA	7: Very high risk, implement change now	6: Medium risk, further investigation

4.4. Exemplarily Workplace with Integrated Exoskeleton

The exoskeleton sample used for this investigation is a passive device that supports the upper-limb region for overhead work and is specifically designed to reduce stresses during prolonged overhead work; the mainframe of these devices is fixed to the body with belts or straps. The bottom side of each upper arm lies on a pad which is connected to the mainframe. The supporting force is assumed to equal the weight of the working tools, which is generally known as a ZeroG compensation approach. Many exoskeleton manufacturers recommend this simplified approach, as the devices' supporting forces can often be adjusted accordingly in reality. In the presented example, the device provides supports of approximately 24 Newton during the described static work.

For applying the assessments methods, the pervious exoskeleton and a workplace description from [3] were assumed. Results are presented in Table 1.

Furthermore, the stress–strain concept must be considered as well: exoskeletons do not change the strain, but exoskeletons do change the individual's perceived stress level. The high interest for exoskeletons for industrial applications in recent times supports the theory that there will be an increasing demand for new ergonomic workplace assessment methods [23]. Without a valid assessment approach that considers the advantages and disadvantages equally, there is no objective basis to help determine whether an exoskeleton should be integrated into a production system. Currently, each combination of a workplace and an exoskeleton needs to be analyzed individually to determine the impact on the workflow. This approach is time-consuming, impractical, and therefore highly cost intensive in an industrial environment [26].

5. Production Impacts of Exoskeletons in Manufacturing, Especially "Time Management"

As mentioned in Section 3.3, in [2], there is an intensive analysis of different impacts of the integration of exoskeletons. After identifying and analyszing the impacts, it is obvious that the time impacts are the most important impacts, with the most experience in research. With this hint, a deeper analysis is started for time management with exoskeletons. Time management means a summary of methods to analyze and integrate tasks and devices in the manufacturing process, under consideration of influences with time aspects. Professional measurements of time aspects are called time measurements, and the results of this analysis are referred to in [4].

5.1. Impact from Time-Management Perspective

Feasibility tests with exoskeletons in the automotive industry lead to the idea that time management is affected during work execution. Scientific work proves this assumption as follows: In [24], an increase in task time from 8 to 9 s was observed during ergonomic investigations; [31] even reports an average productivity increase of 40% for painting and 86% in welding operations; and [32] shows a tripling of the holding capacity (from 3.2 to 9.7 min) through the exoskeleton. Productivity in this context means less time for the same task, but the exoskeleton technology is not to be used as a performance-enhancing tool. However, their benefits can be argued based on ergonomic improvements as an additional side effect. In a further investigation of the applicability of passive upper-limb exoskeletons for the automotive industry [33], several experiments proved that the defined "precision index" and quality increased by 16.7%. This has a relevant impact on time management. In most existing studies, the focus is on ergonomic evaluation, while only a few studies are for time-management purposes.

In the context of exoskeletons integrated into workplaces, the difference between diverse types of the time impacts (Figure 6) has to be considered. The three types are (1) setup times, (2) task-execution impacts and (3) process-related impacts. If exoskeletons are mandatory [34] as required manufacturing tools or employed as prevention measures for the execution of processes, the impact regarding the time management has to be considered in the process planning. In the case of optional use of exoskeletons, these factors of impact are important as well. The focus relies mainly on the integration of this technology in serial production. Additional factors like the initial adjustments, training, briefing and maintenance work are not considered.

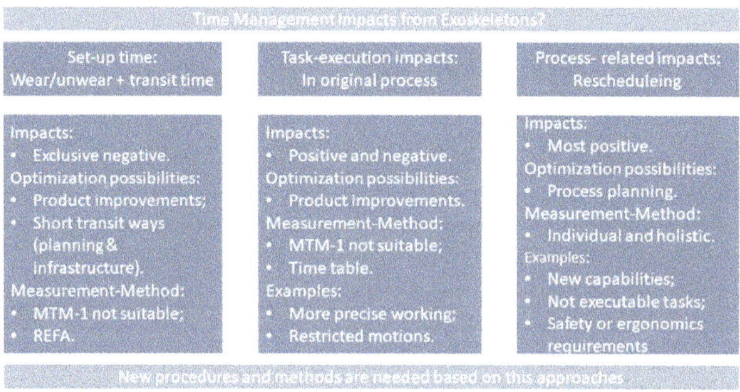

Figure 6. Three categories of time-management impacts for exoskeleton integration [4].

Setup times for exoskeletons in the workplace, can not be avoided. In this phase, there are considered factors which lead to time loss. Due to this reason, the setup times should be decreased (in the interests of user acceptance as well). Wear/unwear tasks, as well as transit times, are frequent examples. Under some circumstances, it is conceivable to use the exoskeleton just for certain tasks or during certain periods of time and not to wear it during the whole process. This creates the so-called "hybrid assembly processes", which are a combination of manual and semiautomated work tasks.

Task-execution impacts are time-related changes due to the integration of exoskeletons into the original process. Once they are introduced, the performance of a certain task is then either slower or faster. The main reasons for performance slowdowns are restrictions of motion [35,36]. By using exoskeletons for better precision [33] and high flexibility, faster movements are possible as well.

Process-related impacts are characterized by the fact that these processes are highly influenced by the integration of exoskeletons. These impacts are connected to rescheduling or emerged abilities due to the use of the device and therefore have a positive impact on time management. Additionally, the embedding of support brackets could optimize gripping processes because of shorter distances (up to a third hand). Likewise, new abilities could be made possible through an additional free hand that was used for support which is now provided by the exoskeleton.

5.2. Exoskeleton "Wear/Unwear" Times Experiments and Evaluation

In this paragraph, the "wear/unwear" times for commercial exoskeletons are presented. A method for establishing and evaluating times for all three different types of passive exoskeletons (overhead, bending and sitting support) is developed. Therefore, six different devices of diverse manufacturers were chosen to perform experiments, as follows: (a) 2x overhead exoskeleton: Exo1 and Exo2; (b) 2x bending exoskeleton: Exo3 and Exo4; and (c) 1x sitting exoskeleton: Exo5 in 2 different versions.

Before the kick-off for the experiments, the exoskeleton was adjusted initially to an optimum for each individual worker. This initial adjustment time is not captured and counted, as in serial operation each worker wears his own exoskeleton and therefore the settings were previously personalized.

Methodologically, first of all, the wear/unwear process was captured with a camera device. The goal was to analyze the particular movements and tasks on the basis of MTM-1 (methods-time measurement) and thereby, to achieve a valid assessment of time management. It became apparent that the setup process was too detailed and diverse to be depicted sufficiently by this method. Therefore, the decision to implement an assessment following the REFA (Reichsausschuß für Arbeitszeitermittlung—Reich Committee for Labour Studies) method was made. For the perspective of work-condition regulations, a measurement of the voluntary workers' performance level is not legitimate. Enough beforehand training confirms this hypothesis of 100% performance level [37].

For this purpose, the test person first picked up the exoskeleton from a tripod. After taking it off the holder, the test person wore the exoskeleton, paused for a short moment in order to signalize the end of the wearing process, subsequently removed the exoskeleton and hung it up properly in its initial tripod. Without another break, once again, a wear/unwear cycle was started. Every exoskeleton was taken thorugh the weared/unweared cycle ten times. The performance of every test person was captured on a video device. A digital signal started and stopped the test. The intermediate results (wearing starts and unwearing starts) were documented on a form. Deviations, for example, resulting from mistakes or twisted straps, were recorded as well. The related table forms, based on validated methods following [38], were used for the measurement process. The respective start of the diverse partial tasks (each wear/unwear) of a whole cycle was documented and enhanced by eventually interfering disturbances of different types. Likewise, the performance level of each partial task was documented at all times (here, 100%, as mentioned before). To simplify the following statistical evaluation, the results were digitalized after 10 cycles of experiments. For evaluation purposes, each time value, t_i, between the partial tasks of each cycle was determined separately. Each cycle's total time was determined through the addition of times within one cycle. The different results are shown in Table 2.

Table 2. Summary of exoskeletons wear/unwear time measurement [4].

Exo	Type	Wear/Unwear	Standard Deviation	Relative Confidence Interval
1	Overhead V1	37 s	1.19 s	2.3
2	Overhead V2	87 s	5.5 s	4.4
3	Bending V1	20.3 s	1.19 s	4
4	Bending V2	66.3 s	4.98 s	7
5a	Sitting V1	52.6 s	3.07 s	4.1
5b	Sitting V2	37.8 s	3.26 s	6.5

In conclusion, the times for wear/unwear were determined via time observation methods used while performing experiments. Depending on the exoskeleton needed, wear/unwear time is between 20.3 and 87 s.

6. Roadmap for Optimized Integration of Exoskeleton

After the knowledge of exoskeleton technology and specific requirements on workplaces (chapter 3) as well as the holistic impact regarding ergonomic (chapter 4) and for the production system (chapter 5), two blocks from Figure 1 are checked. This section closes the gap between the challenges (chapter 2) and the final effectiveness, by presenting a roadmap for optimized integration of exoskeleton, step by step.

In Figure 7, there are four steps of integration. Because in most of the cases not all new innovation are able to be integrated direcly (process experiment, device maturity, habituation, etc.), we need a step-by-step process of integration, starting with "Step 0" by the current state of the art and ending with by "Step 3" as vision fix star.

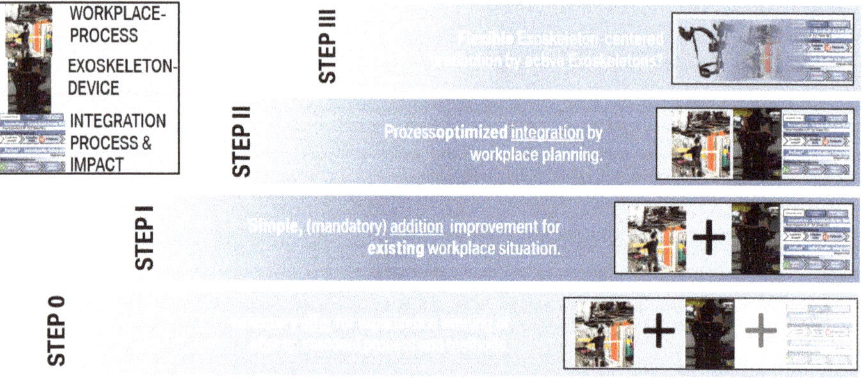

Figure 7. Step-by-step roadmap for optimized integration of exoskeleton.

This roadmap recommends, at first, in "Step 0", a voluntary integration so that the user and the company have time to get more and more experience. The workplace is the same as before, but exoskeleton and its impacts, as well as its needed processes, are learnt.

With this knowledge, in "Step 1" the exoskeleton is integrated as an additional optional working tool. In the end of "Step 1" the exoskeleton could be mandatory but is not necessary (in case of problems).

After good experience from previous steps, "Step 2" is the first step in which the exoskeleton is fully integrated, necessary and mandatory for process execution. The process is planned under known consideration of the exoskeleton process and impacts for optimization.

The last step, "Step 3", is the vision of a fully flexible exoskeleton-centered workplace. All the components (workplace process, exoskeleton device, and integration process and impacts) are merged together.

7. Conclusions

This article present the transformation from the given challenges (industrial and demographic) to an effectiveness production by using exoskeletons for human-centered workplaces as motivation. The challenges—a need of more flexibility and productivity for production systems, but on the other side, demographic change and health issues from a human perspective—are transformed by three identified relevant research questions. The answers are given in the following sections.

The first question, "where and which technology?", is answered by a methodology, called ExoMatch in chapter 3. A deep analaysis of exoskeleton technology and workplace environments is used to formulate specific matching rules by requirements and exclusions based on the knowledge about characteristics and attributes from database.

The second question, "what kind of benefit?", is divided in two sections regarding the challenges (for production and human). Section 4 answers the questions about ergonomic benefits and influences. Given research from studies given an ergonomic improvement from round about 20% but as well an increased discomfort of 56%. This article presents exemplary the ergonomic results by applying five industrial state of the assessments cheats for exemplary workplace and exoskeleton as well. The ergonomic risk indicator is changed to an improved workplace situation, but under consideration of the validity. Section 5 answers the benefits and influences of exoskeletons for the production system. This paper asserts that time-related influences are the main relevant impacts. They are divided into three relevant pillars (setup times, task-execution impacts and process-related impacts). With intense study's by applying the REFA method the result is a wear/unwear time from 20.3 to 52.6 s.

The third question, "when integrate?", for the final challenge-effectiveness transformation, is answered in Section 6 in the form of a integration roadmap. This roadmap gives an overview and outlook of exoskeleton integration.

The exoskeleton has enormous potential to improve the daily life of workers by decreasing work-related health issues. However, in reality, there are some barriers based on legal issues [39], discomfort/acceptence, direct financial or productivity benefit and, therefore, missing long-term experience. This is the reason for increasing device development and research on this topic. As a result, we see some real-life examples for using exoskeletons in the automotive industry. After the community has enough long-time experience with impacts and processes, and the devices are more smart and comfortable, the next big step could be the usage of active (or semi-acitve) exoskeletons when the benefit is more obviously. A next big challenge is to create a realy noticable businesse case, either by transforming ergonomic benefit in payback costs or by a big improvement in the production system.

Author Contributions: Conceptualization, C.D. and C.C.; Investigation, C.D.; Methodology, C.C.; Writing—original draft, C.D. and C.C.; Writing—review & editing, C.C. All authors have read and agreed to the published version of the manuscript.

Funding: This research received no further external funding. Slight expenditure for research, business trips studies, hardware, etc. are supported from BMW as part of a project for predevelopment, in cooperation with Fraunhofer IAO. The APC of this Journal was funded by Fraunhofer IAO.

Acknowledgments: I would like to express my gratitude to my research supervisors. Thank you to Carmen Lucia Consantinescu for her guidance, enthusiastic encouragement and useful critiques of this research work. I would also like to thank to Claudiu-Alin Rusu, Stefan Giosan and Daniele Ippolito for their advice and assistance in developing the article and, overall, their attitude regarding the work done by our scientific team.

Conflicts of Interest: I declare no conflicts of interest.

References

1. Dahmen, C.; Wöllecke, F.; Constantinescu, C. Challenges and Possible Solutions for Enhancing the Workplaces of the Future by Integrating Smart and Adaptive Exoskeletons. In Proceedings of the 11th CIRP Conference on Intelligent Computation in Manufacturing Engineering—CIRP ICME 17, Naples, Italy, 19–21 July 2017.
2. Dahmen, C.; Hölzel, C.; Wöllecke, F.; Constantinescu, C. Approach of Optimized Planning Process for Exoskeleton Centered Workplace Design. In Proceedings of the 51st CIRP Conference on Manufacturing Systems, Stockholm, Sweden, 16–18 May 2018.
3. Dahmen, C.; Hefferle, M. Application of Ergonomic Assessment Methods on an Exoskeleton Centered Workplace. In Proceedings of the XXXth Annual Occupational Ergonomics and Safety Conference, Pittsburgh, PA, USA, 7–8 June 2018.
4. Dahmen, C.; Constantinescu, C. Methodology for Evaluation of the Time-Management impact of Exoskeleton-centred workplaces. *Tech. Univ. Cluj Napoca Acta Tech. Napoc.* **2018**, 55–58.
5. Marinov, B.; Dao, T. Exoskeleton Report. Available online: http://exoskeletonreport.com/ (accessed on 8 March 2017).
6. Bräunig, D.; Kohstall, T. Berechnung des internationalen "Return on Prevention" für Unternehmen—Kosten und Nutzen von Investitionen in den betrieblichen Arbeits- und Gesundheitsschutz. *DGUV Rep.* **2013**, *1*, 34–38.
7. Bungard, S.; Hertle, D.; Kliner, K.; Lüken, F.; Tewes, C.; Trümmer, A. Gesundheit in Bewegung—Gesundheit in Bewegung: Schwerpunkt Muskel- und Skeletterkrankungen. *BKK Gesundh.* **2013**, *1*, 16–22.
8. Bevan, S.; Davies, C. The Impact of Back Pain on Sickness Absence in Europe. *Work Found.* **2012**, *1*, 3–4.
9. Accelopment AG. RoboMate. Available online: http://www.robo-mate.eu/ (accessed on 8 March 2017).
10. Constantinescu, C.; Mureșan, P.C.; Gînța, S.M.; Todorovic, O. Modelling and simulation of advanced factory environments integrating intelligent exoskeleton. *Int. Conf. Prod. Res.* **2014**, *3*, 109–114.
11. Constantinescu, C.; Muresan, P.-C.; Simon, G.-M. JackEx: The New Digital Manufacturing Resource for Optimization of Exoskeleton-based Factory Environments. *Procedia CIRP* **2016**, *26*, 508–511. [CrossRef]
12. Chase, B. The LevitateTM Difference—Analysis of the LevitateTM Personal Lift Assist Device (PLAD) During Various Physical Work-Related Tasks. *White Pap.* **2015**, *1*, 1–11.
13. Constantinescu, C.; Popescu, D.; Muresan, P.-C.; Stana, S.-I. Exoskeleton-centered Process Optimization in Advanced Factory Environments. *Procedia CIRP* **2015**, *48*, 740–745. [CrossRef]
14. Constantinescu, C.; Popescu, D.; Muresan, P.; Todorovic, O. Optimisation of advanced manufacturing environments with integrated intelligent Exoskeletons. *Int. Conf. Prod. Res.* **2016**, *4*, 1–6.
15. Weidner, R.; Redlich, T.; Wulfsberg, J.P. *Technische Unterstützungssysteme*, 1st ed.; Springer: Berlin/Heidelberg, Germany, 2015.
16. Westkämper, E.; Constantinescu, C.; Lentes, J.; Spath, D. *Digitale Produktion*, 1st ed.; Springer: Berlin/Heidelberg, Germany, 2013.
17. Østergaard, E.H. Human-robot collaboration in the age of industry 4.0—How process automation becomes a child's play for SMEs. *Präsentation* **2015**, *1*, 8–32.
18. VDI Wissensforum. Assistenzroboter in der Produktion 2015—Mensch-Roboter-Kollaboration im Industriealltag—Mensch, Roboter & Sicherheit: Wie Industrieroboter gefahrlos mit dem Menschen kollaborieren. *VDI Wissensforum* **2015**, *2*, 1–4.
19. Deutsche Gesetzliche Unfallversicherung e.V. Fragen und Antworten zum Thema Exoskelette. Available online: http://www.dguv.de/fbhl/sachgebiete/physische-belastungen/faq_exo/index.jsp (accessed on 18 April 2017).
20. Becker, T. *Prozesse in Produktion und Supply Chain Optimieren*, 2nd ed.; Springer: Berlin/Heidelberg, Germany, 2008.
21. Levitate Technologies Inc. Levitate Device—Personal-body worn—Performance enhancement. *Technol. Overv.* **2016**, *1*, 1–4.
22. Stadler, K.S.; Elspass, W.J.; van de Venn, H.W. *Mobile Service Robotics: CLAWAR 2014—Robo-Mate: Exoskeleton to Enhance Industrial Production*, 1st ed.; World Scientific Publishing Co. Pte. Ltd.: Singapore, 2014.
23. De Looze, M.P.; Bosch, T.; Krause, F.; Stadler, K.S.; O'Sullivan, L.W. Exoskeletons for industrial application and their potential effects on physical work load. *Ergonomics* **2016**, *59*, 671–681. [CrossRef] [PubMed]

24. Sylla, N.; Bonnet, V.; Colledani, F.; Fraisse, P. Ergonomic contribution of ABLE exoskeleton in automotive industry. *Int. J. Ind. Ergon.* **2014**, *44*, 475–481. [CrossRef]
25. Ulrey, B.L.; Fathallah, F.A. Subject-specific, whole-body models of the stooped posture with a personal weight transfer device. *J. Electromyogr. Kinesiol.* **2013**, *23*, 206–215. [CrossRef]
26. Weston, E.B.; Alizadeh, M.; Knapik, G.G.; Wang, X.; Marras, W.S. Biomechanical evaluation of exoskeleton use on loading of the lumbar spine. *Appl. Ergon.* **2018**, *68*, 101–108. [CrossRef]
27. Spada, S.; Ghibaudo, L.; Gilotta, S.; Gastaldi, L.; Cavatorta, M.P. Analysis of Exoskeleton Introduction in Industrial Reality: Main Issues and EAWS Risk Assessment. *Adv. Phys. Ergon. Hum. Factors* **2017**, *1*, 236–244.
28. Rashedi, E.; Kim, S.; Nussbaum, M.A.; Agnew, M.J. Ergonomic evaluation of a wearable assistive device for overhead work. *Ergonomics* **2014**, *57*, 1864–1874. [CrossRef]
29. Peter, F.; Tropschuh, B.H. Taktgeber für Arbeitsplatzergonomie in der Industrie; UmweltDialog—Wirtschaft; Verantwortung; Nachhaltigkeit. 2015. Available online: https://www.umweltdialog.de/de/wirtschaft/arbeitsbedingungen/2018/Taktgeber-fuer-Arbeitsplatzergonomie-in-der-Industrie.php (accessed on 20 January 2020).
30. Rosemarie, P. Mit klassischen gesundheitsfördernden Maßnahmen und neuesten Technologien ist das BMW Group Werk Steyr für den demografischen Wandel gerüstet. In *Sichere Arbeit: Internationales Fachmagazin für Prävention in der Arbeitswelt*; 2017; Available online: http://www.sicherearbeit.at/cms/X04/X04_1.8.6.a/1342585867168/archiv/html-archiv-2017/sichere-arbeit-6-2017/bmw-vorsorge-heute-fuer-morgen (accessed on 20 January 2020).
31. Butler, T. Exoskeleton Technology—Making Workers Safer and More Productive. *Am. Soc. Saf. Eng. ASSE* **2016**, *61*, 32–36.
32. Bosch, T.; van Eck, J.; Knitel, K.; Looze, M. The effects of a passive exoskeleton on muscle activity, discomfort and endurance time in forward bending work. *Appl. Ergon.* **2016**, *1*, 212–217. [CrossRef]
33. Spada, S.; Ghibaudo, L.; Gilotta, S.; Gastaldi, L.; Cavatorta, M.P. Investigation into the Applicability of a Passive Upper-limb Exoskeleton in Automotive Industry. *Procedia Manuf.* **2017**, *27*, 1255–1262. [CrossRef]
34. BGHM. Einsatz von passiven Exoskeletten an (gewerblichen) Arbeitsplätzen der Automobilindustrie. In *Fach-Information: BGHM FAQ-Liste 05/2017*; BGHM: Mainz, Germany, 2017; pp. 1–5.
35. Abdoli-E, M.; Agnew, M.J.; Stevenson, J.M. An on-body personal lift augmentation device (PLAD) reduces EMG amplitude of erector spinae during lifting tasks. *Clin. Biomech.* **2006**, *5*, 456–465. [CrossRef] [PubMed]
36. Godwin, A.A.; Stevenson, J.M.; Agnew, M.J.; Twiddy, A.L.; Abdoli-Eramaki, M.; Lotz, C.A. Testing the efficacy of an ergonomic lifting aid at diminishing muscular fatigue in women over a prolonged period of lifting. *Int. J. Ind. Ergon.* **2009**, *1*, 121–126. [CrossRef]
37. REFA Bundesverband E.V. *Leistungsgradbeurteilung—Lehrunterlage zu Modul 3210248*, 1st ed.; REFA: Darmstadt, Germany, 2003.
38. REFA. *Durchführen und Auswerten von Zeitaufnahmen—Lehrunterlage zu Modul 3210251*, 1st ed.; REFA-Fachbuchreihe Unternehmensentwicklung: Darmstadt, Germany, 2006.
39. PR—Prävention Aktuell. Das STOP-PRINZIP. Available online: https://praevention-aktuell.de/das-stop-prinzip/ (accessed on 5 July 2018).

Article

Collision-Free Path Planning Method for Robots Based on an Improved Rapidly-Exploring Random Tree Algorithm

Xinda Wang [1], Xiao Luo [2,*], Baoling Han [1], Yuhan Chen [1], Guanhao Liang [3] and Kailin Zheng [1]

1 School of Mechanical Engineering, Beijing Institute of Technology, No. 5 Zhongguancun South Street, Haidian District, Beijing 100081, China; 2120170401@bit.edu.cn (X.W.); hanbl@bit.edu.cn (B.H.); 3120185244@bit.edu.cn (Y.C.); 2120170440@bit.edu.cn (K.Z.)

2 School of Computer Science and Technology, Beijing Institute of Technology, No. 5 Zhongguancun South Street, Haidian District, Beijing 100081, China

3 School of Mechatronical Engineering, Beijing Institute of Technology, No. 5 Zhongguancun South Street, Haidian District, Beijing 100081, China; simonleungbit@hotmail.com

* Correspondence: luox@bit.edu.cn; Tel.: +86-010-6891-8856

Received: 20 January 2020; Accepted: 13 February 2020; Published: 19 February 2020

Featured Application: The new method can be used to guide the path planning of robots with any number of degrees of freedom in complex environments.

Abstract: Sampling-based methods are popular in the motion planning of robots, especially in high-dimensional spaces. Among the many such methods, the Rapidly-exploring Random Tree (RRT) algorithm has been widely used in multi-degree-of-freedom manipulators and has yielded good results. However, existing RRT planners have low exploration efficiency and slow convergence speed and have been unable to meet the requirements of the intelligence level in the Industry 4.0 mode. To solve these problems, a general autonomous path planning algorithm of Node Control (NC-RRT) is proposed in this paper based on the architecture of the RRT algorithm. Firstly, a method of gradually changing the sampling area is proposed to guide exploration, thereby effectively improving the search speed. In addition, the node control mechanism is introduced to constrain the extended nodes of the tree and thus reduce the extension of invalid nodes and extract boundary nodes (or near-boundary nodes). By changing the value of the node control factor, the random tree is prevented from falling into a so-called "local trap" phenomenon, and boundary nodes are selected as extended nodes. The proposed algorithm is simulated in different environments. Results reveal that the algorithm greatly reduces the invalid exploration in the configuration space and significantly improves planning efficiency. In addition, because this method can efficiently use boundary nodes, it has a stronger applicability to narrow environments compared with existing RRT algorithms and can effectively improve the success rate of exploration.

Keywords: Rapidly-exploring Random Tree (RRT); manipulator; motion planning; obstacle avoidance; complex environment

---

1. Introduction

With the development of computer technology and modern manufacturing, particularly in the context of Industry 4.0, the intelligence level requirements continue to increase and the application scenarios of multi-Degree-Of-Freedom (DOF) robots are becoming increasingly complex. These conditions pose challenges for the motion planning of manipulators. Up to now, scholars have extensively researched motion planning in high-dimensional spaces and unstructured complex environments. Sampling-based methods mainly seek a collision-free path from the start point to the goal point by sampling in the Configuration space (C-space). It is unnecessary to model the entire space, and the methods have a probabilistic completeness [1–3]. Therefore, sampling-based methods are widely used in the motion planning of high-dimensional spaces owing to their unique advantages. For example, the Rapidly-exploring Random Tree (RRT) [4] and Probabilistic Roadmap Method (PRM) [5,6] are the most popular and commonly used techniques.

Basic-RRT, as shown in Algorithm 1, has been widely used in many fields, including robot motion planning, as a single-query planner since it was proposed by LaValle et al. in 1998. A large number of algorithms derived from RRT have been proposed to solve different problems. Improvements in RRT have mainly focused on sampling strategies and their guidance for exploring areas, the selection of extension nodes, the directions and step sizes of extensions, selection of metrics, and the collision detection algorithm and local connection method, all of which have many studies available for reference [7]. These improved methods enhance the performance of Basic-RRT from various aspects, but the obtained solutions are still highly suboptimal in most cases. Therefore, various methods of pruning and smoothing [8–11] have been proposed to implement path post processing.

Algorithm 1 Basic-RRT algorithm.

$T \leftarrow$ InitTree(q_{start});
for $i = 1$ **to** n **do**
  $q_{rand} \leftarrow$ RandomSample(i);
  $q_{near} \leftarrow$ NearestNeighbor(q_{rand}, T);
  $q_{new} \leftarrow$ Extend($q_{rand}, q_{near}, \varepsilon$);
  if CollisionFree(q_{near}, q_{new}) **then**
    AddNewNode(T, q_{new});
  end if
  if Distance(q_{new}, q_{goal}) $< \rho_{min}$ **then**
    return T;
  end if
end for
return *Failed*;

Algorithm improvement, regardless of the method, has only two purposes: to reduce the path cost and to reduce the running time of the algorithm. For the former, the typical method is RRT*, which is an optimal planning method based on RRT and proposed by Karaman and Frazzoli [12]. It guarantees the asymptotic optimality of the algorithm by adding two processes to Basic-RRT: "near vertices" and "rewire". However, the planning process is time consuming and thus impractical for planning tasks that have certain time requirements. Many methods [13–15] have been subsequently proposed to accelerate the convergence of RRT*, but the computational time still cannot meet the requirements of the actual application.

This paper focuses on the latter purpose of algorithm improvement, namely to decrease the computational time of the algorithm on the premise that the algorithm can be applied to complex environments. The usual approach involves reducing the number of nodes in the tree and the number of collision detections. RRT-biased [16] is a simple way to improve efficiency. As shown in Algorithm 2, it improves the performance of the algorithm through goal bias sampling with a certain probability (usually 5–10%). However, although the number of nodes in the random tree is appropriately reduced, the computational cost remains large. Gitae Kang et al. [17] proposed a method based on goal-oriented sampling for the motion planning of a manipulator; this method can limit the distribution of sampling points to improve the search speed, but the invalid exploration area is large. Brendan Burns et al. [18] proposed a utility function for selecting the extension node and direction; this function, which determines the maximum expected extension step of the planner on the basis of the obtained state space information, has improved efficiency to some extent. In addition, Dong-Hyung Kim et al. [19] described a method of adaptive body selection based on the complexity of planning; this method provides another description of the variable-dimensional C-space for high-DOF articulated robots, thereby improving efficiency from the perspective of dimensionality reduction.

Algorithm 2 GoalBiasedSample()

$num \leftarrow$ RandomNumber;   RandomNumber $\in [0, 1]$
if $num <$ k **then**
    $q_{rand} = q_{goal}$;
else
    $q_{rand} =$ RandomSample();
end if
return q_{rand};

These methods and many existing RRT variants that guide sampling and expansion have improved efficiency from different aspects, but there is no good general solution for complex environments, such as "narrow" ones. Haojian Zhang et al. [20] proposed an improved method combining the regression and boundary expansion mechanisms; this technique improves efficiency by providing a corresponding solution process for a specific problem. However, although the boundary node can be identified during the exploration process, this method also limits the use of boundary nodes. In addition, some methods [21–23] dedicated to improving the success rate of the algorithm by changing the sampling strategy have been proposed specifically to solve the problem of narrow environments, but their universality is also weak.

It can be seen from the above that an efficient and universal algorithm that is suitable for complex environments is needed to ensure the completion of planning tasks. Therefore, this paper proposes an autonomous path planning algorithm of Node Control based on the architecture of RRT (NC-RRT). Firstly, a method of gradually changing the sampling area is proposed to guide exploration, thereby effectively improving the search speed. In addition, unlike existing methods, which aim to explore new sampling strategies, a node control mechanism is proposed to constrain the extended nodes of the tree and, thus, enhance the applicability of the environment. The results reveal that the algorithm greatly reduces the invalid exploration in the C-space and significantly improves planning efficiency. Moreover, compared with most existing RRT algorithms, NC-RRT is universal for different environments by appropriately adjusting the parameters. For the convenience of explanation, the algorithm will be first described and verified in a two-dimensional space and then applied in a high-dimensional space.

The rest of the paper is organized in the following manner. Section 2 introduces the proposed improved RRT algorithm in two parts. The method of gradually changing the sampling area is proposed in the first part, and we present the node control mechanism in the second part. Section 3 explains the simulation process and results of the algorithm. The proposed algorithm is simulated in a two-dimensional space and then applied to a 6-DOF manipulator. Finally, the conclusions and future work of this paper are provided in Section 4.

2. The NC-RRT Method

2.1. The Method of Gradually Changing the Sampling Area

We propose the method of gradually Changing the Sampling Area based on RRT (CSA-RRT) to guide exploration; the process is shown in Algorithm 3. As in [17], we need to calculate the distance (the Euclidean distance is used in this paper) between the goal configuration point q_{goal} and the configuration point that is initially the farthest from the goal q_f in the C-space, which is represented by

$D_{farthest}$ here, to ensure the completeness of the solution space. Assuming that the dimension of the C-space is s, $D_{farthest}$ is solved as follows in Equation (1):

$$D_{farthest} = \sqrt{(q_{goal(1)} - q_{f(1)})^2 + \ldots + (q_{goal(s)} - q_{f(s)})^2}. \quad (1)$$

Algorithm 3 CSA-RRT algorithm.

$T \leftarrow$ InitTree(q_{start});
$R \leftarrow D_{farthest}$;
for $i = 1$ **to** n **do**
  $q_{rand} \leftarrow$ RandomSample(i);
  if Distance(q_{rand}, q_{goal}) $> R$ **then**
    continue;
  end if
  $q_{near} \leftarrow$ NearestNeighbor(q_{rand}, T);
  $q_{new} \leftarrow$ Extend($q_{rand}, q_{near}, \varepsilon$);
  if CollisionFree(q_{near}, q_{new}) **then**
    AddNewNode(T, q_{new});
    $R =$ Distance(q_{new}, q_{goal});
  else
    $R = R + k \times \varepsilon$;
    continue;
  end if
  if Distance(q_{new}, q_{goal}) $< \rho_{min}$ **then**
    return T;
  end if
end for
return *Failed;*

We initialize the sampling radius R to $D_{farthest}$, and the sampling range in the subsequent sampling process is limited to the area within R of the goal configuration point. If a new node q_{new} is successfully added in a certain iteration, that is there is no collision with obstacles during the generation of q_{new}, then the value of R is changed to the distance between q_{new} and q_{goal}. On the contrary, if there is a collision, then the sampling radius is increased as follows:

$$R = R + k \times \varepsilon, \quad (2)$$

where k is the coefficient used to change the range of the sampling domain; its value is a positive integer and can be adjusted according to the complexity of the environment. ε is the step size of the extension. We can find that, if there is no obstacle in the environment or there are obstacles, but no collision, the sampling domain will be reduced to an area $\| q_{new} - q_{goal} \|$ away from the goal node each time a new node is added to the tree; once obstacles are encountered during expansion, the sampling domain will be expanded to find a new sampling point and the corresponding q_{near} node. When q_{new} is successfully added, the random tree will be restored to the no obstacle state and continue to explore. This gradual change in the sampling area will drive the tree to grow continuously toward the goal node. Figure 1 shows the performance comparison of Basic-RRT and CSA-RRT in the same environment. Obviously, the latter has much fewer nodes. The specific experiment and analysis are provided in Section 3.

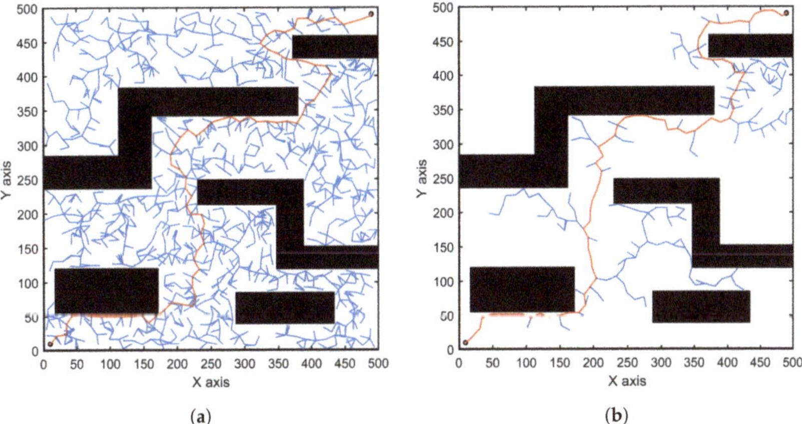

Figure 1. Performance comparison between Basic-RRT and CSA-RRT in the same environment. (**a**) Performance of Basic-RRT. (**b**) Performance of CSA-RRT.

2.2. The Node Control Mechanism

When the CSA-RRT algorithm selects the extended nodes, the nodes in the tree should be traversed to find the node with the smallest distance from the sampling configuration point in the C-space, but this traversal is usually redundant and time consuming. Under the premise of maintaining the performance of the CSA method, a node control mechanism is introduced in this part to further reduce the extension of invalid nodes and extract the boundary nodes (or near-boundary nodes), thereby further improving the speed of the algorithm and enhancing the adaptability of the environment. On the basis of this mechanism, the "local trap" phenomenon in the process of random tree expansion is proposed. We update and record the expansion state of each node when the random tree expands in the C-space. The value of the node control factor is changed according to whether the "local trap" phenomenon occurs. Then, the selection of the extension node is adjusted, such that the tree is expanded by the boundary nodes only or by nodes close to the boundary in most cases. The specific process is as follows.

We represent the extended state value of the node as δ and redefine the following concept that is different from the past: the path that each leaf node of the tree traces back to the initial node q_{start} is called a branch. Then, the change strategy of the node state value is as follows. Its δ value is set to zero each time a new node is successfully added to the tree. Then, starting from the parent node and following the reverse path, one is added to the state value of each node in the branch where this new node is located until it traces back to the initial node. As shown in Figure 2, any node q_n in the process of random tree exploration is regarded as an example. Assuming its δ value is recorded as m at this time, if a child node q_{n+1} is generated with q_n as the parent node in a certain iteration, then the value of δ for q_{n+1} is set to zero; then, starting from the parent node (i.e., q_n) of q_{n+1} and following the reverse path, one is added to the value of δ for all nodes in the branch where q_{n+1} is located until it traces back to the q_{start} node. At this time, the δ value of q_n becomes $m+1$. If the new node q_{n+2} generated in a subsequent iteration is still a child node of q_n, the above process is repeated, and the δ value of q_n becomes $m+2$. However, because the q_{n+1} node is not in the same branch as the q_{n+2} node, the δ value of the q_{n+1} node remains unchanged in this iteration.

According to the state value of each node, a node control factor *control* can be introduced to guide the selection of the expanded node. For each extension, only the node that has a state value less than *control* and lies closest to the sampling node is selected as the extension node. To reduce the invalid exploration in the C-space sufficiently, we set *control* to one by default. However, the introduction of this control factor creates a problem. Because only the δ value of the leaf nodes in the random tree is zero and the value of *control* remains one, it means that the last node in the tree will always be used as the extension node. Combined with the proposed CSA sampling method, this technique will drive the tree to reach the target configuration extremely rapidly if there are no obstacles in the environment. As shown in Figure 3a, we can see that all nodes in the tree exist as "valid nodes" in the final query path. However, once there are obstacles in the environment, the tree easily falls into the "local trap" state, as shown in Figure 3b. The random tree will keep this state, which can be expanded in local areas only until it reaches the set maximum number of failures.

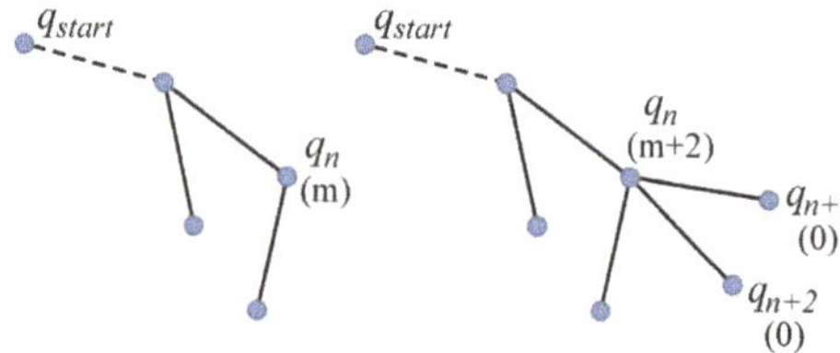

Figure 2. The change strategy of node state value δ in the node control mechanism. q_{start} is the initial node, and q_n is any node in the tree. q_{n+1} and q_{n+2} are child nodes of q_n. The values in parentheses are the node state values δ we recorded.

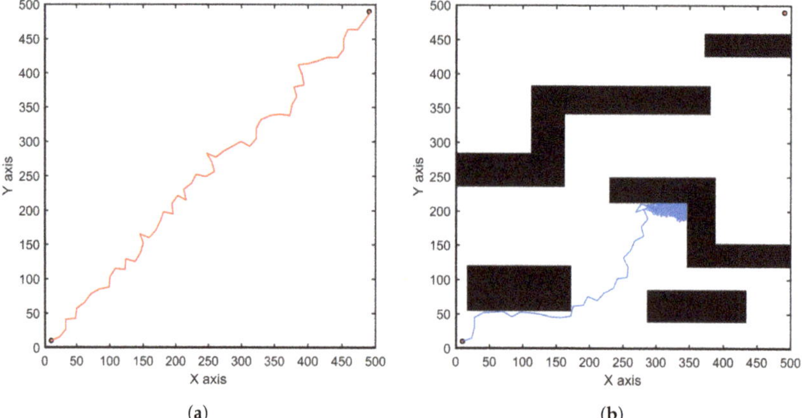

Figure 3. Under a *control* value of one and the combination of the exploration with the CSA method, the random tree exhibits different behaviors when it does and does not encounter obstacles. (**a**) Efficient exploration in an environment without obstacles. (**b**) The "local trap" phenomenon during exploration in an environment full of obstacles.

Therefore, we need to change the value of *control* at the appropriate time to increase the number of optional extension nodes to prevent this problem. This phenomenon is likely to occur when the tree encounters obstacles. Thus, we consider the collision occurrence in the expansion process approximately as the judgment condition for the occurrence of the "local trap" phenomenon, that is the condition for changing the value of the node control factor, similar to the change condition of the sampling radius in the CSA-RRT method. When collision occurs, the *control* value is increased, and the tree can effectively escape from this area. After the new node is successfully added, the *control* value is restored to one to continue the exploration. At this point, the NC-RRT algorithm proposed in this paper is obtained, and its pseudocode is shown in Algorithm 4.

Algorithm 4 NC-RRT algorithm.

```
T ← InitTree(q_start);
Δ_T ← {δ(q_start) = 0};
R ← D_farthest;
control ← 1;
for i = 1 to n do
    q_rand ← RandomSample(i);
    if Distance(q_rand, q_goal) > R then
        continue;
    end if
    T_ctrl ← LessThanControl(Δ_T, control, T);
    q_near ← NearestNeighbor(q_rand, T_ctrl);
    q_new ← Extend(q_rand, q_near, ε);
    if CollisionFree(q_near, q_new) then
        AddNewNode(T, q_new);
        Δ_T ← {δ(q_new) = 0};
        Δ_T ← BranchNodeUpdate(Δ_T, q_new, T);
        R = Distance(q_new, q_goal);
        control = 1;
    else
        R = R + k × ε;
        control = c;    (c = 2, 3, ...)
        continue;
    end if
    if Distance(q_new, q_goal) < ρ_min then
        return T;
    end if
end for
return Failed;
```

3. Simulation and Analysis

Computation in a two-dimensional space is small and yields an intuitive effect. Therefore, several algorithms were firstly simulated in a two-dimensional environment in this section to evaluate the performance of the proposed algorithm. Basic-RRT, CSA-RRT, and NC-RRT were analyzed and evaluated by comparing their simulation results in various environments. Subsequently, each algorithm was applied to a 6-DOF serial manipulator and simulated in an environment full of obstacles. Simulations of all algorithms were performed using MATLAB 2015b on a Windows 10 system with an Intel Core i7-8750H 2.2 GHz CPU, and 8 GB of RAM. In addition, a virtual prototype experiment of the 6-DOF manipulator was completed by using ADAMS. The simulation results in the two-dimensional space and the virtual prototype experiments of the 6-DOF manipulator were the average values of 50 runs.

3.1. Simulations in a Two-Dimensional Environment

Each algorithm was simulated in three different environments in a two-dimensional space. As shown in Figure 4, three typical maps were used: a map cluttered with obstacles, a trapped environment, and a narrow passage. The environment dimensions were 500×500 in all cases. The black areas in the map indicate obstacles, and the start and goal points were set separately in each map. The step size ε was set to 15, and the maximum number of failures nwas set to 2000. In addition, the k and c values in all the algorithms could be adjusted appropriately according to the different environments.

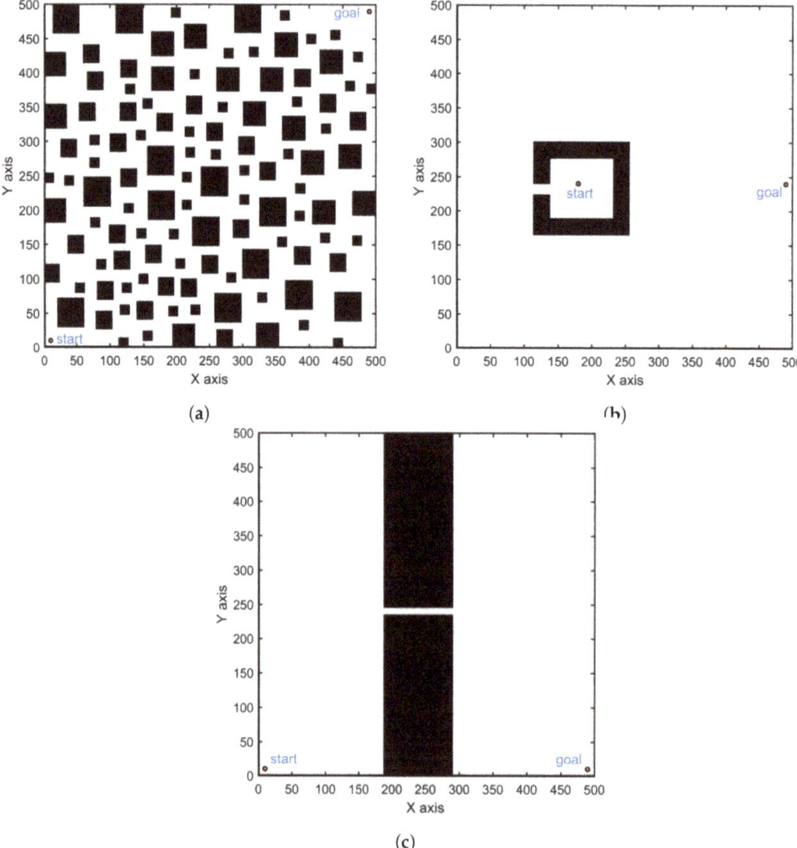

Figure 4. Three typical environments in a two-dimensional space. (**a**) The cluttered environment. (**b**) The trapped environment. (**c**) The narrow environment.

Figures 5–7 present the performance of the three algorithms in the cluttered, trapped, and narrow environments, respectively. The entire exploration process is denoted by blue lines, and the resulting query path is denoted by red lines. When the Basic-RRT algorithm was simulated, the nodes of the random tree almost filled the entire free space in each map. By contrast, the number of nodes was greatly reduced after the proposed CSA-RRT and NC-RRT algorithms were used. In addition, as seen in Figures 6c and 7c, the extended nodes in the tree were distributed more around the obstacles because of the node control mechanism in the NC-RRT algorithm.

Tables 1–3 present the simulation results of the three algorithms with respect to the average computational time, average number of nodes in the tree, average number of collision detections, average path length, and success rate in each environment. The results revealed that the proposed CSA-RRT and NC-RRT had good effects in each environment. Compared with that of Basic-RRT, the running times of the proposed algorithm were greatly reduced, and the numbers of nodes in the random tree and the numbers of collision detections were decreased.

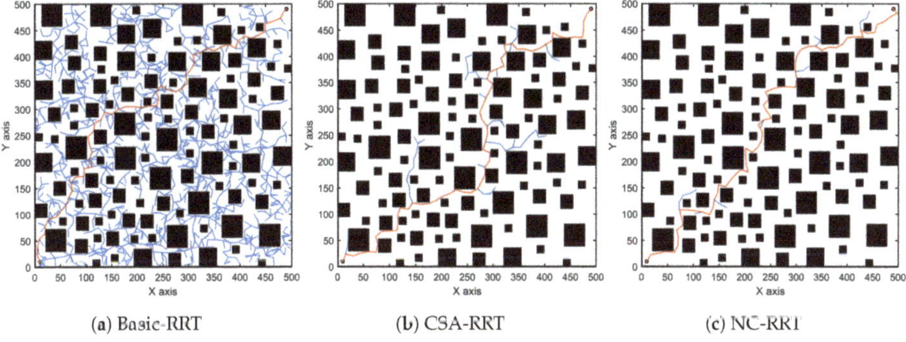

(a) Basic-RRT (b) CSA-RRT (c) NC-RRT

Figure 5. Performance of the three algorithms in the cluttered environment. ($k = 1,\ c = 2$).

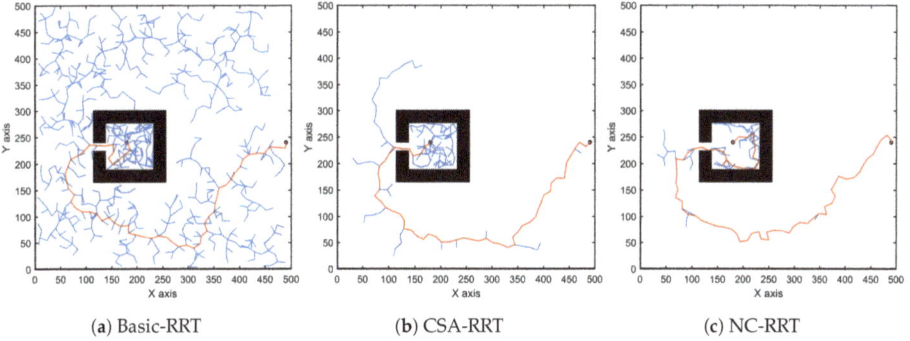

(a) Basic-RRT (b) CSA-RRT (c) NC-RRT

Figure 6. Performance of the three algorithms in the trapped environment. ($k = 3,\ c = 2$).

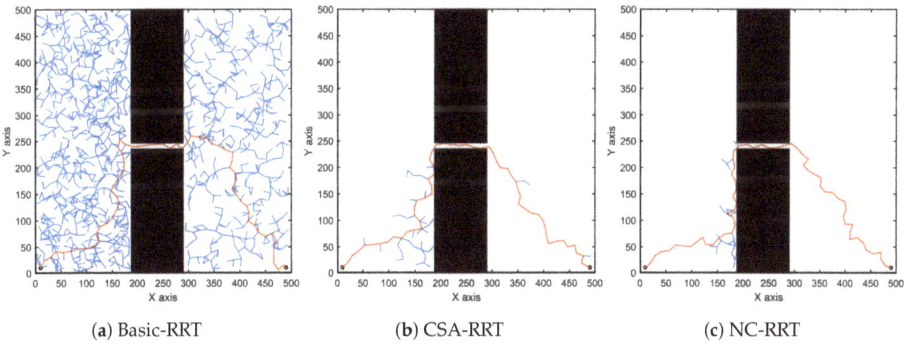

(a) Basic-RRT (b) CSA-RRT (c) NC-RRT

Figure 7. Performance of the three algorithms in the narrow environment. ($k = 1,\ c = 2$).

However, as can be seen from Tables 1–3, the average computational time of CSA-RRT and NC-RRT in the cluttered environment was 0.058 and 0.055 s, respectively; the average computational time in the trapped environment was 0.118 and 0.159 s, respectively; and the average computational time in the narrow environment was 0.106 and 0.117 s, respectively. Thus, the computing efficiency of the two algorithms in each environment was nearly the same. Moreover, the CSA method could reduce the path length to some extent, but the path length increased slightly after the node control mechanism was added. The reason was that the random tree in the NC-RRT method usually expanded along the boundary of obstacles. However, it was irrelevant for our purposes because the node control mechanism proposed in this paper was concerned more about how to improve the environmental adaptability of the algorithm by using boundary nodes than about the path cost. Furthermore, all the above-mentioned algorithms would eventually need to complete path pruning and smoothing when used in practice, so this cost could be ignored. In addition, we could see that the success rates of NC-RRT in the cluttered and trapped environments were not significantly different compared with those of the other two algorithms. However, in the narrow environment, the success rate was effectively increased, and the occurrence of pathological cases was reduced due to the excellent boundary property of the NC-RRT algorithm. Therefore, the proposed node control mechanism was necessary. In other words, the NC-RRT algorithm, which combined the CSA method and the node control mechanism, could effectively improve the efficiency of planning and was more suitable for complex environments than other algorithms.

Table 1. Simulation results of the three algorithms in the cluttered environment. $(k = 1,\ c = 2)$.

Algorithm	Average Computational Time (s)	Average Number of Nodes in the Tree	Average Number of Collision Detections	Average Path Length	The Success Rate of the Algorithm
Basic-RRT	0.463	844.520	38,026	954.369	96%
CSA-RRT	0.058	92.900	4860.1	915.982	100%
NC-RRT	0.055	82.860	3941.7	977.667	100%

Table 2. Simulation results of the three algorithms in the trapped environment. $(k = 3,\ c = 2)$.

Algorithm	Average Computational Time (s)	Average Number of Nodes in the Tree	Average Number of Collision Detections	Average Path Length	The Success Rate of the Algorithm
Basic-RRT	0.471	928.680	35,821	865.236	84%
CSA-RRT	0.118	122.909	9705.3	805.751	88%
NC-RRT	0.159	126.817	9214.6	992.595	88%

Table 3. Simulation results of the three algorithms in the narrow environment. $(k = 1,\ c = 2)$.

Algorithm	Average Computational Time (s)	Average Number of Nodes in the Tree	Average Number of Collision Detections	Average Path Length	The Success Rate of the Algorithm
Basic-RRT	0.550	1145.600	40,159	871.825	78%
CSA-RRT	0.106	172.732	8867.2	868.609	82%
NC-RRT	0.117	126.245	6746.4	942.615	98%

3.2. Simulation of the 6-DOF Manipulator

In this section, the above algorithms are applied to a 6-DOF serial manipulator. As presented in Figure 8, the manipulator was surrounded by obstacles (blue solid blocks) mainly distributed in the upper and lower parts of the manipulator workspace. These two parts restricted the path of the manipulator through a tunnel. Each algorithm was implemented and applied to the manipulator to complete the task of passing through the tunnel from the initial configuration to the target configuration. Because the success rate of using uniform sampling for planning in high-dimensional spaces was almost zero, all the algorithms in this experiment adopted the sampling strategy with a 10% goal bias to maintain the consistency of the conditions. The step size ε was set to 2°, and the maximum number of failures n was set to 2000. $k = 15,\ c = 2$ were set in the NC-RRT and CSA-RRT algorithms, and the average time obtained after 50 runs of each algorithm is presented in Table 4. It could be seen that compared with RRT with the 10% goal bias, the algorithm proposed in this paper significantly enhanced the efficiency.

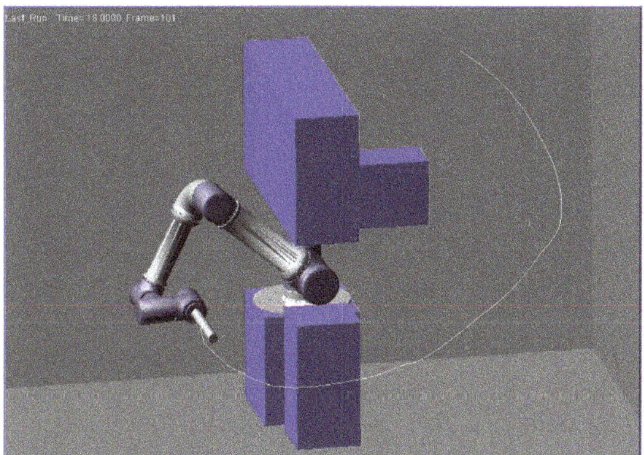

Figure 8. Simulation of a 6-DOF manipulator. The three algorithms are separately applied to the manipulator to allow it to pass through a tunnel and reach the target configuration.

Table 4. Average computational time for each algorithm applied to the 6-DOF manipulator (10% goal bias, 100% success rate).

Algorithm	Basic-RRT	CSA-RRT	NC-RRT
Average computational time (s)	3.658	1.541	1.469

4. Conclusions and Future Work

To address the problem of existing sampling-based planners having low exploration efficiency and poor environmental adaptability and the increasingly sophisticated level of intelligence requirements in the Industry 4.0 era not being met, this study proposed a path planning algorithm that was based on the architecture of the RRT algorithm and suited complex environments. The algorithm included a method of gradually changing the sampling area and a node control mechanism, which were used to guide the random tree exploration and reduce the expansion of invalid nodes, respectively. Furthermore, the node control mechanism could extract boundary nodes to improve the environmental adaptability. The algorithm was tested in three scenarios in two-dimensional space and was applied to a 6-DOF manipulator. The results revealed that the algorithm was effective and universal. It could significantly improve the planning efficiency and had a stronger applicability to complex environments, particularly narrow environments, compared with the traditional RRT algorithm.

However, the proposed method had certain limitations. The selection of the parameters k and c in the algorithm was a problem. Although the algorithm was universal, these parameters sometimes needed to be adjusted appropriately for different environments in order to obtain the best results. In addition, the path quality needed to be improved. This issue will be further researched by attempting an adaptive adjustment of parameters and considering the introduction of curvature constraints and other kinematic and dynamic constraints to improve the performance of the algorithm.

Author Contributions: Conceptualization, X.W. and X.L.; methodology, X.W.; software, Y.C.; validation, G.L. and K.Z.; formal analysis, X.L.; investigation, G.L. and K.Z.; data curation, X.L.; writing, original draft preparation, X.W.; writing, review and editing, X.W.; visualization, Y.C.; supervision, B.H.; project administration, B.H. All authors read and agreed to the published version of the manuscript.

Funding: This research was funded by National Key R & D Program of China (2016YFC0803000, 2016YFC0803005).

Acknowledgments: Thanks to Professor Qingsheng Luo for providing some suggestions to improve this manuscript. Additionally, thanks to Shanda Wang, Yan Jia and Lei Wang for the language help.

Conflicts of Interest: The authors declare no conflict of interest.

Abbreviations

The following abbreviations are used in this manuscript:

C-space	Configuration space
RRT	Rapidly-exploring Random Tree
PRM	Probabilistic Roadmap Method
CSA-RRT	Gradually Changing the Sampling Area-RRT
NC-RRT	Node Control-RRT
DOF	Degree-Of-Freedom

References

1. Barraquand, J.; Kavraki, L.; Latombe, J.C.; Motwani, R.; Li, T.Y.; Raghavan, P. A random sampling scheme for path planning. *Int. J. Robot. Res.* **1997**, *16*, 759–774. [CrossRef]
2. Hsu, D.; Latombe, J.C.; Kurniawati, H. On the probabilistic foundations of probabilistic roadmap planning. *Int. J. Robot. Res.* **2006**, *25*, 627–643. [CrossRef]
3. LaValle, S.M.; Kuffner, J.J., Jr. Randomized kinodynamic planning. *Int. J. Robot. Res.* **2001**, *20*, 378–400. [CrossRef]
4. LaValle, S.M. Rapidly-exploring random trees: A new tool for path planning. *Comput. Sci. Dept. Oct.* **1998**, *98*. Available online: http://citeseerx.ist.psu.edu/viewdoc/summary?doi=10.1.1.35.1853 (accessed on 19 February 2020).
5. Amato, N.M.; Wu, Y. A randomized roadmap method for path and manipulation planning. In Proceedings of the IEEE International Conference on Robotics and Automation, Minneapolis, MN, USA, 22–28 April 1996; Volume 1, pp. 113–120.
6. Kavraki, L.E.; Svestka, P.; Latombe, J.C.; Overmars, M.H. Probabilistic roadmaps for path planning in high-dimensional configuration spaces. *IEEE Trans. Robot. Autom.* **1996**, *12*, 566–580. [CrossRef]
7. Elbanhawi, M.; Simic, M. Sampling-based robot motion planning: A review. *IEEE Access* **2014**, *2*, 56–77. [CrossRef]
8. Thrun, S.; Montemerlo, M.; Dahlkamp, H.; Stavens, D.; Aron, A.; Diebel, J.; Fong, P.; Gale, J.; Halpenny, M.; Hoffmann, G.; et al. Stanley: The robot that won the DARPA Grand Challenge. *J. Field Robot.* **2006**, *23*, 661–692. [CrossRef]
9. Urmson, C.; Anhalt, J.; Bagnell, D.; Baker, C.; Bittner, R.; Clark, M.; Dolan, J.; Duggins, D.; Galatali, T.; Geyer, C.; et al. Autonomous driving in urban environments: Boss and the urban challenge. In *The DARPA Urban Challenge*; Springer: Berlin, Germany, 2009; pp. 1–59.
10. Yang, K.; Sukkarieh, S. An analytical continuous-curvature path-smoothing algorithm. *IEEE Trans. Robot.* **2010**, *26*, 561–568. [CrossRef]
11. Wei, K.; Ren, B. A method on dynamic path planning for robotic manipulator autonomous obstacle avoidance based on an improved RRT algorithm. *Sensors* **2018**, *18*, 571. [CrossRef] [PubMed]
12. Karaman, S.; Frazzoli, E. Sampling-based algorithms for optimal motion planning. *Int. J. Robot. Res.* **2011**, *30*, 846–894. [CrossRef]
13. Nasir, J.; Islam, F.; Malik, U.; Ayaz, Y.; Hasan, O.; Khan, M.; Muhammad, M.S. RRT*-SMART: A rapid convergence implementation of RRT. *Int. J. Adv. Robot. Syst.* **2013**, *10*, 299. [CrossRef]
14. Qureshi, A.H.; Iqbal, K.F.; Qamar, S.M.; Islam, F.; Ayaz, Y.; Muhammad, N. Potential guided directional-RRT* for accelerated motion planning in cluttered environments. In Proceedings of the 2013 IEEE International Conference on Mechatronics and Automation, Takamatsu, Japan, 4–7 August 2013; pp. 519–524.
15. Weghe, M.V.; Ferguson, D.; Srinivasa, S.S. Randomized path planning for redundant manipulators without inverse kinematics. In Proceedings of the 2007 7th IEEE-RAS International Conference on Humanoid Robots, Pittsburgh, PA, USA, 29 November–1 December 2007; pp. 477–482.
16. Kuffner, J.J.; LaValle, S.M. RRT-connect: An efficient approach to single-query path planning. In Proceedings of the 2000 ICRA, Millennium Conference, IEEE International Conference on Robotics and Automation, Symposia Proceedings (Cat. No.00CH37065), San Francisco, CA, USA, 24–28 April 2000; Volume 2, pp. 995–1001.

17. Kang, G.; Kim, Y.B.; Lee, Y.H.; Oh, H.S.; You, W.S.; Choi, H.R. Sampling-based motion planning of manipulator with goal-oriented sampling. *Intell. Serv. Robot.* **2019**, *12*, 265–273. [CrossRef]
18. Burns, B.; Brock, O. Single-query motion planning with utility-guided random trees. In Proceedings of the 2007 IEEE International Conference on Robotics and Automation, Roma, Italy, 10–14 April 2007; pp. 3307–3312.
19. Kim, D.H.; Choi, Y.S.; Kim, S.H.; Wu, J.; Yuan, C.; Luo, L.P.; Lee, J.Y.; Han, C.S. Adaptive rapidly-exploring random tree for efficient path planning of high-degree-of-freedom articulated robots. *Proc. Inst. Mech. Eng. Part C J. Mech. Eng. Sci.* **2015**, *229*, 3361–3367. [CrossRef]
20. Zhang, H.; Wang, Y.; Zheng, J.; Yu, J. Path Planning of Industrial Robot Based on Improved RRT Algorithm in Complex Environments. *IEEE Access* **2018**, *6*, 53296–53306. [CrossRef]
21. Boor, V.; Overmars, M.H.; Van Der Stappen, A.F. The Gaussian sampling strategy for probabilistic roadmap planners. In Proceedings of the 1999 IEEE International Conference on Robotics and Automation (Cat. No. 99CH36288C), Detroit, MI, USA, 10–15 May 1999; pp. 1018–1023.
22. Sun, Z.; Hsu, D.; Jiang, T.; Kurniawati, H.; Reif, J.H. Narrow passage sampling for probabilistic roadmap planning. *IEEE Trans. Robot.* **2005**, *21*, 1105–1115.
23. Zhong, C.; Liu, H. A region-specific hybrid sampling method for optimal path planning. *Int. J. Adv. Robot. Syst.* **2016**, *13*, 71. [CrossRef]

Article

Life Cycle Engineering 4.0: A Proposal to Conceive Manufacturing Systems for Industry 4.0 Centred on the Human Factor (DfHFinI4.0)

Susana Suarez-Fernandez de Miranda [1,*], Francisco Aguayo-González [1], Jorge Salguero-Gómez [2] and María Jesús Ávila-Gutiérrez

[1] Design Engineering Department, University of Seville, Polytechnic School, 41011 Seville, Spain; faguayo@us.es (F.A.-G.); mavila@us.es (M.J.Á.-G.)

[2] Mechanical Engineering and Industrial Design Department, University of Cádiz, Avenida de la Universidad nº 10, 11519 Puerto Real, Cádiz, Spain; jorge.salguero@uca.es

* Correspondence: ssuarez1@us.es

Received: 28 May 2020; Accepted: 25 June 2020; Published: 27 June 2020

Abstract: Engineering 4.0 environments are characterised by the digitisation, virtualisation, and connectivity of products, processes, and facilities composed of reconfigurable and adaptive socio-technical cyber-physical manufacturing systems (SCMS), in which Operator 4.0 works in real time in VUCA (volatile, uncertain, complex and ambiguous) contexts and markets. This situation gives rise to the interest in developing a framework for the conception of SCMS that allows the integration of the human factor, management, training, and development of the competencies of Operator 4.0 as fundamental aspects of the aforementioned system. The present paper is focused on answering how to conceive the adaptive manufacturing systems of Industry 4.0 through the operation, growth, and development of human talent in VUCA contexts. With this objective, exploratory research is carried, out whose contribution is specified in a framework called Design for the Human Factor in Industry 4.0 (DfHFinI4.0). From among the conceptual frameworks employed therein, the connectivist paradigm, Ashby's law of requisite variety and Vigotsky's activity theory are taken into consideration, in order to enable the affective-cognitive and timeless integration of the human factor within the SCMS. DfHFinI4.0 can be integrated into the life cycle engineering of the enterprise reference architectures, thereby obtaining manufacturing systems for Industry 4.0 focused on the human factor. The suggested framework is illustrated as a case study for the Purdue Enterprise Reference Architecture (PERA) methodology, which transforms it into PERA 4.0.

Keywords: life cycle; knowledge- and technology-intensive industry (KTI); manufacturing; VUCA; key enabling technology (KET); Operator 4.0; cyber-physical system (CPS); DfHFinI4.0; PERA 4.0

1. Introduction

Manufacturing processes are becoming increasingly automated and connected within companies. This means that the engineer has to acquire new competencies related to planning [1] and process management in VUCA (volatile, uncertain, complex and ambiguous) contexts [2], in both the manufacturing field and in the market. Due to these challenges in Industry 4.0, it is necessary to develop new competencies and improve existing competencies, for engineering students, as well as for professionals who have to design, manufacture, and manage interconnected smart products and processes. This requires the identification and development of training activities associated to the complex and creative characteristics of environments 4.0. These training activities are made possible by the potential of the interactions of digital enablers, and are integrated into the tasks to be carried out, whose main feature is the exchange of knowledge in real time.

Manufacturing systems 4.0 [3] have to be provided, not only with adaptability to VUCA contexts in their technological, economic, environmental, and social aspects, but also in a sustainable manner, so that, depending on the purpose established by the market [4], the system can co-evolve in a stable way. This entails a continuous evolution of the competencies associated to Operators 4.0, in order to deal successfully with increasingly complex and creative problems [5]. The aforementioned evolution of the competencies associated with Operators 4.0 gives rise to the interest in conceiving socio-technical cyber-physical manufacturing systems (SCMS), in which the processes and relationships between human and technological factors are integrated and can co-evolve, which is crucial in the management, development, and growth of smart factories and learning factories. This interaction of the human factor with machines and robots acquires major importance in these factories through interfaces based on cognitive and neurocognitive technologies.

The adjustment in the field of design and production management between manufacturing systems and Operators 4.0 [6] can be supported by cyber-physical systems, and from the possibilities offered by key enabling technologies (KETs) and new frameworks. The aforementioned lack of adjustment provokes the need for engineers and technicians to be trained as managers of the digital transformation by updating their knowledge and competencies with online support through interfaces that enable connection in workflows. The digital transformation process constitutes a key element that places the human factor at the centre of Industry 4.0, by creating integrated and co-evolutionary systems that take into account the work environment and the marketplace [7]; these systems are herein labelled SCMS.

The foregoing constitutes a research potential in which contributions have been made that refer to general aspects of the organisation of Industrial 4.0 [8–10], and others aspects related to the integration of Operator 4.0 within the socio-technical systems of Industry 4.0 [11–21].

This paper focuses on the life cycle engineering of manufacturing systems for Industry 4.0 [22–24], and the potential of KETs and the variety required thereof for the integration of Operators 4.0 towards the growth and development of such systems. In this paper, a new framework called Design for the Human Factor in Industry 4.0 (DfHFinI4.0) is therefore proposed, which allows the human factor to be placed at the core of Industry 4.0 and is based on the conceptual frameworks of the connectivist paradigm, the law of requisite variety, and on activity theory. Its main contribution lies in the modelling of SCMS from the consideration of the relationships between the human and technological factors (equipment and information system). This consideration enables the transfer of routine and smart competencies from human operators to the technical systems and takes full advantage of the engineer´s talent by encouraging competencies of greater scope derived from the incorporation of KETs. All this brings added value to the management of the engineering processes, technologies, and competencies required in the different phases of the life cycle of smart and learning factories, and ensures the adaptation of products and processes to market dynamics [25].

The goal of this paper involves responding to the problem of conceiving cyber-physical socio-technical manufacturing systems from socio-cognitive conceptual frameworks under the perspective of life cycle engineering, in which the growth and development of the talent of Operators 4.0 is made possible. This research objective, yet to be put into practice, constitutes exploratory research, with a qualitative approach, and uses deductive methodology, in which the DfHFinI4.0 framework is formulated from conceptual frameworks identified with bibliographic review techniques [26], whose feasibility is explored in the inclusion proposal for the Purdue Enterprise Reference Architecture (PERA) architecture and methodology.

In the design and application of the proposed methodology, the following steps can be distinguished:

- First, the state of the art is analysed to identify the gap.
- The conceptual frameworks are then presented and analysed to value their usefulness in the resolution of the formulated gap.
- The proposed framework is designed while taking the inclusion of the previous conceptual frameworks into consideration.
- Finally, the method is applied to a case study.

The conception of SCMS considering the perspective of life cycle engineering enables the work systems of the engineer and technicians to be configured as Operators 4.0, through the integration of all aspects associated with professional competence for their operational and effective efficiency [27], which permits the gap between human and technological factors to be bridged. Regarding these SCMS, the navigation by the Operator 4.0, through the cyber-physical space, is managed by the interfaces, and the associated training actions are adjusted for the acquisition of higher levels of experience and professionalism. All this fosters progressive adaptation from the academic to the professional field and career development. The proposed DfHFinI4.0 framework can be integrated into various life cycle engineering methodologies of smart and learning factories. A case study for the PERA methodology has been implemented in cyber-physical systems and is configured as PERA 4.0.

The organisation of the paper is structured as follows: Section 2 contextualises and considers the aspects related to smart manufacturing, while paying particular attention to the cyber-physical systems (CPS), for its projection in the conceptualisation of SCMS. Section 3 describes the conceptual domains that are employed to obtain SCMS. In Section 4, the conceptual domains are articulated in a DfHFinI4.0 framework, which allows the configuration of integrated and co-evolutive SCMS, and establishes the relationships between its elements. Section 5 applies the proposed framework to the PERA methodology, thereby transforming it into PERA 4.0. Section 6 lays out the discussion and proposes future work. Finally, Section 7 presents the conclusions.

2. Background of the Literature

In this section, from among the possible types of reviews characterised by Mayer [26], a status quo review is carried out that entails a description of the state of knowledge in smart manufacturing, especially regarding cyber-physical systems in Industry 4.0. The content of the review has been graphically represented, as shown in Figure 1. The result of the review will allow us to characterise different aspects of SCMS and the gap associated to the human factor.

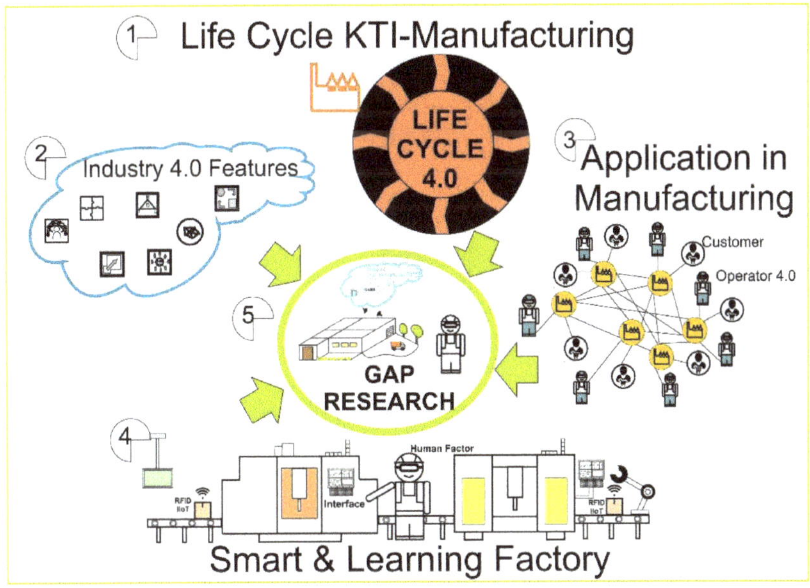

Figure 1. Background of the literature organisation.

2.1. Life Cycle Knowledge- and Technology-Intensive Industry (KTI) Manufacturing

The Organisation for Economic Co-operation and Development (OECD) taxonomy classifies the industries into five groups (high, medium-high, medium, medium-low, and low), and includes all the manufacturing industries in the high and medium categories [28]. Knowledge- and technology-intensive industries (KTIs) are those that have a particularly strong link to science and technology, and are classified by the OECD taxonomy as high- and medium-intensive R&D industries. Aerospace, computers and office machinery, testing instruments, pharmaceuticals, motor vehicles, chemicals, machinery and equipment, business, communications, and education constitute knowledge- and technology-intensive industries (KTIs) [29].

The evolution of automation, given the possibilities presented by digital enablers, connectivity and artificial intelligence, has made possible the inclusion of knowledge in a wide variety of cyber-physical elements [15], by proposing areas of research in which the distribution of knowledge and intelligence between Operators 4.0 and technological solutions is produced, not only with the assistance of KETs [30], but also by cognitive capabilities [9]. Another area of research [31] of great significance for the present work is related to the personalisation of technology and occupational environments that use subrogate models of Operators 4.0 to parameterise the adaptation of the technology, thanks to their ability to conceptualise and consider them as another cyber-physical system of the SCMS.

Within the framework of Industry 4.0, the life cycle engineering of manufacturing systems, together with the concept of cyber-physical systems of technical equipment, includes the concept of Operator 4.0 [32]. Its differential features include creative intelligence and expertise in the domain of knowledge that constitutes the field of responsibility. Their modes of operation in Industry 4.0 environments are performed cooperatively with robots and machines, and with cyber-physical resources, and employ advanced human-machine interaction technologies and adaptive automation to achieve a suitable degree of symbiosis [11–33]. Collaborative robots enable the creation of shared work environments where productivity can increase while minimising delivery response time [34], and developing tasks cooperatively to solve open or complex problems under creative approaches that are representative of VUCA contexts. This provides evidence that environments based on the combination of human and technological factors can successfully tackle such contexts [35].

On the one hand, all the challenges that characterise the smart manufacturing process require continuous innovation and learning [36]. Distributed manufacturing is supported by a real-time operation planning system that controls manufacturing networks [37]. Manufacturing processes are becoming increasingly automated and connected within organisations [1]. Complexity and flexibility in manufacturing require analytics, efficient problem-solving, and process improvements. Business intelligence (BI) analysis graphs represent expert knowledge on analysis processes [38]. Manufacturing involves collaborative information exchange from several sources under different working conditions [39]. The concept of collective intelligence has been applied in engineering within the field of cyber-physical systems (CPS). Knowledge is used by automated problem-solving methods to coordinate and supervise manufacturing systems. Ontology can thereby play a major role in the process of creating and managing knowledge [40]. On the other hand, lean manufacturing is a management model that focuses on minimising losses and optimising the creation of value for the client. Enke et al. [14] take into consideration the combination of lean manufacturing and Industry 4.0. Lean-based methodologies can improve organisational capabilities and tools to facilitate the transformation of a company into Industry 4.0 [41]. The concepts of lean manufacturing and Industry 4.0 can be developed in an end-to-end value chain for the Learning Factory to learn how to carry out a digital transformation [42].

2.2. Industry 4.0 Features

Among the characteristics of Industry 4.0 systems, boundaries between operation technologies (OT) and information technologies (IT) are disappearing [43]. Visualisation technologies, and fundamentally augmented [16], virtual and mixed reality are incorporated into production processes and training, since it has been proven that these constitute useful tools for Industry 4.0 [17,35]. Previously established immersion technologies [44], as well as brain-computer interfaces and brain-machine interfaces, improve manufacturing systems [45] and Operator 4.0 performance [35].

Several models and definitions, such as reconfigurable manufacturing systems, smart factories, and ubiquitous factories, are associated with Industry 4.0 [46]. The complexity and flexibility in which companies and engineers have to operate in these environments require analysis [10], efficient problem-solving, and process improvement. The use of business intelligence (BI) analysis charts represents expert knowledge for analysis processes and provides support in the work carried out by the engineer [38], from the engineering perspective of the life cycle 4.0, in cyber-physical systems design, development, and management environments [47].

In Industry 4.0 systems, the customisation of products, processes, and services requires flexibility in the manufacturing and intelligence for smart products. It is also possible to enable the integration of ontology-based web services for flexible manufacturing systems [48]. Semantic web technologies can be used with cybernetic systems to integrate the decision-making process into smart machinery [49]. This allows automated decisions to be made, to help in the configuration of the manufacturing system from its representation of a virtual or digital twin. Therefore, ontology can play a major role in the process of creating and managing knowledge of cyber-physical systems [40].

One critical aspect of Industry 4.0 relative to advanced manufacturing is that of the availability of real-time information to optimally program the objectives of manufacturing systems throughout systems that have edge, fog, and cloud architecture [50]. Qu et al. [51] propose an ontology-based framework to represent a synchronised and station-based flow workshop, and develop a multi-agent reinforcement learning approach for optimal programming.

In order to apply simulation solutions that improve the efficiency and profitability of Industry 4.0 systems, digital twins are created to describe the behaviour of the system. Stark, Kind and Neumeyer [47] consider the digital twin as the digital representation of a product, machine, service, product service system, or other intangible assets, that alters their properties, conditions, and behaviour, through models, information, and data. This concept is not only restricted to the operational part, however: it is also transposed to the human component [18] by generating a digital twin for the Operator 4.0 [31]. Recent developments in machine learning and other big data techniques offer new possibilities in conjunction with the concept of a digital twin [52] and subrogated models.

2.3. Application in Manufacturing

Relevant themes that have emerged as a result of flexibility, customisation, optimisation (saving time and costs), and smartisation, and those that require increasing connectivity include: predictive maintenance, which serves the objectives of sustainable manufacturing in the three dimensions of 3E; virtual commissioning [53] and crowdsourced manufacturing organisations [54]; real-time online support for production operators; new opportunities for servitisation; co-design; and co-manufacturing, cloud manufacturing and social manufacturing [55].

The role of Industry 4.0 maintenance, especially that of predictive maintenance, presents a strategic factor in manufacturing [56]. Techniques, such as forecasting, health and safety management (PHM), and condition-based maintenance (CBM), create a demand for Operators 4.0 with adaptive interfaces that allow suitable characterisation and development of the required maintenance, with the necessary connectivity and the appropriate decision support [19]. The implementation of such techniques requires the use of the Industrial Internet of Things (IIoT) [40], cloud computing [57], big data [58], machine learning, and augmented reality [59,60]. In this respect, Cachada et al. [56] describe the architecture of intelligent and predictive maintenance to support Operators 4.0 by providing guided intelligent decision support.

Virtual commissioning, through the creation of a simulation model in a virtual environment of a manufacturing plant, allows Operator 4.0 to propose the necessary changes for its subsequent implementation in the real plant. However, today´s lack of competencies and associated experiences hinder the full integration of this tool in manufacturing [53].

Crowdsourced manufacturing organisations share their manufacturing resources based on their demand and capacity. Kaihara et al. [54] have developed a manufacturing simulation model in collaboration with a resource model and a negotiation algorithm based on cyber-physical systems to evaluate the effectiveness of manufacturing. This enables Operator 4.0 to cover the task of resource manager and requires interfaces capable of providing real-time feedback on shared resources and on any needs that may arise.

2.4. *Smart and Learning Factories*

The new professional engineer profiles that have emerged in Industry 4.0 environments require a suitable characterisation of the competencies to be acquired [20,61], in order to interact with smart manufacturing agents. To this end, the learning factories have been developed and are employed to instruct and train engineers through an approach between learning and professional practice, thereby contextualising Industry 4.0 environments. Furthermore, learning factories enable applied research to be carried out, both in engineering areas and in other areas of interest [8,21], and also foster collaboration between companies, students, and universities [62], with dual training models.

For successful operations in smart and learning factories, not only must training in technical competencies be taken into account, but also training in solving complex problems with uncertainty and decision-making in real time, under time pressure [63]. Special attention should be given to the acquisition of creative competencies, innovation, multicultural teamwork, and the ability to solve complex problems, all as enablers for their operation in VUCA contexts [61]. Research studies are currently being carried out in which transformation processes are developed in a manufacturing workshop, led by training in a learning factory, which involves instruction, integration, and engineering. This instruction is related to training strategies and objectives [64], in which serious gaming instructional techniques are incorporated, which can be employed to develop competencies related to new technologies in Industry 4.0 [65].

A smart factory [66] is made up of cyber-physical systems consisting of a physical part and an associated digital twin, with great possibilities for connectivity, intelligence, and data processing in the cloud and in the fog, and for operation with subrogate models [50]. These operational environments determine the need for tools that integrate the potential of the competencies acquired by the engineer with the use of technologies 4.0 [64]. In this context, Operator 4.0 constitutes a cyber-physical system with the possibility of multiscale and multilevel connectivity, with great analytical, calculation, and simulation capacity.

The development of learning factories, which allows the introduction of the concept of a digital twin and its applications, is necessary not only at the level of large companies, but also for SMEs (small and medium-sized enterprises) [13,67]. This leads to the integration of all the possible business fabric within the concept of Industry 4.0, whatever the size of the company.

2.5. *Research Gap*

Once the review of the most specific research parameters of Industry 4.0 and its manufacturing systems has been carried out, these parameters can then be characterised in a more detailed way as shown in Table 1, in order to establish a way in which to conceive the adaptive manufacturing systems of Industry 4.0, through the operation, growth, and development of human talent in VUCA environments. In this table, those papers that are most relevant for their contribution towards adaptive manufacturing through synergistic actions of processes, technology, and human factors in SCMS, characterise the research gap.

The contributions of those various studies whose objective involves the integration of technological enablers and their potential for adaptive manufacturing have been examined, both for exogenous changes (external client and contextual factors) and for internal changes (internal client and contextual factors). Identification has been made of the need for: conceptual resources and frames that allow the analysis of the productive activity located at any level of granularity; the search for allostasis (dynamic balance) between technology, processes, and Operator 4.0; and for the leverage of human talent and its affective coupling to manufacturing systems. In the following section, conceptual frameworks are presented that lead to the synergistic and adaptive manufacturing system framework proposed in the paper.

Table 1. Specific research parameters of Industry 4.0 and its manufacturing systems.

	Adaptative Manufacturing					
	D1	D2	D3	D4	D5	D6
[8]	✗	-	✓	✗	-	✓
[9]	-	✓	-	✗	✓	✗
[10]	✗	✓	-	✗	-	-
[11]	✗	✓	✓	-	-	-
[12]	✗	-	✓	✗	✓	✗
[13]	✗	✓	-	✗	✗	✗
[14]	✗	-	-	✗	✗	✓
[15]	✗	-	-	-	-	✗
[16]	✗	-	✗	-	✗	✗
[17]	✗	✓	-	✓	✗	✗
[18]	✗	-	✗	✓	-	✓
[19]	✗	-	✗	✓	-	✗
[20]	✗	✓	-	✗	-	✗
[21]	✗	-	✗	-	✓	✗

The levels considered are: (✗) minor level, (-) medium level and (✓) mayor level. D1: Modelling and simulation of SCMS. D2: Assistance in decision-making and navigation strategies. D3: Soft and hard skills. D4: Affective interaction. D5: Cognitive, socio-cognitive and cultural ergonomics. D6: Competence management and talent development throughout the professional life cycle.

After this review of the literature, observation is made of the effort being carried out in the field of digital transformation for the application of key enabling technologies (KETs), as shown in Figure 2, and of the implementation of SCMS, which enables successful operations in VUCA environments. The preceding situation modifies the bases of productivity and the competencies required to adapt to this new situation, in which the life cycles of products, manufacturing technology, and knowledge are becoming shorter and more volatile.

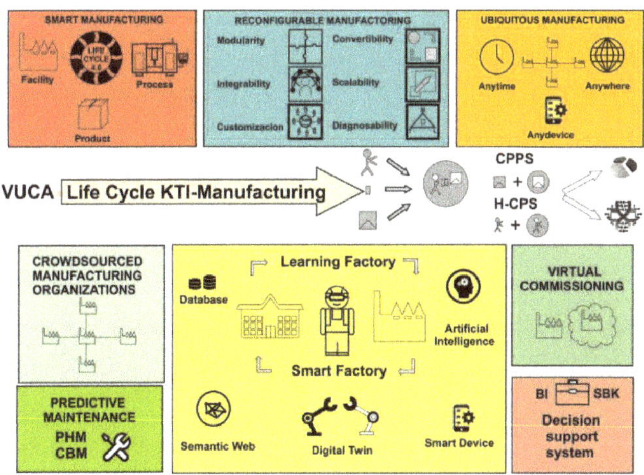

Figure 2. Application context for DfHFinI4.0.

Life cycle engineering 4.0 must be equipped with tools to integrate experience and to project human talent in the growth and development of SCMS from the opportunities of the operational environment. To this end, it is necessary to identify a set of conceptual frameworks that enable tools to be derived: for the conception of the work in the SCMS environment, in the most complete and possible way; to integrate the various elements therein, both individually and socially; and to assign tasks to the human and technological factors so that they work collaboratively by establishing adaptive interfaces. For this purpose, Vigosky's activity theory [68,69] has been considered.

Complementary to the decision to identify a conceptual framework for the formalisation of work in all dimensions of its complexity, its conception is required to integrate elements and solutions that allow its adaptation and self-regulation, depending on the operational, cognitive and affective variety of the Operators 4.0 that can undertake the job, thereby allowing its coupling to the established work system. Ashby's law of requisite variety has therefore been selected [67,70], since it is integrated in the activity theory, which enables the self-regulation of tasks in accordance with the Operator 4.0 who develops it and with the affective connection required.

Finally, the further aspect of the human factor integration that requires special attention is that of the co-evolutionary development of the joint action of the socio-technical system which enables growth and development. This aspect provides online support strategies in the development of tasks and assistance in the growth and structuring of the experience acquired in the form of lessons learned, for their subsequent reuse in analogous situations, and consequently enables navigation strategies in the information system under semantic websites that allow the innovation and growth at both the individual and collective levels. To this end, the instructional framework of connectivism has been selected to derive strategies and tools [71,72].

Subsequently, these conceptual areas are structured in the proposed framework which is oriented towards the integration of the human factor within the environment of a smart company [73] which configures an SCMS, based on the search for the best available techniques, with the aim of seeking manufacturing excellence. All these techniques integrate, on the one hand, the potential of emerging conceptual frameworks; on the other hand, they integrate key digital and technological enablers, by merging the human and technological factors that correspond to the environment that produces them.

In the following section, the conceptual approaches and the KETs that will integrate the framework are presented, and an analysis is given of the most characteristic features that make them value drivers for the conception of the DfHFinI4.0 framework.

3. Conceptual Frameworks

The adaptation by companies to the requirements of Industry 4.0 entails the appropriate training of both engineers and technicians [42] that is focused on handling problems of a greater complexity and open situations of a wider scope, as depicted in Figure 3. The acquisition, training, and improvement of these competencies, defined in accordance with the updating of the tasks to be carried out Operator 4.0 and their relationship with the other elements of the SCMS, enable professional profiles to be generated in accordance with the requirements in the smart and learning factories [74]. This represents a major training challenge, both in the academic and professional fields [75]. The engineer, as Operator 4.0, has to be capable of handling interoperability, virtualisation, decentralisation, service orientation, modularity, and key technology enablers (KETs) [76].

Along with the new demands of competencies required for workers, it is necessary to conceive the technology and associated processes of interaction with operators in Industry 4.0, as cognitive and socio-cognitive systems with affective connectivity. These characteristics make possible the integration and co-evolution of the SCMS formed by the human and technological factors under an organisational framework in the cultural context of a given company. Both factors, human and technological, are coupled [77] through adaptive interfaces that co-evolve as a cyber-physical system, together with Operator 4.0 and industrial equipment.

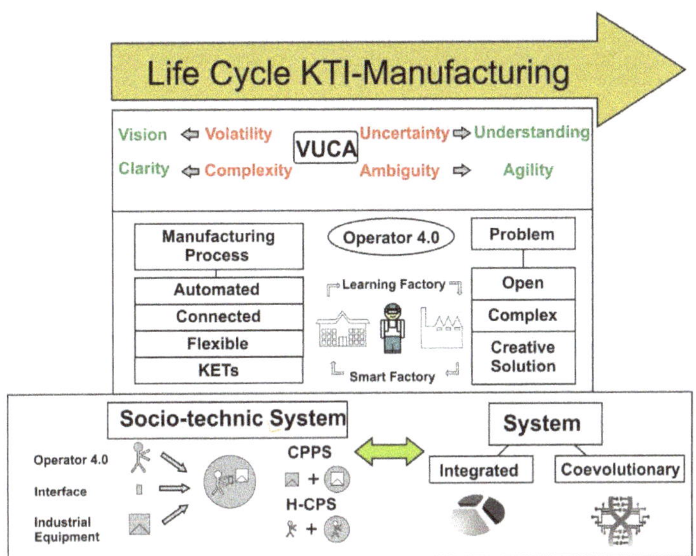

Figure 3. Smart Manufacturing requirements.

As shown in Figure 4., the potential from KETs triggers the search for a framework articulated in a toolbox that enables affect-socio-cognitive SCMS to be obtained, that respond to the different situations in which the engineer as an Operator 4.0 must be competent within an Industry 4.0 environment. This potential is articulated in conceptual frameworks derived from other areas of research: the theory of action [78]; the activity theory [79]; the required variety of Ashby for affective integration and coupling; and support in obtaining objectiveness and intentionality in the domains of cyber-physical entities through the connectivist paradigm:

- For the division of the labour to be performed by the engineer and technicians as Operators 4.0, the formalisation and analysis of its elements is carried out by Vigotsky's activity theory, as a tool that supports the elements of work, their variety, and the social fabric in which they are developed.
- The work to be carried out by engineers requires adaptation to their cognitive and affective characteristics, as well as to the particular characteristics of the task to be performed. Consequently, Ashby's law of requisite variety is employed, which is articulated in different elements and relationships of the activity theory.
- The establishment of the network workflow in real time, as well as the training required depending on the type of situation requested, are carried out by applying the connectivist methodology, which provides the supports and strategies of online navigation.
- The potential of these conceptual frameworks is implemented under the DfHFinI4.0 framework with KETs.

The proposed DfHFinI4.0 framework requires the use of high-frequency trading concepts and technologies (HFTs) that are useful in modelling cyber-physical systems 4.0. A descriptive review of their state of knowledge based on Squires [80] is therefore proposed, which generates a proposal adapted to the field of smart manufacturing. In the following subsections, these various conceptual approaches are developed and articulated within this field.

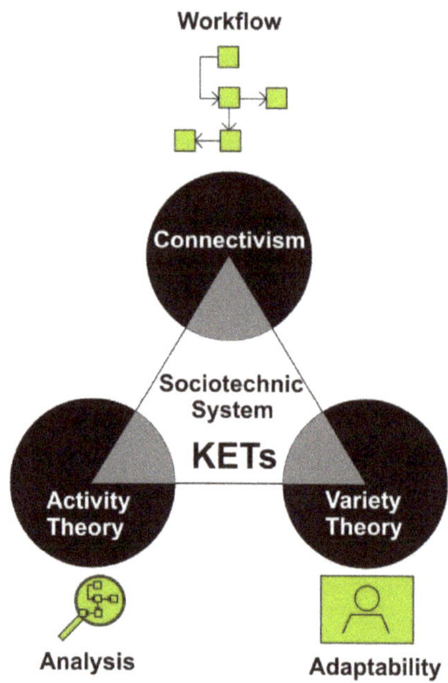

Figure 4. Conceptual frameworks that configure DfHFinI4.0.

3.1. Activity Theory

According to Vigotsky's activity theory, the action of the engineer as an Operator 4.0, which is represented in Figure 5, takes place in a SCMS [81], where the engineer with the interface solves problems or, with the help of a tool, transforms the raw material or the product [82] as a result of its application. The interaction between the human factor and machines and robots is mediated through interfaces [12], whereby the latter constitute the tools within the activity system [83]. The adoption of Vigotsky's activity theory as a unit of analysis for containing all its elements involved in work situations was reformulated by Yrjo Engeström [84]. This author has considered its evolution through three generations of research. Its use is especially significant as an instrument to model the activity and knowledge, that, in Industry 4.0, is largely transferred from technology to the human factor.

In the second generation of activity theory, Engeström expands the Vigotsky triangle to represent the collective elements of the activity system. From the Engeström model [85], it is possible to develop a general structure of the activity based on Operator 4.0, machines, robots, work equipment, interfaces, rules, and division of labour, as well as on the knowledge implicit in the technology (tool) and in the operator, and even in the evolution of their competencies as a consequence of the development of work. The rules and regulations, explicit and implicit, define the course of action to accomplish the task. The tasks are carried out according to the organisational structure of the company, through a division of labour. In the third generation of activity theory, tools are developed for dialogue, diversity of perspectives and networks of interacting activity systems. Modelling operational activity with activity theory enables the analysis to be performed with different granularity [86], and also the incorporation of the analysis of the different dimensions of the elements for the study of their integration and of the effect that arises from articulated solutions.

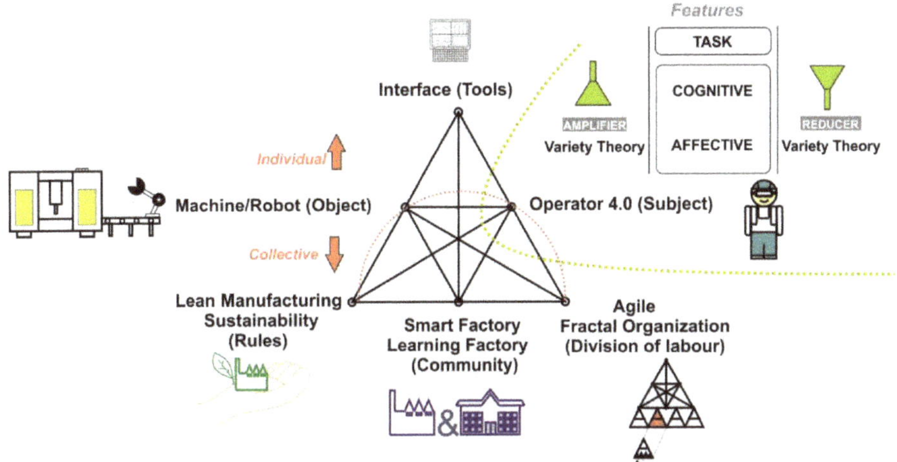

Figure 5. Activity theory and Law of requisite variety to model characteristics of Operator 4.0.

This theory has been applied, among other areas, in training organisation [87], human-computer interaction (HCI) [88], and information systems (IS) [89], since it has been proven to be a highly useful tool in the establishment of the way in which ICTs and other technologies interact with their context [90]. In this system of analysis of the proposed activity, the operator and the technological and organisational elements are interrelated [91], thereby enabling a holistic analysis of the socio-technical system. Its articulation in engineering has great potential for engineers to adapt to the requirements of smart and learning factories [92], when it comes to modelling and responding to training and co-evolution needs [93,94]. All this reduces the static and dynamic complexity in manufacturing systems regarding cyber-physical systems.

Activity theory [95] facilitates interactivity and assertive navigation, and encourages the engineer's creativity as an Operator 4.0, in the resolution of open and complex problems. This is made possible by enhancing not only cognitive competencies [96], but also affective coupling, since the latter enables better responses to be made to possible feedbacks, thereby facilitating the resolution of problems in collaborative work. In this respect, the same objective as that which the video game industry [97] demands from designers can be applied to smart manufacturing; an innovative design that attracts the attention of the internal customer (Operator 4.0), where there is constant feedback at the cognitive and affective level, which encourages constant engagement. This can be employed to design interfaces and the work system as a whole, with the aim of promoting its use through motivation [98], affectivity, dependency, and feedback generated by the engineer.

3.2. Law of Requisite Variety

The work analysis requirements focused on the explicit and implicit knowledge that cyber-physical manufacturing systems must support, both for human and technological factors, to justify the incorporation of Vigotsky's activity theory as a tool to formalise the elements that integrate a work system, both in its individual and social dimension, with the aim of obtaining socio-cognitive manufacturing systems. By taking this theory into consideration, it is possible to identify various elements of the work and their features. Such is the case of Operators 4.0 and their associated competencies, the tools, products and problems on which they take part, the organisation and specialisation required for the work accomplished, in addition to the associated operational culture. Nevertheless, the aforementioned analysis suffers from the inclusion of mechanisms to identify, among others, sensory, cognitive, and affective capacities, and experience of the Operators 4.0 that will be assigned to a workstation and the required demands of the work system. This situation considers the need to identify a conceptual framework that enables the characterisation of the variety of competencies of workers and the work system demand with which they interact, and to determine their discrepancy and variety adaptation mechanisms. The aforementioned mechanisms and the affection that comes from their implementation must be represented in the activity theory for the articulation of mechanisms that allow the adaptation of the required variety of the technological system to the variety established in the worker.

In relation to this, given the need for a conceptual framework to adapt variety on the structure of the elements of the activity theory, it is worth considering Ashby's law of requisite variety [67]. This law establishes that a regulator-regulated system is one that is made up of a regulatory subsystem that exercises its action based on the information collected from the regulated subsystem [70]. In this kind of system, the regulatory part (work station, machine, robot, process) must have at least the same variety as the regulated part (Operator 4.0 or work equipment 4.0) so that the system reaches stability, which necessitates the establishment of a one-to-one correspondence between the varieties on each side. In order to achieve this adjustment, adapters are employed to reduce or amplify the variety, depending on what is required [99]. This adapter is itself a system or part of a system, and can act in either direction, increasing the variety by means of amplifiers or reducing it by means of reducers, until the regulator and regulated subsystems reach the same variety and determine an affective occupational use experience.

The adjustment of the collaborative environment to the engineers as Operators 4.0 [100] can be carried out by taking into account the law of variety (law of requisite variety), and will depend both on the constraints of the tasks to be performed [101] and on the cognitive [102] and affective [103] features possessed by the engineer and the technicians who operate in Industry 4.0, as illustrated in Figure 5 above.

Regarding the adaptation elements of the variety, in the context of Industry 4.0, it is necessary to consider the characteristics of the Operator 4.0 tasks, limited to situations of complex, creative problem-solving, with uncertainty and deadlines. This requires online support regarding knowledge, embodied in Operators 4.0, from the company's knowledge bases and other stakeholders in the project. All this determines that, as adaptation strategies of the variety of technology 4.0 (through interfaces, mobile devices, tablets, and wearables), the potential of KETs and the management of the variety required are utilised, so that the navigation strategies provide the online support required by the Operator 4.0.

As mentioned earlier, this enables the personalised occupational activity of Operators 4.0 to be modelled, who, in their implementation of achievements, determine the support of navigation strategies and of the dynamic management of the operational requirements, of their learning and improvement of competencies and of the reconfiguration of the technological environment 4.0 for the variety required according to the designated worker. In order to answer these questions, it is considered as a conceptual framework to manage the variety in the navigation strategies, learning support, and systematisation of lessons learnt in the connectivist paradigm [104], which will be presented with KETs.

3.3. Connectivist Paradigm

In the field of smart manufacturing, connectivity acquires major importance in a cyber-physical environment, with the hybridisation of the physical and digital world, and with artificial intelligence and knowledge, not only in the cloud, but also integrated into production, equipment and tools in the edge. This has resulted in Operator 4.0 competencies moving towards complex and open creative, social, and problem-solving competencies, with opportunities for continuous learning. Support in the navigation strategies of Operator 4.0 is necessary for the resolution of problems; it is here where the connectivist theory of Stephen Downes and George Siemens intervenes, which affects the process of connecting specialised knowledge [105,106], under the theoretical principles of the Clark and Chalmers concept of the extended mind [107].

A crucial component of this connectivity falls on the semantic web as the informational dimension of the extended mind [108]. On the one hand, the semantic web is an intelligent entity that generates, shares, and connects content, capable of being interpreted by operators, machines and robots, to work collaboratively through specific languages such as XML, RDF, RDFS, OWL. It enables the articulation of connectivist schemes for the online support of navigation strategies in problem-solving, learning, collaboration, and systematisation of the experience. This semantic web enables the meaning of this data to be interpreted in a similar way to that of the semantic analysis of the operator's language and therefore to support the workers in their various tasks. It provides communication content between Operator 4.0 and machines and robots with semantic content, and allows the information to be processed based on a semantic assessment of its content, which permits its best coupling through interfaces so that it can be optimally interpreted by the engineer [49]. On the other hand, the use of ontologies in the semantic web enables the correct identification of the meaning of instructions according to a given situation and context [109]. Together, these facilitate the establishment of interactive and navigation strategies to support problem-solving and to enable the learning process and the improvement of competencies [110].

Connectivism integrates the IIoT, cloud computing, and virtual reality technologies, among others, which enable connection, accessibility, and data sharing [111], as depicted in Figure 6. The engineer has to perform tasks, both training and professional, by collaboratively exchanging information in real time with machines and robots, in order to adapt to different working conditions [39]. Furthermore, connectivism seeks to create collaborative environments, connected not only between Operator 4.0 and machines and robots, but also between end customers, vendors, suppliers, and all those agents involved [112]. These strategies, in the form of instructional knowledge and subrogate instructional models, will be carried out from the cloud.

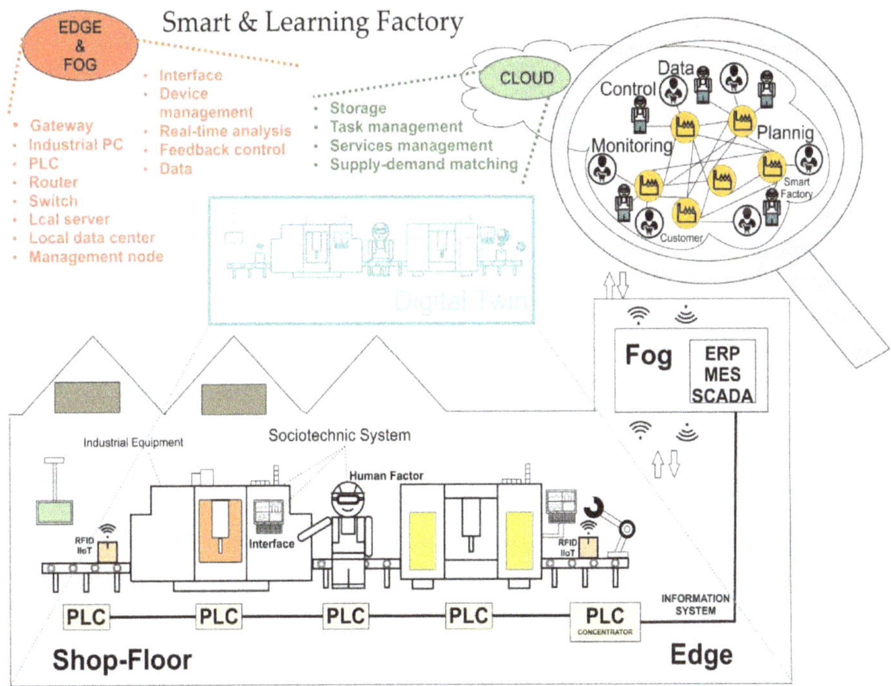

Figure 6. Connectivity in the Smart and Learning Factories.

Connectivism is based on theories of neural networks, chaos, complexity, self-organisation, and non-linear systems, from which navigation strategies are structured. Due to these influences, it integrates tools that increase, from connectivity and virtuality, the ability to facilitate interaction between the elements that shape the smart and learning factories. Connectivism has been described as the amplification of knowledge and understanding through the extension of the network, and is called the theory of knowledge for the digital age [106].

Thanks to these characteristics, the connectivist paradigm has been established as the basis for the cyber-physical socio-technical system, since it facilitates the real-time network support of cyber-physical systems as part of DfHFinI4.0, which allows constant and dynamic network workflow and training.

The navigation strategies, in conjunction with a variety of filters and amplifiers, are established in an integrated manner regarding the activity theory, which in turn, as support for activities of less granularity, constitute fractal elements [113] of top-down and button-up analysis of the SCMS for Industry 4.0, and, through the KETs, determine its greater significance by acting as lever arms.

3.4. KETs

Key enabling technologies have been defined as knowledge-intensive technologies associated with high-intensity research and development, rapid innovation cycles, high capital expenditure, and high-skilled employment [114]. They can also be classified as advanced manufacturing technologies, advanced materials and nanotechnologies, life-science technologies, artificial intelligence, micro/nano electronics and photonics, and security and connectivity tools.

Examples where KETS have a place include those of an organisational type (fractal, holonic, or bionic organisation), digital (cloud, big data, AR&VR, wearables, and mobile devices) and technological type (collaborative robots, additive manufacturing, etc.) [115] Data generation, data analytics and decision-making technologies are examples of KET-artificial intelligence. Human-machine interfaces and cyber-physical systems provide examples of KET-security and connectivity [116,117].

Information and communication technology (ICT) and KETs challenge traditional production structures and require the search for alternative and innovative solutions to those proposed so far, through the use of disruptive activities and compentencies by Operator 4.0. This requires engineers with the ability to solve problems based on different outlooks rather than those already raised by cultural heritage. At this point, the third generation of activity theory becomes useful by allowing the transcending of established frameworks and by generating creative and innovative solutions that break with the mental structures of established systems, thereby providing solutions to the new challenges that arise.

KETs are used as variety filters and amplifiers to support DfHFinI4.0. Connectivism, together with the law of requisite variety, activity theory and KETs, help SCMS modelling, and enable real-time support and assistance for the engineer, through the design of dynamic interfaces that co-evolve according to needs considered in engineering environments 4.0, which are characterised by the digitisation and virtualisation of products and processes. This involves an impact on improvement in the acquisition and training of competencies, as well as on the affective connectivity established between human and technological factors.

4. DfHFinI4.0 Framework

As shown above, the technological environment promoted by Industry 4.0 determines that the tasks of Operator 4.0 are mainly focused on solving complex problems, decision-making, and the ability to adapt to new scenarios and situations in which human and technological factors work collaboratively. It is therefore important to define, design, and build SCMS in a way that allows part of the knowledge of the workers to be supported, and to respond and support, giving response and support to articulate competencies of a wider scope at the cognitive, operational, affective, and co-evolutive levels with the best working conditions. The acquisition of the competencies required in Industry 4.0 for the development of day-to-day tasks and the accumulation of know-how from the lessons learned at the workstation must also be supported. The context in which the analysis of work must be placed is within that of the value chain as an instrument of analysis of added value that incorporates the activity into products and is demanded by customers. One example of this is related to lean manufacturing systems [41], in the form of a map of the value stream, otherwise known as value stream mapping (VSM).

The analysis of the value chain of smart factories must be carried out, while considering their primary and support activities in operating conditions (professional domain), and in learning factories (academic domain). To this end, the potential of digital enablers and their projection in the integration of the human factor, machines and robots for collaborative work performance must be considered, by interacting jointly, through interfaces for both primary and support activities. The latter are elements of the SCMS that enable the interaction and navigation of Operator 4.0, not only internally, but also externally, with other socio-technical systems from other value chains, such as suppliers, distributors, customers, and plants of reverse manufacturing. This interaction takes place at a physical and virtual level, and hence the interfaces must be capable of acting as links between both physical and virtual realities. Their study is required at the level of sustainable smart manufacturing value chain design, and of associated chains of suppliers and distributors.

The productive activity in the manufacturing systems is carried out from a set of activities represented in its value chain, or VSM, based on customer value maps, which are specified downstream with various lean manufacturing tools. Subsequently, the requirements of processes, material and information flows, scheduling, tasks, equipment, operators, and their associated competencies are established [118]. As indicated before, the value chain contains primary activities and support activities that host the KETs as a means of improving their efficiency and sustainability. The process of incorporating the KETs into the activities of the value chain, focused on the human factor, requires their analysis, the potential of the KETs, the distribution of activities between those carried out by the human factor and the machines and collaborative robots, the knowledge required for this analysis, and strategies that support the entire manufacturing system, whose priority involves the development of strategies to empower the human factor and its affective coupling to the environment.

Based on the analysis of the added value, the analysis of the potential of the KETs and of the objective of SCMS focused on the human factor, a set of tools is proposed for the of the analysis. As illustrated in Figure 7, this analysis of the value creation activity is based on the conceptual frameworks, for the configuration of the value chain as an integrated and co-evolving SCMS to the highest degree of abstraction.

The set of proposed tools are related to the activity theory model proposed in a fractal way in terms of various degrees of granularity [119], based on the analysis of the levels of company, departments, activities and workstations. Subsequently, the variety required of Operator 4.0 in workstations is characterised, as is the variety required for the successful and satisfactory development of the task, by means of the addition of the necessary filters and amplifiers through adaptive interfaces, and by establishing the connectivist navigation strategies that allow the operator, assisted by the competencies, to manage the lessons learned in day-to-day work. This ensures that the process, machines and robots have at least the same variety as Operator 4.0, so that the cyber-physical socio-technical system reaches stability. The modelling of the SCMS employs the duality established by the digital twin by integrating the human and technological factors through adaptive communication interfaces, thereby allowing the generation of the variety of use required for the sustainable production process, through subrogate models that are dynamically built and managed from the cloud with tools of big data and artificial intelligence, such as machine learning, classification techniques, and deep learning.

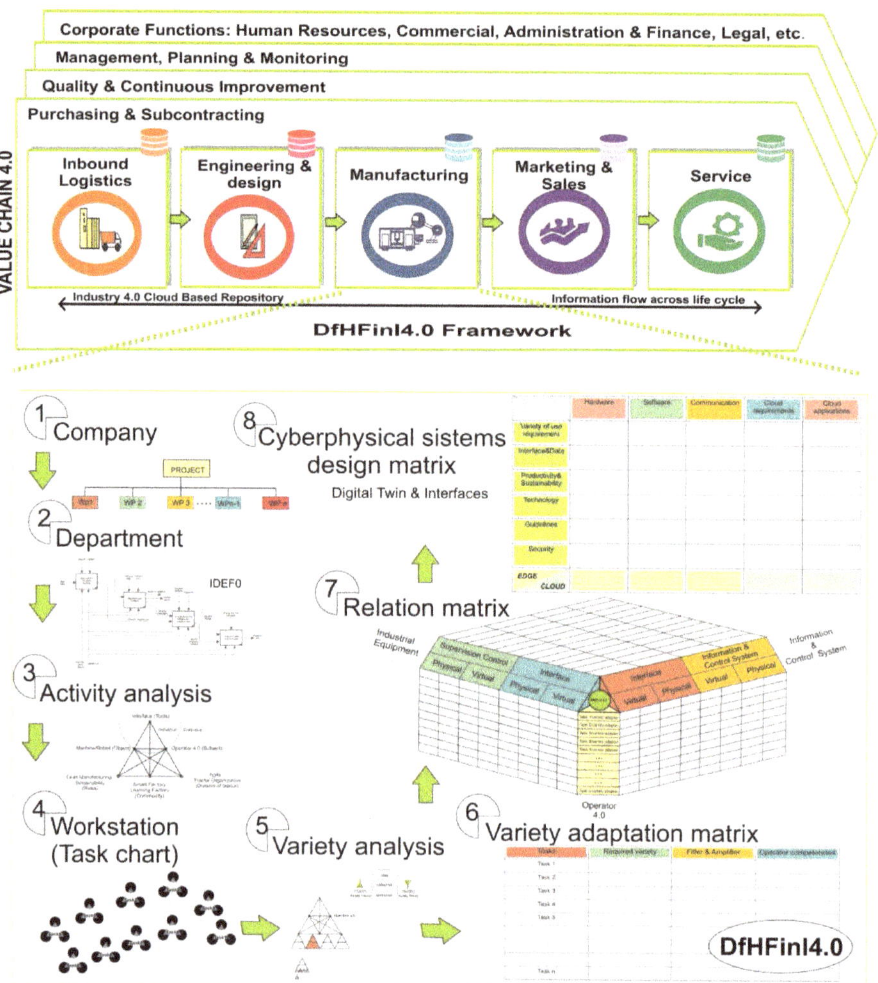

Figure 7. DfHFinI4.0 for the configuration of the integrated cyber-physical and co-evolutionary system.

Given the potential of the digitisation of Industry 4.0, the company can establish new forms of relationships with suppliers, consumers and other value chains, by forming the global value chain through horizontal and vertical clustering with great innovative potential. This generates a horizontal integration that involves real-time cooperation between human and technological factors, as well as vertical integration between partners, suppliers, and customers, which, when brought to the field of cyber-physical systems, can be developed in a fractal way in easily replicable structures, throughout the DfHFinI4.0 framework that is proposed for the value chain of the company.

In the framework developed in this paper, the workflow is analysed through activity theory, and forms an activity system that produces and develops actions based on said theory, by breaking it down into elements at the individual level and elements at the collective level, whose explicit and implicit knowledge can be ascertained, modelled and divided between technology and the human factor. The contradictions that occur between the elements of the activity system appear once the forces that drive creativity and innovation within the smart and learning factories are resolved. This study of provoked interactions facilitates the modelling of the system itself, and explains the relationships that lead to the co-evolutive and affective coupling between the human and technological factors, which enables continuous improvement in the adaptive process between the two factors. This modelling results in a dynamic and multilateral flow of data associated with the information systems necessary for the adaptation of intelligent manufacturing to customer requirements.

Connectivism and the law of requisite variety enable the SCMS to be modelled, and characterize the particularity of the activity and the specific profile of the engineer or technician as the Operator 4.0 that carries out this activity, regarding experience, and cognitive and affective level, and adapts it through filters and variety amplifiers. Connectivism allows the establishment of navigation strategies and online assistance in problem-solving processes, through the information system, from subrogated models from the cloud, fog, and edge. This necessary connection is held on the semantic web that causes, shares, and connects content capable of being interpreted by all the elements of the cyber-physical socio-technical system. The information is collected through IIoT technologies that allow data sharing between smart devices that configure smart and learning factories, thereby fostering collaborative affective environments [120].

By establishing a cognitive design appropriate to the required variety through adaptive interfaces, its adaptive reconfiguration can take place based on the task specification and the competency model of Operator 4.0 [100]. In order to make this possible, variety adapters employed as either amplifiers or reducers are introduced in accordance with the requirements, so that the regulatory part has at least the same variety as the regulated subsystem, and the SCMS can therefore achieve stability. The design and assembly of the system, within the fractalized context of the company [121], is made up of the tasks carried out by Operator 4.0, the equipment with which it interacts, and the associated information system.

5. Case Study: DfHFinI4.0 in PERA 4.0

Companies, and their associated manufacturing systems 4.0, are becoming increasingly complex and dynamic. In order to reduce this complexity, the management of knowledge and operational information is needed. Business architectures, such as PERA, GERAM, and CIMOSA, have hitherto been used, while more recently, different architectures have been proposed for Industry 4.0, such as Holonic, RAMI 4.0, IIRA, SME, and IVI [122–124], which correspond to the different ways of implementing the informational requirements of primary activities and support the smart and sustainable value chain. Among the proposed architectures, PERA is formed by the ecosystem of business entities [125], as shown in Figure 8, in which this methodology, which constitutes its life cycle engineering, can be implemented for each of its entities.

From among its characteristics, it is worth mentioning its orientation to the life cycle of the entities that make up the architecture, and considering the interaction and interfaces between technology and the human factors within its methodology. Li and Williams [126] highlighted the importance of considering good design in the communication interfaces between the diagrams that constitute the PERA model, in order to guarantee the correct exchange of information and the integration of the company, both vertically and horizontally. These characteristics determine that PERA constitutes a model of reference architecture and methodology on which to integrate the DfHFinI4.0 framework proposed under a cyber-physical conception of the entities contained therein, which evolves towards PERA 4.0. This enables research questions to be answered in life cycle engineering 4.0 formulated by Romero et al. [11], which establishes the need for the reference architecture to focus on the human factor.

The basic methodology of a PERA entity is illustrated in Figure 9. In the following, manufacturing PERA entity 3 of the architecture is employed, on which it is illustrated how the DFHinI4.0 framework proposed should be integrated together with the associated tools, which enable the cyber-physical manufacturing systems to be conceived based on the human factor.

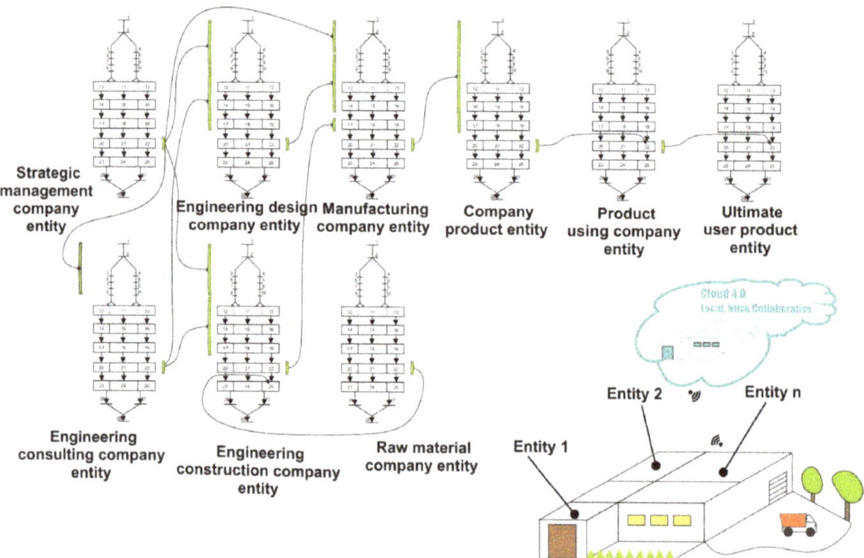

Figure 8. Entities and processes for the company in Purdue Enterprise Reference Architecture (PERA) 4.0 architecture.

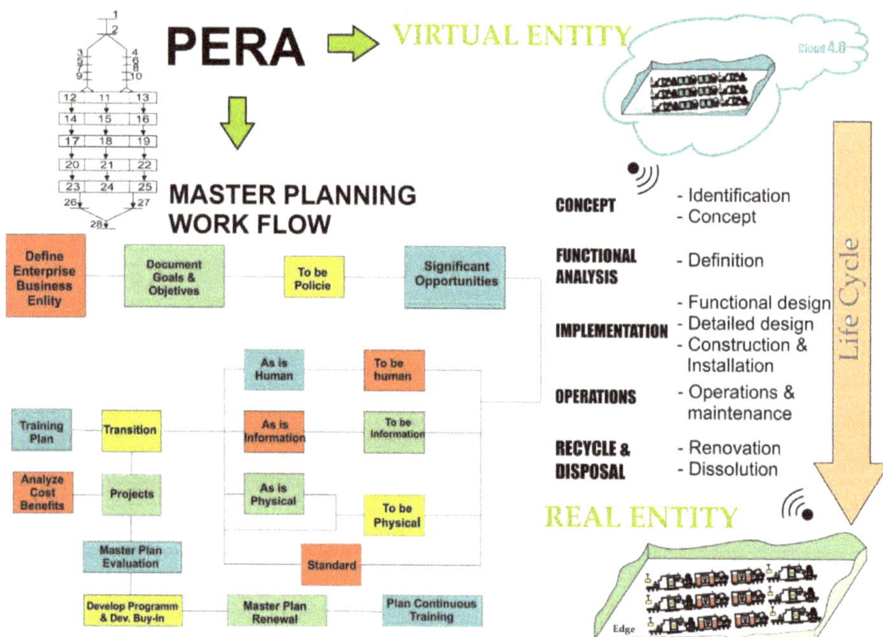

Figure 9. Basic methodology of a PERA 4.0 entity.

As indicated, PERA [127], in its methodological aspect as life cycle engineering, establishes the various regions, phases, and layers, into which the entities that constitute the company can be decomposed throughout its life cycle, while taking into account that the production equipment, the human factor, and the information and control system are involved in each element. Three separate elements are established in the design and implementation for entity three of PERA that correspond to the manufacturing system:

- The architecture of the information system.
- Human and organisational architecture.
- The architecture of the manufacturing team.

Certain vertical lines of great significance for the integration of Operator 4.0 can be observed among these elements, within which the DfHFinI4.0 framework is integrated:

- The line related to automation the PERA diagram is limited, since many tasks and functions require human innovation.
- The line related to human factors is limited by human competencies.
- The extent of the automation line represents the actual degree of automation carried out, and defines the boundaries between the three elements.

As depicted in Figure 9, the PERA methodology, in its initial proposal, contemplated the human factor, the technology, and the interfaces for their operation on industrial equipment and the information system, hitherto with no set of tools derived from conceptual frameworks of other areas of research that would allow the integration and empowerment of the human factor into digital transformation processes characterised by the connectivity and smartisation of technology.

In the proposal regarding PERA 4.0, which integrates the human factor into SCMS, as illustrated in Figure 10, the aforementioned elements are maintained. The elements and tools belonging to the DFHinI4.0 framework are incorporated into the design of socio-technical systems for the integration of the human factor, the industrial equipment, and the information and control system, thereby giving rise to two types of interfaces. In the same model, Operator 4.0 is configured as one more cyber-physical system, whose subrogate model obtained in the cloud will serve to adapt the interfaces and technology of the occupational environment to the operator positioned at the workstation.

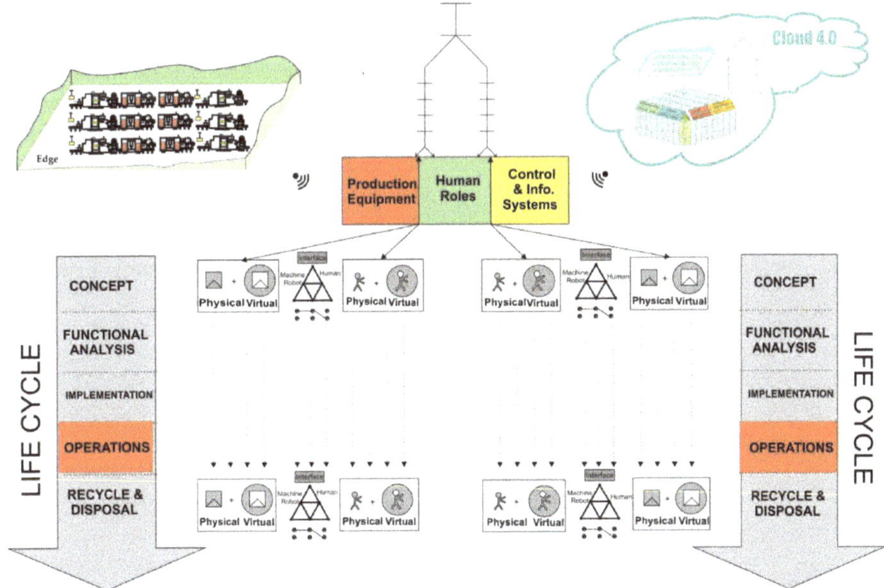

Figure 10. Integrating DfHFinI4.0 in PERA 4.0 operations phase.

The evolution of the PERA model of manufacturing entity three (analogously for the rest of the PERA entities) towards a PERA 4.0 entity three as a cyber-physical system under the possibilities of KETs, determines that the hierarchical architecture of the control system (analogous to ISA-85 and ISA-99) associated with manufacturing devices has been modified by distributed intelligent cyber-physical systems architecture, with real-time connectivity for monitoring and control in the edge and cloud. For this reason, the engineering or re-engineering of PERA towards PERA 4.0 architecture not only implies the integration and empowerment of the human factor with the DFHinI4 framework proposed, but also indicates the transformation of the hierarchical architecture, initially proposed under the PERA methodology of entity three, into a distributed architecture of cyber-physical systems based on micro-services under the PERA 4.0 methodology.

Entity three of the PERA manufacturing systems has established certain levels, characteristics and interrelations [128], which include:

- Level 0: Process. In this level, the real physical processes are defined by means of sensors, actuators, and other devices of the manufacturing process, and perform the functions of the automation and industrial control system for the measurement of the variables of the machines and the control of the process outputs. The devices communicate with each other, with the operator, and with top-level control devices.
- Level 1: Basic control. This level employs programmable automation controllers (PAC), which control and manipulate the manufacturing process, and act according to the feedback offered by the level-0 devices. The operator programs, configures, and manages these devices from the workstation through the human machine interface (HMI). In turn, the PACs (which for discrete manufacturing are called PLCs, and for process manufacturing are more specifically called DCSs) communicate with the specific information and control elements of levels 2 and 3, and also with other PACs.
- Level 2: Supervision control area. At this level, the supervision of the execution time and the operation of an area of the production facility are carried out using HMI, alert systems, batch-processing management systems, and the control of workstations. This level 2 communicates with PACs of level 1 and shares data with business systems and the applications of levels 4 and 5.
- Level 3: Manufacturing and control operations. This represents the highest level of the industrial automation and control system. This level includes the functions involved in managing workflows.
- Level 4: Business planning and site logistics. This level includes programming systems, material flow applications, manufacturing execution systems (MES), and information technology services (ITS).
- Level 5: Company. Residing at this level are the business resource management services, company-company through ERP and company-client through CRM for the PLM product, and BIM for the facility.

Figure 11 shows the incorporation of the DFHinI4.0 framework in a fractalized way, in the design and development of the various entities of the PERA 4.0 architecture for all the phases of the life cycle engineering that integrates the PERA methodology. This framework enables interactions between the human and technological factors to be modelled, and establishes their integration and development dynamically for each of the stages of the life cycle. This situation, together with the conception of the manufacturing entity as an intelligent and distributed cyber-physical system, gives rise to PERA 4.0 as a distributed system and life cycle engineering 4.0 methodology, which empowers the human factor, as shown.

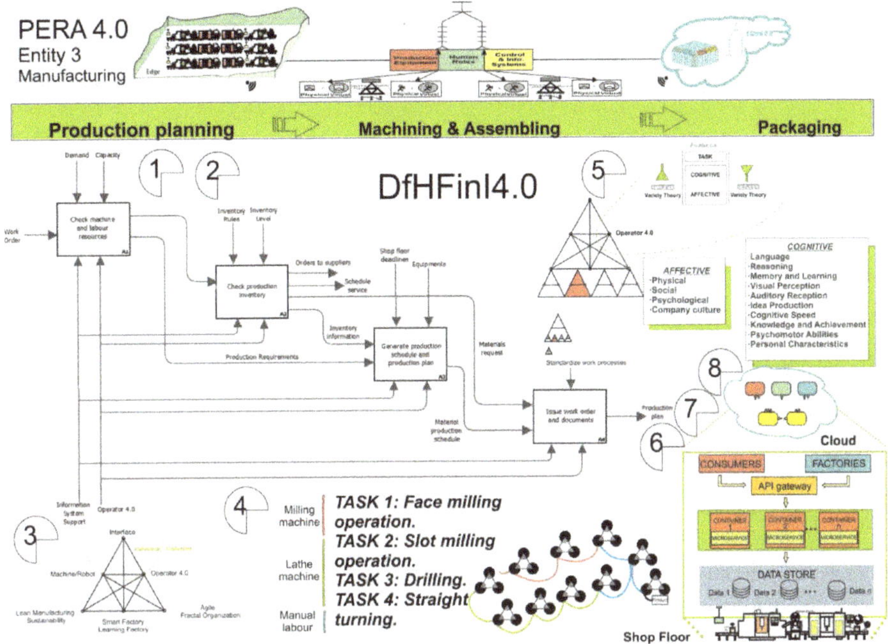

Figure 11. Application of the DfHFinI4.0 framework to the design and development process of entity three of PERA in the manufacturing system.

For the management of the reconfiguration of the technological occupational environment, accordance with the characteristics and competence of the Operators 4.0, who can be interacting with the system at any given moment, it is proposed that the operator, as a cyber-physical system, possess a cloud model of its operational singularity. This model refers to experience, knowledge, capabilities, competences, and other characteristics, which, as parameters of a subrogated model, allow the technology to be adapted, as represented in Figure 11. To this end, the operator model is sent from the cloud to the edge when necessary. The operators are sensed and assisted by the KETs, which transmit the data of the operators to the cloud to configure a more refined subrogate model, that in turn, is sent to the edge, which leads to the adaptation of the interface and technologies to Operator 4.0, thereby empowering this operator and enabling affective coupling. Big data techniques, learning machines, and deep learning will be employed in the preparation of the subrogate model.

Under the PERA 4.0 approach of manufacturing system entity three, the hierarchical levels of control established by PERA must be embedded in a distributed system of its cyber-physical entities, with intelligence and local connectivity in the edge, and global intelligence and connectivity in the cloud through IIoT. Figure 12 presents a schematic of the way in which the information system of the manufacturing system can evolve for its integration into any of the distributed architectures of Industry 4.0. In our proposal, we opted for holonic architecture [122], and for the modelling of the different holons, the Arrowhead methodology is proposed, both locally and globally [129]. This approach can be carried out by using blockchain technology, through the open-source container orchestration and choreography software tool called Kubernetes, and the creation of container images by using Docker [130].

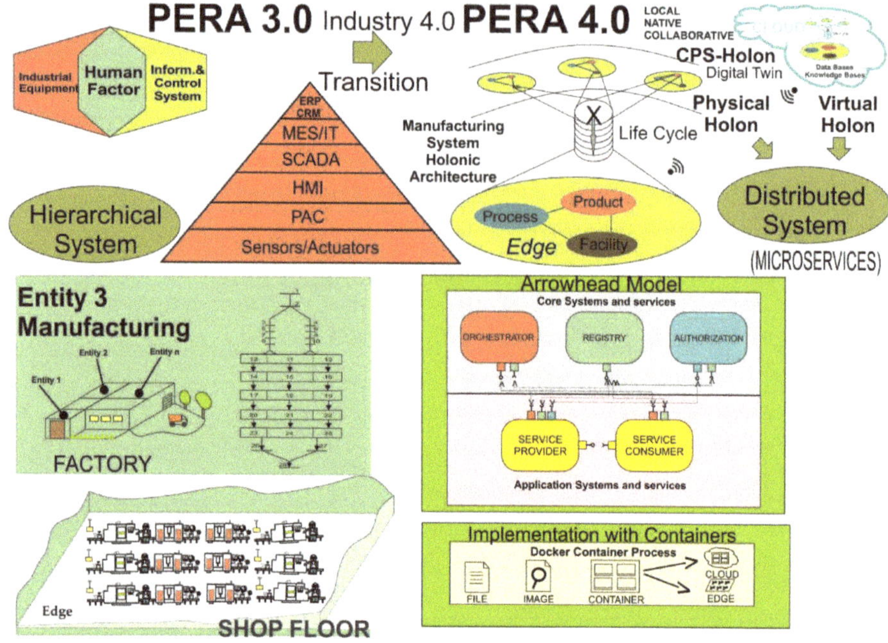

Figure 12. Evolution of the PERA control architecture to the PERA 4.0 architecture.

6. Discussion

This work proposes a framework (DfHfinI4.0) for the conception of manufacturing systems as cyber-physical socio-technical systems, from the perspective of Life Cycle Engineering 4.0. In this framework, the symbiotic conception of linked processes, teams, and people is established through adaptive interfaces that enable the operation, growth, and development of Operator 4.0. The proposed model enables the processes, their activities, and tasks to be studied, in order to search for potential synergies through its reconfiguration and customisation in accordance with Operator 4.0, thereby providing the online support required.

The most significant limitation of the proposed model involves the necessary sensitisation of the engineering teams in the socio-cognitive approach to the conception of socio-technical systems.

Regarding the relationship of the results obtained from the proposed DfHfinI4.0 model and from the application of the case study to the PERA 4.0 architecture, it should be borne in mind that it is impossible to establish a comparative relationship of results, due to the current lack of investigations regarding Operator 4.0 with the proposed scope and reference architectures that support techniques for the conception synergy of socio-technical systems.

7. Conclusions

The objectives of this work include: the identification of the degree of incorporation of the human factor in the different areas of digital and technological transformation of Industry 4.0, and the formulation of a model that the value of this factor in engineering and re-engineering processes to be evaluated, while taking into account Industry 4.0, which incorporates knowledge from other areas of research. A proposal is made for the application of the DfHFinI4.0 framework to one of the most significant reference architectures and methodologies, with the incorporation of the characteristics of Industry 4.0 as a reference. All of the above constitutes a solid proposal that enables the development of a framework for SCMS Life Cycle 4.0 Engineering.

With regard to the first objective, it should be considered that the human factor has not received the same attention as technology has in digital transformation processes. As a consequence of the previous consideration, it is necessary to investigate proposals that constitute an empowerment of the situation in which, in most cases, the human factor has been installed in digital transformation processes, thereby wasting human talent and its wider scope with the hybridisation of the technology.

In relation to the identification and selection of conceptual frameworks on which to build a SCMS design methodology focused on the human factor, after a bibliographic analysis and due to the importance of interfaces and navigation processes in Industry 4.0, we have opted for the conceptual fields of Vigotsky's activity theory (AT), Ashby's law of required variety (LRV), and the connectivist paradigm (CP). The latter two methologies are integrated into the activity theory model, thereby establishing an integrated formalism for the distribution of work, competencies, and analysis of the required variety, as well as establishing the filters and amplifiers of a variety of navigation strategies of Operators 4.0.

The proposed framework is called DFHinI4.0, within which AT, LRV, and CP are articulated in eight integrated steps that lead to the configuration of SCMS, focused on the human factor. This framework enables affective and timeless occupational experiences in the development of work, through the interfaces of hybridised equipment, both in its physical and virtual dimensions in the cloud. This framework is conceived with a degree of generality that allows its fractal implementation at the various levels of granularity and phases of Life Cycle Engineering 4.0.

Finally, both the DFHinI4.0 framework and the perspective of Life Cycle Engineering 4.0 are implemented in PERA, thereby configuring it as PERA 4.0 in the case of entity 3 of SCMS. A proposal is made for a distributed approach to hierarchical control architectures over PERA manufacturing systems, by means of a holonic architecture based on the model of connectors that considers micro-services for their development with blockchain technology. In this case, Section 5 enables the proposed framework to be validated, as shown by the improvement in dimension 1 related to the modelling and simulation of SCMS, which appear in Section 2.5 Research Gap.

Author Contributions: Conceptualisation, methodology, investigation and writing—original draft, S.S.-F.d.M. and F.A.-G.; writing—review and editing, S.S.-F.d.M., F.A.-G., J.S.-G., and M.J.Á.-G. All authors have read and agreed to the published version of the manuscript.

Funding: This research has received no external funding.

Conflicts of Interest: The authors declare there to be no conflict of interest.

References

1. Trstenjak, M.; Cosic, P. Process Planning in Industry 4.0 Environment. *Procedia Manuf.* **2017**, *11*. [CrossRef]
2. Bennett, N.; Lemoine, G.J. What a difference a word makes: Understanding threats to performance in a VUCA world. *Bus. Horiz.* **2014**, *57*, 311–317. [CrossRef]
3. Tao, F.; Cheng, Y.; Xu, L.D.; Zhang, L.; Li, B.H. CCIoT-CMfg: Cloud computing and internet of things-based cloud manufacturing service system. *IEEE Trans. Ind. Inform.* **2014**, *10*, 1435–1442.
4. Pandit, D.; Joshi, M.P.; Sahay, A.; Gupta, R.K. Disruptive innovation and dynamic capabilities in emerging economies: Evidence from the Indian automotive sector. *Technol. Forecast. Soc. Chang.* **2018**, *129*, 323–329. [CrossRef]
5. Suárez Fernández-Miranda, S.; Marcos, M.; Peralta, M.E.; Aguayo, F. The challenge of integrating Industry 4.0 in the degree of Mechanical Engineering. *Procedia Manuf.* **2017**, *13*. [CrossRef]
6. Yan, H.; Wan, J.; Zhang, C.; Tang, S.; Hua, Q.; Wang, Z. Industrial Big Data Analytics for Prediction of Remaining Useful Life Based on Deep Learning. *IEEE Access* **2018**, *6*, 17190–17197. [CrossRef]
7. Plumanns, L.; Printz, S.; Vossen, R.; Jeschke, S. Strategic Management of Personnel Development in the Industry 4.0. In Proceedings of the 14th International Conference on Intellectual Capital, Knowledge Management & Organisational Learning: ICICKM 2017, Hong Kong, China, China, 7–8 December 2017; pp. 179–186.

8. Tirabeni, L.; De Bernardi, P.; Forliano, C.; Franco, M. How Can Organisations and Business Models Lead to a More Sustainable Society? A Framework from a Systematic Review of the Industry 4.0. *Sustainability* **2019**, *11*, 23. [CrossRef]
9. Tran, N.-H.; Park, H.-S.; Nguyen, Q.-V.; Hoang, T.-D. Development of a Smart Cyber-Physical Manufacturing System in the Industry 4.0 Context. *Appl. Sci.* **2019**, *9*, 24. [CrossRef]
10. Vrchota, J.; Pech, M. Readiness of Enterprises in Czech Republic to Implement Industry 4.0: Index of Industry 4.0. *Appl. Sci.* **2019**, *9*, 25. [CrossRef]
11. Romero, D.; Stahre, J.; Wuest, T.; Noran, O.; Bernus, P.; Fast-Berglund, Å.; Gorecky, D. Towards an Operator 4.0 Typology: A Human-Centric Perspective on the Fourth Industrial Revolution Technologies. In Proceedings of the 46th International Conference on Computers & Industrial Engineering, Tianjin, China, 29–31 October 2016; pp. 1–11.
12. Romero, D.; Bernus, P.; Noran, O.; Stahre, J.; Berglund, Å.F. The operator 4.0: Human cyber-physical systems & adaptive automation towards human-automation symbiosis work systems. In Proceedings of the. IFIP International Conference on Advances in Production Management Systems, Iguassu Falls, Brazil, 3–7 September 2016; Springer LLC: New York, NY, USA, 2016; Volume 488, pp. 677–686. [CrossRef]
13. Taylor, M.P.; Boxall, P.; Chen, J.J.J.; Xu, X.; Liew, A.; Adenijib, A. Operator 4.0 or Maker 1.0? Exploring the implications of Industrie 4.0 for innovation, safety and quality of work in small economies and enterprises. *Comput. Ind. Eng.* **2020**, *139*, 5. [CrossRef]
14. Enke, J.; Glass, R.; Kreß, A.; Hambach, J.; Tisch, M.; Metternich, J. Industrie 4.0-Competencies for a modem production system A curriculum for Learning Factories. *Procedia Manuf.* **2018**, *23*, 267–272. [CrossRef]
15. Emmanouilidis, C.; Pistofidis, P.; Bertoncelj, L.; Katsouros, V.; Fournaris, A.; Koulamas, C.; Ruiz-Carcel, C. Enabling the human in the loop: Linked data and knowledge in industrial cyber-physical systems. *Annu. Rev. Control.* **2019**, *47*, 249–265. [CrossRef]
16. Zakoldaev, D.A.; Gurjanov, A.V.; Shukalov, A.V.; Zharinov, I.O. Implementation of H2M technology and augmented reality for operation of cyber-physical production of the Industry 4.0. *J. Phys. Conf. Ser.* **2019**, *1353*, 5. [CrossRef]
17. Segura, A.; Diez, H.V.; Barandiaran, I.; Arbelaiz, A.; Álvarez, H.; Simões, B.; Posada, J.; García-Alonso, A.; Ugarte, R. Visual computing technologies to support the Operator 4.0. *Comput. Ind. Eng.* **2020**, *139*, 9. [CrossRef]
18. Ruppert, T.; Jaskó, S.; Holczinger, T.; Abonyi, A. Enabling Technologies for Operator 4.0: A Survey. *Appl. Sci.* **2018**, *8*, 19. [CrossRef]
19. Zolotová, I.; Papcun, P.; Kajáti, E.; Miškuf, M.; Mocnej, J. Smart and cognitive solutions for Operator 4.0: Laboratory H-CPPS case studies. *Comput. Ind. Eng.* **2020**, *139*, 15. [CrossRef]
20. Fantini, P.; Pinzone, M.; Taisch, M. Placing the operator at the centre of Industry 4.0 design: Modelling and assessing human activities within cyber-physical systems. *Comput. Ind. Eng.* **2020**, *139*, 11. [CrossRef]
21. Peruzzinia, M.; Fabio Grandia, M.P. Exploring the potential of Operator 4.0 interface and monitoring. *Comput. Ind. Eng.* **2020**, *139*, 19. [CrossRef]
22. Umeda, Y.; Takata, S.; Kimura, F.; Tomiyama, T.; Sutherland, J.W.; Kara, S.; Herrmann, C.; Duflou, J.R. Toward integrated product and process life cycle planning-An environmental perspective. *CIRP Ann. Manuf. Technol.* **2012**, *61*, 681–702. [CrossRef]
23. Yan, P.; Zhou, M.C. A life cycle engineering approach to development of flexible manufacturing systems. *IEEE Int. Conf. Robot. Autom.* **2003**, *19*, 465–473. [CrossRef]
24. Wanyama, W.; Ertas, A.; Zhang, H.C.; Ekwaro-Osire, S. Life-cycle engineering: Issues, tools and research. *Int. J. Comput. Integr. Manuf.* **2003**, *16*, 307–316. [CrossRef]
25. Würtz, G.; Kölmel, B. Integrated Engineering—A SME-Suitable Model for Business and Information Systems Engineering (BISE) towards the Smart Factory. *IFIP Adv. Inf. Commun. Technol.* **2012**, *380*, 494–502. [CrossRef]
26. Mayer, P. Guidelines for writing a review article. *Zurich-Basel Plant Sci. Cent.* **2009**, *82*, 1–10.
27. Coelho, D. A growing concept of ergonomics including pleasure. comfort and cognitive engineering: An engineering design perspective. Ph.D. Thesis, The University of Beira Interior, Covilhã, Portugal, 2002.
28. Galindo-Rueda, F.; Verger, F. OECD taxonomy of economic activities based on R&D intensity. OECD Publishing, Paris. *OECD Sci. Technol. Ind. Work. Pap.* **2016**, *4*. [CrossRef]
29. National Science Board, N.S.F. *Science and Engineering Indicators 2020: The State of U.S. Science and Engineering*; NSB-2020-1; National Science Board, N.S.F.: Alexandria, VA, USA, 2020.

30. Lee, J.; Bagheri, B.; Kao, H.A. A Cyber-Physical Systems architecture for Industry 4.0-based manufacturing systems. *Manuf. Lett.* **2015**. [CrossRef]
31. Zhou, J.; Zhou, Y.; Wang, B.; Zang, J. Human–Cyber–Physical Systems (HCPSs) in the Context of New-Generation Intelligent Manufacturing. *Engineering* **2019**, *5*, 624–636. [CrossRef]
32. Rødseth, H.; Eleftheriadis, R.; Lodgaard, E.; Fordal, J.M. Operator 4.0—Emerging job categories in manufacturing. *Lect. Notes Electr. Eng.* **2019**, *484*, 114–121. [CrossRef]
33. Krugh, M.; McGee, E.; McGee, S.; Mears, L.; Ivanco, A.; Podd, K.C.; Watkins, B. Measurement of Operator-machine Interaction on a Chaku-chaku Assembly Line. *Procedia Manuf.* **2017**. [CrossRef]
34. Zamora, M.; Caldwell, E.; Garcia-Rodriguez, J.; Azorin-Lopez, J.; Cazorla, M. Machine Learning Improves Human-Robot Interaction in Productive Environments: A Review. In Proceedings of the International Work-Conference on Artificial Neural Networks, IWANN 2017, Cadiz, Spain, 14–16 June 2017; Proc. Lect. Notes Comput. Sci. (including Subser. Lect. Notes 812 Artif. Intell. Lect. Notes Bioinformatics). Springer: Berlin, Germany, 2017; Volume 10306, pp. 283–293.
35. Frynas, J.G.; Mol, M.J.; Mellahi, K. Management Innovation Made in China: Haier's Rendanheyi. *Calif. Manag. Rev.* **2018**, *61*, 71–93. [CrossRef]
36. Shamim, S.; Cang, S.; Yu, H.; Li, Y. Management approaches for Industry 4.0: A human resource management perspective. In Proceedings of the 2016 IEEE Congress on Evolutionary Computation (CEC), Vancouver, BC, Canada, 24–29 July 2016; pp. 5309–5316. [CrossRef]
37. Lv, Y.; Lin, D. Design an intelligent real-time operation planning system in distributed manufacturing network. *Ind. Manag. Data Syst.* **2017**, *117*, 742–753. [CrossRef]
38. Neuböck, T.; Schrefl, M. Modelling Knowledge about Data Analysis Processes in Manufacturing. In Proceedings of the IFAC Symposium on Information Control in Manufacturing Ottawa, ON, Canada, 11–13 May 2015; Volume 48, pp. 277–282. [CrossRef]
39. Sanin, C.; Shafiq, I.; Waris, M.M.; Toro, C.; Szczerbicki, E. Manufacturing collective intelligence by the means of Decisional DNA and virtual engineering objects, process and factory. *J. Intell. Fuzzy Syst.* **2017**, *32*, 1585–1599. [CrossRef]
40. Chen, Y.; Lee, G.M.; Shu, L.; Crespi, N. Industrial Internet of Things-based collaborative sensing intelligence: Framework and research challenges. *Sensors* **2016**, *16*, 215. [CrossRef] [PubMed]
41. Synnes, E.L.; Welo, T. Enhancing Integrative Capabilities through Lean Product and Process Development. *Procedia CIRP* **2016**, *54*, 221–226. [CrossRef]
42. Küsters, D.; Praß, N.; Gloy, Y.S. Textile Learning Factory 4.0-Preparing Germany's Textile Industry for the Digital Future. *Procedia Manuf.* **2017**, *9*, 214–221. [CrossRef]
43. Mehta, P.; Rao, P.; Wu, Z.D.; Jovanović, V.; Wodo, O.; Kuttolamadom, M. Smart manufacturing: State-of-The-Art reviewin context of conventional & modern manufacturing. In Proceedings of the ASME 2018 13th International Manufacturing Science and Engineering Conference, College Station, TX, USA, 18–22 June 2018; 2018; Volume 3, pp. 1–21. [CrossRef]
44. Büth, L.; Juraschek, M.; Posselt, G.; Herrmann, C. Supporting SMEs towards adopting mixed reality A training concept to bring the reality-virtuality continuum into application. In Proceedings of the 2018 IEEE 16th International Conference on Industrial Informatics, Porto, Portugal, 18–20 July 2018; pp. 544–549. [CrossRef]
45. Govindarajan, U.H.; Trappey, A.J.C.; Trappey, C.V. Immersive Technology for Human-Centric Cyberphysical Systems in Complex ManufacturingProcesses: A Comprehensive Overview of the Global Patent Profile Using Collective Intelligence. *Complexity* **2018**, *17*. [CrossRef]
46. Cimini, C.; Pinto, R.; Cavalieri, S. The business transformation towards smartmanufacturing: A literature overview about reference models and research agenda. *IFAC-PapersOnLine* **2017**, *50*, 14952–14957. [CrossRef]
47. Stark, R.; Kind, S.; Neumeyer, S. Innovations in digital modelling for next generation manufacturing system design. *CIRP Ann-Manuf. Technol.* **2017**, *66*, 169–172. [CrossRef]
48. Cheng, H.; Xue, L.; Wang, P.; Zeng, P.; Yu, H. Ontology-Based Web Service Integration for FlexibleManufacturing Systems. In Proceedings of the 2017 IEEE 15th International Conference on Industrial Informatics, Emden, Germany, 24–26 July 2017; pp. 351–356. [CrossRef]
49. Klöber-Koch, J.; Pielmeier, J.; Grimm, S.; Brandt, M.M.; Schneider, M.; Reinhart, G. Knowledge-Based Decision Making in a Cyber-Physical Production Scenario. *Procedia Manuf.* **2017**, *9*, 167–174. [CrossRef]

50. Qi, Q.; Tao, F. A Smart Manufacturing Service System Based on Edge Computing, Fog, Computing, and Cloud Computing. *IEEE Access* **2019**, *7*, 86769–86777. [CrossRef]
51. Qu, S.; Wang, J.; Govil, S.; Leckie, J.O. Optimized Adaptive Scheduling of a Manufacturing Process System with Multi-Skill Workforce and Multiple Machine Types: An Ontology-Based, Multi-Agent Reinforcement Learning Approach. *Procedia CIRP* **2016**, *57*, 55–60. [CrossRef]
52. Jaensch, F.; Csiszar, A.; Scheifele, C.; Verl, A. Digital Twins of Manufacturing Systems as a Base for Machine Learning. In Proceedings of the 25th International Conference on Mechatronics and Machine Vision in Practice (M2VIP), Stuttgart, Germany, 20–22 November 2018; pp. 1–6. [CrossRef]
53. Mortensen, S.T.; Madsen, O. A Virtual Commissioning Learning Platform. *Procedia Manuf.* **2018**, *23*, 93–98. [CrossRef]
54. Kaihara, T.; Katsumura, Y.; Suginishi, Y.; Kadar, B. Simulation model study for manufacturingeffectiveness evaluation in crowdsourcedmanufacturing. *CIRP Ann. Manuf. Technol.* **2017**, *66*, 445–448. [CrossRef]
55. Jiang, P.; Leng, J. The Configuration of Social Manufacturing: A Social Intelligence Way Toward Service-Oriented. *Int. J. Manuf. Res.* **2016**, *12*. [CrossRef]
56. Cachada, A.; Barbosa, J.; Leitão, P.; Gcraldcs, C.A.S.; Deusdado, L.; Costa, J.; Teixeira, J.; Moreira, A.H.J.; Miguel, P.; Romero, L.; et al. Maintenance 4.0: Intelligent and Predictive Maintenance System Architecture. In Proceedings of the 2018 IEEE 23rd International Conference on Emerging Technologies and Factory Automation (ETFA), Turin, Italy, 4–7 September 2018; pp. 139–146. [CrossRef]
57. Fisher, O.; Watson, N.; Porcu, L.; Baco, D.; Rigley, M.; Gomes, R.L. Cloud manufacturing as a sustainable process manufacturing route. *J. Manuf. Syst.* **2018**, *47*, 53–68. [CrossRef]
58. Zhang, Y.; Ren, S.; Liu, Y.; Si, S. A big data analytics architecture for cleaner manufacturing and maintenance processes of complex products. *J. Clean. Prod.* **2017**. [CrossRef]
59. Berg, L.P.; Vance, J.M. Industry use of virtual reality in product design and manufacturing: A survey. *Virtual Real.* **2017**, *21*, 1–17. [CrossRef]
60. Pai, Y.S.; Yap, H.J.; Zawiah, S.; Dawal, S.Z.; Ramesh, S.; Phoon, S.Y. Virtual Planning, Control., and Machining for a Modular-Based Automated Factory Operation in an Augmented Reality Environment. *Sci. Rep.* **2016**. [CrossRef]
61. Lawrence, K. *Developing Leaders in a VUCA Environtment*; UNC Kenan-Flagler Bussines School: Chapel Hill, NC, USA, 2013.
62. Centea, D.; Elbestawi, M.; Singh, I.; Wanyama, T. SEPT Learning Factory Framework. In *Smart Industry & Smart Education, Proceedings of the 15th International Conference on Remote Engineering and Virtual Instrumentation, Duesseldorf, Germany, 21–23 March 2018*; Lecture Notes in Networks and Systems; Auer, M., Langmann, R., Eds.; Springer: Cham, Switzerland, 2019; Volume 47. [CrossRef]
63. Schallock, B.; Rybski, C.; Jochem, R.; Kohl, H. Learning Factory for Industry 4.0 to provide future skills beyond technical training. *Procedia Manuf.* **2018**, *23*, 27–32. [CrossRef]
64. Baena, F.; Guarin, A.; Mora, J.; Sauza, J.; Retat, S. Learning Factory: The Path to Industry 4.0. *Procedia Manuf.* **2017**, *9*, 73–80. [CrossRef]
65. Duin, H.; Gorldt, C.; Thoben, K.D.; Pawar, K. Learning In Ports With Serious Gaming. In Proceedings of the International Conference on Engineering, Technology and Innovation (ICE/ITMC), Funchal, Portugal, 27–29 June 2017; pp. 431–438. [CrossRef]
66. Papazoglou, M.P.; Andreou, A. Smart connected digital factories: Unleashing the power of industry 4.0 and the industrial internet. In *Cloud Computing and Services Science*; Springer: Berlin, Germany, 2019; Volume 1073, pp. 77–101. [CrossRef]
67. Ashby, W.R. Requisite variety and its implications for the control of complex systems. In *Facets of Systems Science*; Part of the International Series in Systems Science and Systems Engineering; Springer: Berlin, Germany, 1991; Volume 7, pp. 405–417. [CrossRef]
68. Engeström, Y. Activity theory and individual and social transformation. In *Perspectives on Activity Theory*; Cambridge University Press: Cambridge, UK, 1999; ISBN 0-521-43127-1.
69. Foot, K.A. Cultural-Historical Activity Theory as Practical Theory: Illuminating the Development of a Conflict Monitoring Network. *Publ. Commun. Theory* **2001**. [CrossRef]
70. Ashby, W.R. *Variety, Constraint, and the Law of Requisite Variety*; Wiley: Hoboken, NJ, USA, 2017; ISBN 15327000.

71. Clinton, G.; Lee, E.; Logan, R. Connectivism as a framework for creative productivity in instructional technology. In Proceedings of the 2011 IEEE 11th International Conference on Advanced Learning Technologies, Athens, GA, USA, 6–8 July 2011; pp. 166–170. [CrossRef]
72. Rodríguez, A.J.; Molero de Martins, D.M. Conectivismo como gestión del conocimiento. *REDHECS Rev. Electrónica Humanidades, Educ. y Comun. Soc.* **2009**, *4*, 73–85.
73. Vitali, I.; Arquilla, V.; Tolino, U. A Design perspective for IoT products. A case study of the Design of a Smart Product and a Smart Company following a crowdfunding campaign. *Des. J.* **2017**, *20*, S2592–S2604. [CrossRef]
74. Rajnai, Z.; Kocsis, I. Labor Market Risks of Industry 4.0, Digitization, Robots and AI. In Proceedings of the IEEE 15th International Symposium on Intelligent Systems and Informatics (SISY), Subotica, Serbia, 14–16 September 2017; pp. 000343–000346. [CrossRef]
75. Gualtieri, L.; Rojas, R.; Carabin, G.; Palomba, I.; Rauch, E.; Vidoni, R.; Matt, D.T. Advanced Automation for SMEs in the I4.0 Revolution: Engineering Education and Employees Training in the Smart Mini Factory Laboratory. In Proceedings of the IEEE International Conference on Industrial Engineering and Engineering Management (IEEM), Bangkok, Thailand, 16–19 December 2018; pp. 1111–1115. [CrossRef]
76. Jeganathan, L.; Khan, A.N.; Kannan Raju, J.; Narayanasamy, S. On a Frame Work of Curriculum for Engineering Education 4.0. In Proceedings of the 2018 World Engineering Education Forum-Global Engineering Deans Council (WEEF-GEDC), Albuquerque, NM, USA, 12–16 November 2018; pp. 1–6.
77. Tzafestas, S. Concerning human-automation symbiosis in the society and the nature. *Int. J. Fact. Autom. Robot. Soft Comput.* **2006**, *1*, 16–24.
78. Norman, D.A. *El Diseño de los Objetos del Futuro. La Interacción Entre el Hombre y la Máquina*; Ediciones Paidós: Barcelona, Spain, 2010; ISBN 9788449323881.
79. Engeström, Y. The future of activity theory. In *Learning and Expanding with Activity Theory*; Cambridge University Press: Cambridge, UK, 2009; pp. 303–328. [CrossRef]
80. Squires, J.E.; Estabrooks, C.A.; Gustavsson, P.; Wallin, L. Individual determinants of research utilization by nurses: A systematic review update. *Implement. Sci.* **2011**, *6*, 43. [CrossRef] [PubMed]
81. Allen, D.K.; Brown, A.; Karanasios, S.; Norman, A. How should technology-mediated organizational change be explained? A comparison of the contributions of critical realism and activity theory. *MIS Quart.* **2013**, *37*, 835–854. [CrossRef]
82. Hyysalo, S. *Health Technology Development and Use: From Practice-Bound Imagination to Evolving Impacts*; Routledge, Taylor & Francis Group: New York, NY, USA, 2010; ISBN 0-203-84915-9.
83. Crawford, K.; Hasan, H. Demonstrations of the activity theory framework for research in information systems. *Australas. J. Inf. Syst.* **2006**, *13*, 49–67. [CrossRef]
84. Engeström, Y. Expansive Learning at Work: Toward an activity theoretical reconceptualization. *J. Educ. Work* **2001**, *14*, 133–156. [CrossRef]
85. Sannino, A.; Engestrom, Y. Cultural-historical activity theory: Founding insights and new challenges. *Cult. Hist. Psychol.* **2018**, *14*, 43–56. [CrossRef]
86. Henric-Coll, M. *La Organización Fractal: El Futuro del Management*; Fractal Teams: Navarra, Spain, 2014; ISBN 978-8461696628.
87. Jarzabkowski, P. Strategic practices: An activity theory perspective on continuity and change. *J. Manag. Stud.* **2003**, *40*, 23–56. [CrossRef]
88. Kuutti, K. Activity theory as a potential framework for human-computer interaction research. In *Context and Consciousness: Activity Theory and Human-Computer Interaction*; The MIT Press: Cambridge, MA, USA, 1995; pp. 17–44.
89. Wilson, T.D. Activity theory and information seeking. *Annu. Rev. Inf. Sci. Technol.* **2009**, *42*, 119–161. [CrossRef]
90. Issroff, K.; Scanlon, E. Using technology in higher education: An activity theory perspective. *J. Comput. Assist. Learn* **2002**, *18*, 77–83. [CrossRef]
91. Benson, A.; Lawler, C.; Whitworth, A. Rules, roles and tools: Activity theory and the comparative study of e-learning. *Br. J. Educ. Technol.* **2008**, *39*. [CrossRef]
92. Barab, S.; Schatz, S.; Scheckler, R. Using activity theory to conceptualize online community and using online community to conceptualize activity theory. *Mind Cult. Act.* **2004**, *11*, 25–47. [CrossRef]

93. Brine, J.; Franken, M. Students' perceptions of a selected aspect of a computer mediated academic writing program: An activity theory analysis. *Australas. J. Educ. Technol.* **2006**, *22*, 21–38. [CrossRef]
94. Blin, F. CALL and the development of learner autonomy: Towards an activity-theoretical perspective. *ReCALL Camb. Univ.* **2004**, *16*, 377–395. [CrossRef]
95. Abdullah, Z. Activity Theory as Analytical Tool: A Case Study of Developing Student Teachers' Creativity in Design. *Procedia-Soc. Behav. Sci.* **2014**, *131*, 70–84. [CrossRef]
96. Hannah, J.; Hinson, L. Development of Propositions on Human Cognitive Abilities Matching Impacts on Accounting Job Performance. *UF J. Undergrad. Res.* **2019**, *21*. [CrossRef]
97. Beard-Gunter, A.; Ellis, D.G.; Found, P.A. TQM, games design and the implications of integration in industry 4.0 systems. *Int. J. Qual. Serv. Sci.* **2019**, *11*, 235–247. [CrossRef]
98. Rodríguez, R.L. *La Gestión del Tiempo Personal y Colectivo*; Graó: Barcelona, Spain, 2010; ISBN 978-84-9980-406-4.
99. Arenas, T.; Martínez, M.Á.; Honggang, X.; Morales, O.; Chávez, M. Integrating VSM and Network Analysis for Tourism Strategies–Case: Mexico and the Chinese Outbound Market. *Syst. Pract. Action Res.* **2019**, *32*, 315–333. [CrossRef]
100. Hancock, P.A.; Jagacinski, R.J.; Parasuraman, R.; Wickens, C.D.; Wilson, G.F.; Kaber, D.B. Human-automation interaction research: Past, present and future. *Ergon. Des. Q. Hum. Factors Appl.* **2013**, *21*, 9–14. [CrossRef]
101. Sun, S.; Zheng, X.; Gong, B.; García, J.; Ordieres-Meré, J. Healthy Operator 4.0: A Human Cyber–Physical System Architecture for SmartWorkplaces. *Sensors* **2020**, *20*, 2011. [CrossRef] [PubMed]
102. Carrol, J. *Human Cognitive Abilities: A Survey of Factor-Analytic Studies*; Cambridge University Press: New York, NY, USA, 1993; ISBN 9780511571312.
103. Jordan, P.W. *Designing Pleasurable Products: An Introduction to the New Human Factors*; CRC Press: Boca Raton, FL, USA, 2000; ISBN 9780203305683.
104. Saritas, M.T. The Emergent Technological and Theoretical Paradigsn in Education: The Interrelations of Cloud Computing (CC), Conectivism and Internet of things (IoT). *Proc. Acta Polytech. Hungarica* **2015**, *12*, 161–179. [CrossRef]
105. Downes, S. *Connectivism and Connective Knowledge: Essays on Meaning and Learning Networks*; National Research Council Canada: Ottawa, ON, Canada, 2012; ISBN 9781105778469.
106. Salmon, G.; Siemens, G.; Ally, M. A Learning Theory for the Digital Age. *Instr. Technol. Distance Educ.* **2004**.
107. Menary, R. The Extended Mind. In *A Bradford Book*; The MIT Press: Cambridge, MA, USA, 2010; ISBN 978-0-262-01403-8.
108. Patel, P.; Ali, M.I.; Sheth, A. From Raw Data to Smart Manufacturing: AI and Semantic Web of Things for Industry 4.0. *IEEE Intell. Syst.* **2018**, *33*, 79–86. [CrossRef]
109. Cheng, Y.-J.; Chen, M.-H.; Cheng, F.-C.; Cheng, Y.-C.; Lin, Y.-S.; Yang, C.-J. Developing a Decision Support System (DSS) for a Dental Manufacturing Production Line based on Data Mining. In Proceedings of the IEEE International Conference on Applied System Invention (ICASI), Tokyo, Japan, 13–17 April 2018; pp. 638–641. [CrossRef]
110. Wang, S.; Ouyang, J.; Li, D.; Liu, C. An Integrated Industrial Ethernet Solution for the Implementation of Smart Factory. *IEEE Access* **2017**, *5*, 25455–25462. [CrossRef]
111. Cagnin, R.L.; Guilherme, I.R.; Queiroz, J.; Paulo, B.; Neto, M.F.O. A Multi-agent System Approach for Management of Industrial IoT Devices in Manufacturing Processes. In Proceedings of the INDIN 2018: IEEE 16th International Conference on Industrial Informatics, Porto, Portugal, 18–20 July 2018; pp. 31–36. [CrossRef]
112. Madsen, O.; Møller, C. The AAU Smart Production Laboratory for Teaching and Research in Emerging Digital Manufacturing Technologies. *Procedia Manuf.* **2017**, *9*, 106–112. [CrossRef]
113. Lampón, J.F.; Cabanelas, P.; González-Benito, J. The impact of modular platforms on automobile manufacturing networks. *Prod. Plan. Control.* **2017**, *28*, 335–348. [CrossRef]
114. European Commission. Directorate-General for Research and Innovation. In *Defining Innovation. Report of the independent High. Level Group on Industrial Technologies*; Directorate D–Industrial Technologies: Luxembourg, 2018; ISBN 978-92-79-85271-8.
115. Ryu, K.; Jung, M. Agent-based fractal architecture and modeling for developing distributed manufacturing systems. *Int. J. Prod. Res.* **2003**, *41*, 4233–4255. [CrossRef]
116. Lee, J.; Jin, C.; Bagheri, B. Cyber physical systems for predictive production systems. *Prod. Eng. Res. Dev.* **2017**, *11*, 155–165. [CrossRef]

117. Wu, D.; Ren, A.; Zhang, W.; Fan, F.; Liu, P.; Fu, X.; Terpenny, J. Cybersecurity for digital manufacturing. *J. Manuf. Syst.* **2018**, *48*, 3–12. [CrossRef]
118. Suaily, S.; Zubaidah, S. Development of Product Service System Modelling in SMED: The Case of Inventory Control. *J. Mod. Manuf. Syst. Technol.* **2018**, *1*, 94–106. [CrossRef]
119. Shin, M.; Mun, J.; Jung, M. Self-evolution framework of manufacturing systems based on fractal organization. *Comput. Ind. Eng.* **2009**. [CrossRef]
120. Wiltshire, T.; Fiore, S.M. Social Cognitive and Affective Neuroscience in Human-Machine Systems: A Roadmap for Improving Trainig, Human-Robot Interaction and Team Performance. *IEEE Trans. Human Mach. Syst.* **2014**, *44*, 779–787. [CrossRef]
121. Warneke, H.-J. *The Fractal Company: A Revolution in Corporate Culture*; Springer: Berlin/Heidelberg, Germany, 1993; ISBN 978-3-642-78126-1.
122. Ávila-Gutiérrez, M.J.; Aguayo-González, F.; Marcos-Bárcena, M.; Lama-Ruiz, J.R.P.-Á. Reference holonic architecture for sustainable manufacturing enterprises distributed. *DYNA* **2017**, *84*, 160. [CrossRef]
123. Hübner, I. RAMI 4.0 und die Industrie-4.0-Komponente. *Open Autom.* **2015**, 24–29.
124. Yao, X.; Lin, Y. Emerging manufacturing paradigm shifts for the incoming industrial revolution. *Int. J. Adv. Manuf. Technol.* **2016**. [CrossRef]
125. Johannessen, J.-A. Knowledge Management and Organizational Learning. In *Knowledge Management as a Strategic Asset*; Emerald Publishing: Bingley, UK, 2018; ISBN 978-1-4419-0007-4.
126. Li, H.; Williams, T.J. Interface design for the Purdue enterprise reference architecture (PERA) and methodology in e-Work. *Prod. Plan. Control.* **2003**, *14*, 704–719. [CrossRef]
127. Williams, T. *The Purdue Enterprise Reference Architecture and Methodology (PERA)*; Kluwer Academic: Dordrecht, The Netherlands, 1998; ISBN 412812509.
128. Odewale, A. Implementing secure architecture for industrial control systems. In Proceedings of the 27th COREN Engineering Assembly, Abuja, Nigera, 6–8 August 2018; p. 17.
129. Plósz, S.; Hegedűs, C.; Varga, P. Advanced security considerations in the arrowhead framework. In Proceedings of the Intelligent Tutoring Systems, Trondheim, Norway, 20–23 September 2016; Springer Science and Business Media LLC: Berlin, Germany; Volume 9923, pp. 234–245.
130. Larrinaga, F.; Aldalur, I.; Illarramendi, M.; Iturbe, M.; Perez, T.; Unamuno, G.; Lazkanoiturburu, I. Analysis of technological architectures for the new paradigm of the Industry 4.0. *Dyna* **2019**, *94*, 267–271. [CrossRef]

Article

Real-Time Remote Maintenance Support Based on Augmented Reality (AR)

Dimitris Mourtzis *, Vasileios Siatras and John Angelopoulos

Laboratory for Manufacturing Systems and Automation, Department of Mechanical Engineering and Aeronautics, University of Patras, 26504 Patras, Greece; siatras@lms.mech.upatras.gr (V.S.); angelopoulos@lms.mech.upatras.gr (J.A.)

* Correspondence: mourtzis@lms.mech.upatras.gr; Tel.: +30-2610-910-160

Received: 31 January 2020; Accepted: 5 March 2020; Published: 8 March 2020

Abstract: In the realm of the current industrial revolution, interesting innovations as well as new techniques are constantly being introduced by offering fertile ground for further investigation and improvement in the industrial engineering domain. More specifically, cutting-edge digital technologies in the field of Extended Reality (XR) have become mainstream including Augmented Reality (AR). Furthermore, Cloud Computing has enabled the provision of high-quality services, especially in the controversial field of maintenance. However, since modern machines are becoming more complex, maintenance must be carried out from experienced and well-trained personnel, while overseas support is timely and financially costly. Although AR is a back-bone technology facilitating the development of robust maintenance support tools, they are limited to the provision of predefined scenarios, covering only a limited number of scenarios. This research work aims to address this emerging challenge with the design and development of a framework, for the support of remote maintenance and repair operation based on AR, by creating suitable communication channels between the shop-floor technicians and the expert engineers who are utilizing real-time feedback from the operator's field of view. The applicability of the developed framework is tested in vitro in a lab-based machine shop and in a real-life industrial scenario.

Keywords: augmented reality; maintenance; real-time

1. Introduction

Equipment maintenance is one of the key elements consisting of manufacturing systems. In their research work, Mourtzis et al. in Reference [1] highlight the importance equipment maintenance as part of the production lifecycle, which reaches 60%–70% of the total production cost. Therefore, being able to forecast machine maintenance operations and perform them in a short time period, can lead to successful troubleshooting and simultaneously increase machine tools availability. Additionally, since the replacement of damaged components can be as high as 70% of the total maintenance cost [1], it is one of the top priorities among manufacturing firms to discover alternative policies for cutting maintenance costs as a means of increasing their revenues. As a result, the overall production performance is optimized [2].

Besides the failure forecast, a major issue identified is the completion of accurate and error-free maintenance operations, and, thus, the reassurance of fully functional machines in the earliest time possible. Based on that concept, a considerable amount of research effort has been put on the design and development of real-time maintenance support tools and applications based on mobile devices to prevent unnecessary errors from happening [3–6]. With the evolution of Information and Communication Technologies (ICT) as well as the massive break out of augmented reality (AR), the research community extended that approach by exploiting the advantages of AR for data projection during the maintenance operations [6–11]. Generally, under the framework of Industry 4.0 in an attempt to bridge the physical world with virtual worlds, new technologies have emerged, such as

Augmented Reality (AR), Mixed Reality (MR), and Virtual Reality (VR) [12]. These digital technologies lie under the umbrella term Extended Reality (XR). The term XR is often used by practitioners and scientists in order to describe the three immersive technologies mentioned above [13].

However, most of the research works available on the web cover basic aspects of Maintenance and Repair Operations (MRO) and provide flexible and accurate troubleshooting [3–6]. The constant innovation of cutting-edge technologies unveils new opportunities for remote maintenance support. Therefore, there is an apparent need for the development of real-time AR frameworks for the support of MRO.

More specifically, the proposed research deals with the design and development of a novel framework for real-time remote maintenance support based on AR. Inspired by the concepts of Cloud manufacturing and Product Service Systems (PSS), the framework was built in a cloud-enabled environment aimed at bridging the different working groups involved in the production lifecycle. Considering that companies seek increased operational availability, also known as availability (A) of their machines, engineers are constantly developing new maintenance policies that can ensure decreased Mean Time To Repair ($MTTR$) and increased Actual Machining Time (AMT). The following formula reflects the correlation between the above-mentioned variables and characteristics of actual machines.

$$ROTBF = MTBF - AMT \tag{1}$$

where $ROTBF$ is the Remaining Operating Time Between Failures. $MTBF$ is the Mean Time Between Failures. AMT is the Actual Machining Time.

$$A = \frac{MTBF}{MTTR + MTBF} \tag{2}$$

where A is the machine availability. $MTTR$ is the Mean Time To Repair.

From Equation (2), it is clear that, in order to increase the operational availability of the machine, represented by A in Equation (2), there are two possible alternatives. The first alternative is to increase the numerator, i.e., the $MTBF$, and the second alternative is to decrease the denominator, i.e., $(MTTR + MTBF)$. However, under the scope of this research work, the aim is to develop a framework that will facilitate the maintenance process of machine tools and, by extension, it will result in the minimization of $MTTR$. Therefore, if $MTTR$ is decreased, then the denominator of Equation (2) will also decrease and, thereby, the operational availability of the machine tool will eventually increase.

It is stressed out that the above-mentioned indexes are the cornerstones of maintenance contracts offered either as a service by maintenance contractors or OEMs (Original Equipment Manufacturers) as part of their sales program. Although, a maintenance plan may seem attractive due to the low $MTTR$ offered by the maintenance contractor, e.g., 24 h. The Mean Time To Recovery is another index that is highly affecting the quality of the service, as it directly affects Operational Availability of the machine. The Mean Time To Recovery is an aspect of technical maintenance, dependent on a variety of factors, such as part lead times, administrative delays, and technician transportation delays. Aimed at minimizing the $MTTR$ and eliminating the factors mentioned above, this research work is dedicated to the modeling, development, and practical implementation of a real-time, AR-based framework for remote maintenance support.

The reminder of the paper is structured as follows. In Section 2, an extensive literature review is carried out, focusing on topics relevant to technical maintenance and AR. Then, in Section 3, the proposed system architecture is presented and, in continuation in Section 4, the steps leading to the practical implementation of the framework are discussed. Section 5 deals with the discussion of a real-life industrial scenario used for validating the developed framework. Lastly, the paper concludes in Section 6, arguing the results and the future work of this research work.

2. Literature Review

2.1. Literature Review Methodology

In order to conduct a substantial literature review, a research methodology has been adopted by Reference [14], restructured so as to meet the requirements of the current research work. For the research, the most well renowned publication databases, namely Google Scholar, ResearchGate, ScienceDirect, Scopus, and Web of Science, were used. a total of 34 research papers has been tracked and examined. The investigation of the available literature was performed in two levels based on the year of publication. More specifically, publications older than a decade have been examined in order to track down the evolution of AR throughout past years and the newer publications were examined so as to conclude on the current trends in the field of Remote Maintenance Support and, by extension, to compare similar frameworks for the proposed framework.

2.2. Maintenance as Part of the Modern Manufacturing System

As stated in the previous paragraphs, it is clear that maintenance besides being a crucial part of the production line, as it ensures the continuous and flawless operation of the physical assets, is also a very complex process. Therefore, there is a considerable amount of research work trying to cover all of its aspects. However, maintenance activities can be further categorized regarding the status of the physical assets. Concretely, the most common types of technical maintenance found are: (i) Corrective Maintenance, (ii) Preventive Maintenance, (iii) Risked-based Maintenance, and (iv) Condition-based Maintenance [15]. Throughout recent decades, many types of technical maintenance policies have been developed in an aim to cover the majority of the industrial needs. However, with the big diversity in the general policies that firms adopt, there is no optimal solution regarding the equipment maintenance plan. With that said, the maintenance policy adopted a company, and is highly dependent on the general business model and, by extension, to the general goals to be achieved. As a result, all of the above-mentioned policies are equally important for the research community and the industrial domain as well. Regardless, the scope of this research work is to present the design and development of a framework that supports field technicians mainly in corrective scenarios, where a malfunction has already occurred and, in many cases, standard procedures cannot be applied. Therefore, the research work of this paper could be classified as an enhanced corrective maintenance policy, which is based on the provision of service for on-demand guidance.

The current industrial era is characterized by the immense decentralization of the manufacturing networks. Thus, the actual manufacturing plant may be in a different region than the firm's headquarters, which makes communication of the two departments more difficult [3,16]. However, the ongoing technological advances in the ICT have enabled the remote monitoring, remote operation, and the remote maintenance of the modern manufacturing systems [17,18]. Modernized maintenance under the term of Smart Maintenance have occurred over the last decade.

2.3. Augmented Reality Based Remote Maintenance

The concept of Digital Reality has been under the research scope during recent decades. With the advent of the fourth industrial revolution, such cutting-edge digital technologies have met an increasing advance, as a result of the technological advances in the ICT. More specifically, Augmented Reality (AR) is a very popular digital technology, which offers the advantage of enhancing the user's perception by augmenting their physical environment with computer-generated information. In the field of AR, Azuma is considered a pioneer, with the first survey on AR being published two decades ago [19], concluding that AR back then was a very immature technology. a huge potential for future development was expected. In a more recent publication [20], the challenges that must be overcome by researchers in order to make AR a mainstream form of media are discussed. Among others, the most important challenge the research and development community have to overcome, is the creation of highly intuitive AR applications and tools.

In the available literature, there can be found many publications covering the AR-based aspect of maintenance. The majority of the works is focused on the remote assistance of technical maintenance operations through the various AR-based frameworks. a typical application of remote maintenance support is presented in Reference [21], where the technicians are capable of performing maintenance operations in robotic arm manipulators with the use of AR-projected instructions. Another unique AR framework for the support of MRO is presented in Reference [22]. In Reference [23], the authors cover another aspect of AR-based remote maintenance under the framework of Product Service Systems (PSS). Therefore, the AR tools developed can facilitate manufacturers to add value to their line of products by transitioning to the PSS philosophy.

In the research work presented in Reference [4], the authors have carried out research on the recent trends around the Smart Maintenance paradigm by investigating recent publications as well as by extracting useful empirical knowledge. In Reference [5], a collaborative remote maintenance framework has been suggested. Although the research work is inspiring and of good quality, the framework implementation is based on technologies and equipment that currently can be considered as obsolete, as the technician has limited mobility since a PC is required by their side so as to visualize the AR instructions. Moreover, the instructions are based mostly on the projection of textual information. Lastly, in this research work, the use of frame markers is not suitable for the maintenance of bigger and more complex machines. Therefore, what is needed is more robust and more compact framework enhancing the technician's mobility when working on the machine. In a more recent paper, a similar remote maintenance support system based on AR is presented [24]. It is remarkable that the authors have put a considerable amount of effort on providing a mobile tool, based on the utilization of tablet PC. However, similar to the previous work presented, the use of frame markers is inevitable, which can highly affect the overall performance of the tool and, by extension, the performance of the shop-floor/field technician.

Although there is a continuous development of the AR-based tools proposed by the research community during the last two decades, there is a lack of added functionality. The above-mentioned development as well as the practical implementation of these frameworks in real-life industrial scenarios and the in-vitro experimentation in laboratory environments has led to the conclusion that AR has to treated as a back-bone technology. By extension, in the near future, the development of AR tools has to be further enhanced by the addition of more functionalities. a representative example is the framework presented in Reference [25], where an adaptive AR framework for machine operation as well as technical maintenance is presented. The adaptivity of the framework is based on the skill level of the technicians. Similarly, in Reference [26], an innovative framework based on a wrist-haptic tracking methodology is introduced.

2.4. Maintenance under the Framework of Industry 4.0

The modern manufacturing world is undergoing a digitalization phase under the framework of Industry 4.0. During the current industrial revolution, digital technologies have tremendously advanced, which enables the improvement of other scientific fields, including that of technical maintenance. With the recent development in the ICT, and the integration of Internet of Things (IoT), the manufacturing domain has passed in a whole new phase by relying on the utilization of data, i.e., Big Data Analytics. For the successful completion of MRO besides the technical knowledge/expertise of the stuff carrying out the operations, it is crucial that all the needed components are acquired. Therefore, companies are maintaining an inventory of components. However, under the framework of Industry 4.0, and with the utilization of digital technologies, the existing inventory practices can be further improved through proper classification of the MRO components while additive manufacturing can facilitate in reducing lead times to produce several components [6–10,26–28].

For Industry 4.0 applications, two key objectives are to ensure maximum uptime throughout the production chain, and to increase productivity while reducing production costs. In Reference [29], a systematic approach to analyze the strengths and weaknesses of current open-source Big Data and stream processing technologies and assess their use for Industry 4.0 use cases is presented. In addition, the model of Industry 4.0 is gradually being implemented in worldwide development, distribution, and marketing chains. As such, the design and integration of a stainless-steel predictive maintenance system is presented in Reference [29]. This case study uses data from machinery involved in producing high-quality steel sheets. Following the predictive maintenance field, the use of real-time detection and prediction algorithms regarding future failures has significantly benefited from the technological advances of Industry 4.0 era. There's also growing interest in decision-making algorithms triggered by predictions of failure over the past few years. a literature review on decision making in predictive maintenance in the context of smart manufacturing is done in References [30,31]. Additionally, a review on predictive maintenance as an integrated predictive platform for production systems, focusing on maintenance approaches, methods, and tools is described in Reference [32].

Another significant issue is the estimation of the maintenance time for a new maintenance project, which is among the main maintenance offerings, is based solely on the experience and knowledge of the engineer. The research work presented in Reference [33] proposes a framework for knowledge-based calculation of maintenance time based on Key Performance Indicators (KPI) monitoring to support the capture and reuse of information in maintenance activities as well as to improve the performance of the given maintenance PSS. Next, an extendable and reusable scheduling approach in the context of predictive maintenance, which supports multiple heterogeneous inputs and outputs, is described in Reference [33]. In accordance with the current schedule, predictive maintenance indicators from the tracked equipment are used for scheduling maintenance operations. a web service architecture is implemented to accommodate highly different use scenarios such as suppliers of equipment.

3. System Architecture

3.1. General System Architecture

As discussed in the previous paragraphs, the scope of this research work is focused on the design and development of a real-time, remote maintenance assistance framework based on AR by enabling the creation of new communication channels between the expert engineers and the shop-floor technicians. Concretely, there is an opportunity to eliminate the need for preparing AR scenes, which considerably decreases the time and effort needed for creating such content. Moreover, certain limitations can be overcome as expert engineers, through the use of the proposed framework, are able to guide shop-floor technicians in less likely MRO scenarios. In order to meet the above-mentioned technical requirements, the methodology adopted by aiming at addressing the identified scientific gaps will be presented in detail in the following paragraphs. The general architecture flowchart is depicted in Figure 1.

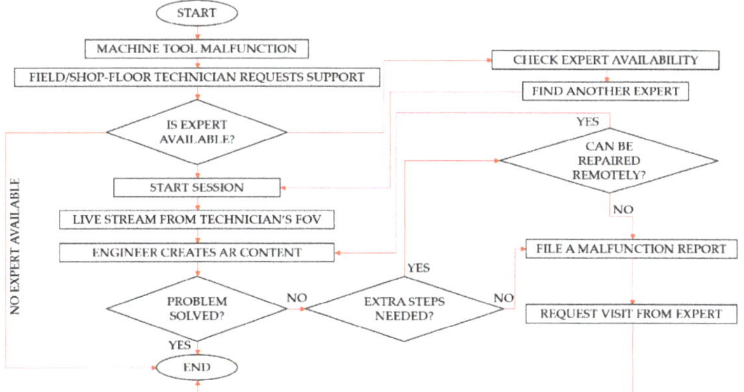

Figure 1. Flowchart of the proposed system.

The sequence of actions begins as long as a malfunctioning machine tool has been noticed. Then, the shop-floor technician who is carrying a device, a Head Mounted Display (HMD), contacts the expert engineer for assistance. It is stressed out that the coexistence of the two involved parties under the same premises is not required. More specifically, through the use of dedicated communication protocols, the shop-floor technicians can be remotely connected to the technical support department of the machine tool manufacturer. As long as the connection between the two parties has been established, a live video stream from the shop-floor technicians' device is broadcast to the expert engineer. This process can be realized as a live teleconference between the two parties. Therefore, while the video stream is broadcast to the expert engineer, the shop-floor technician is capable of making annotations so as to ease the engineer to get a better understanding of the machine tool malfunction. As soon as the engineer is fully aware of the malfunction, they can create on-demand AR content, which will be projected in the shop-floor technician's FOV (Field Of View). It is stressed out that, during the live-video casting from the technician, the device is also capable to 3D scan the space around the technician. As such, the algorithm running in the background is fully aware of the user's surroundings. Therefore, the expert engineer can register 3D content in the technician's FOV more easily and more successfully. The framework architecture can be realized through a platform, where the shop-floor technicians can connect and request assistance from expert engineers in real-time. From the expert engineer point-of-view, the platform serves as a tool enabling the expert to guide the shop-floor technician with the use of simple 3D tools. Therefore, it is of great importance to initially present the platform architecture. In Figure 2, the architecture of the proposed framework is depicted. From the figure, the structure of the Cloud Platform is visible and consists of services and the toolbox.

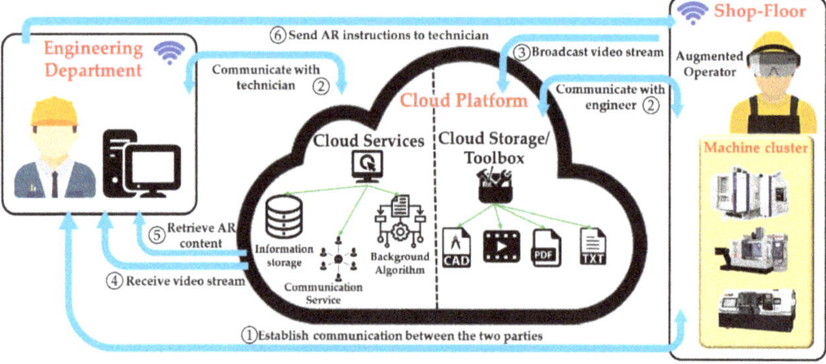

Figure 2. Proposed system architecture and steps sequence.

The most important component of the platform architecture is the online connection and user interconnection framework. More specifically, the platform acts as a cloud platform, where users, either expert engineers or field/shop-floor technicians, can connect to and, most importantly, they can exchange digital information. It is stressed out that there are no limitations in the type and volume of the exchanged information. Additionally, the user experience can be further improved by the support of live teleconference of the involved actors. In order to enhance the communication experience between the involved parties, the framework supports a teleconference feature. When this feature is enabled, then the user is capable of having a live video feed from the expert engineer, projected in his FoV. As such, it is expected that the communication between the two parties is facilitated, as the technician can perform MRO simultaneously. As far as the information exchange is concerned, the platform is connected with a cloud database. In the database, there is dedicated space where certain 3D geometries are saved. These geometries are imported in the AR scene only by the expert engineers. The collection of these tools can be realized as a digital toolbox of the most common tools used in

MRO. However, due to increasing machine tool complexity, specialty tools might be needed. As such, the expert engineer is capable of adding a digital copy of the specialty tool through the "Create tool" Graphical User Interface (GUI).

In Figure 3, the GUI for joining an online maintenance session is depicted. On the top left corner of the GUI, the user through the provided dialog can select a session to join. In case the session has reached its end, then the user has to leave the session. On the lower left corner, the provided buttons assist the master engineer to retrieve from the Cloud platform the required tools by performing a "drag and drop" action to position them in the technician's FOV. However, if the desired tool does not exist in the Cloud toolbox, then the engineer is capable of uploading a 3D geometry. More specifically, in Figure 3, the "plus" icon when hit triggers the "Create Tool" function, so that the engineer adds the geometry in a suitable filetype along with relevant info and a description of the tool. The filetypes accepted by the Cloud Database have been limited to the filetypes accepted by the AR tool. Therefore, since the framework is developed in the Unity 3D game engine, the supported model file formats for 3D geometries are, .fbx, .dae (Collada), .3ds, .dxf, .obj. Then the algorithm running in the background is responsible for uploading the geometry on the Cloud Database as well as to create a new record in the SQL Database. The SQL database is created and used in an ontological manner, so that a smart-searching algorithm can automatically generate suggestions to the engineer during the AR scene creation phase. In Figure 4, the set of the initial tools used for populating the Cloud Toolbox is presented. For this manner, the most common technical tools have been utilized.

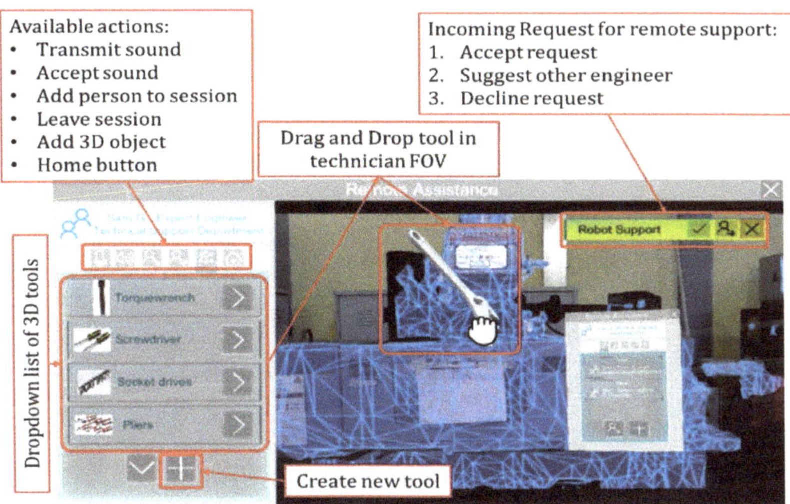

Figure 3. GUI for online maintenance session.

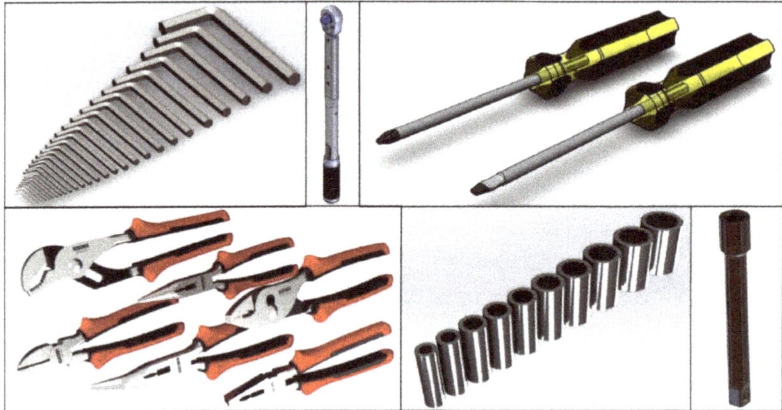

Figure 4. 3D CAD geometries of tools for AR instructions.

3.2. Cloud Platform

The proposed remote maintenance service is developed as a cloud-based platform. The Cloud platform enables the communication of the shop-floor/field technicians and the expert engineer as well as the distribution of information between the involved parties. The Cloud platform is comprised of two domains: (i) the communication domain and (ii) the data storage domain. As discussed in the previous paragraphs, the communication domain is used for matching shop floor technicians for expert engineers and for establishing the communication channels between them. On the other hand, the data storage domain in essence is the Cloud Database where all the data are saved in SQL tables as well as the 3D geometries uploaded.

3.3. Remote Maintenance Sessions

As long as the shop-floor/field technicians identify a malfunctioning machine, then, with the use of a mobile device, they use the cloud platform so as to communicate with an expert engineer. In the cloud platform, this form of communication is realized through the creation of online sessions. For each session created, a communication domain is created. Upon the establishment of the communication domain, the two parties are capable of bilaterally exchanging information.

4. Implementation

As far as software is concerned, the first step for the development of the framework is the development of the cloud platform. From a hardware point of view, the architecture of the framework has been developed so that a variety of platforms can be supported including PCs, HMDs, and handheld devices. In its current form, the framework has been developed so that the technician uses a Microsoft Hololens HMD and the expert engineer operates from a desktop PC.

For the platform, a remote server is utilized. The server has a dedicated domain for the storage of files, where 3D geometries are also uploaded. The supported filetypes for the 3D geometries are .fbx, .dae (Collada), .3ds, .dxf, .obj, and .skp. Besides the 3D geometries, the storage domain supports any form of files, providing enough degrees of freedom to the engineers in exchanging files with the technicians during a remote maintenance session.

As far as the development of the software is concerned, the Unity 3D game engine has been utilized. Regarding the AR application for the technicians, Unity 3D in conjunction with the Vuforia and MRTK (Mixed Reality Tool Kit) libraries were used. The Vuforia library is useful for the creation and handling of the AR content, whereas the MRTK library supports advanced user interaction functionalities for the Microsoft Hololens HMD. Regarding the expert engineer application for the remote support, a Windows Form Application has been developed. In order to enable the communication of the two applications, the integrated Unity UNet API (Application Programming Interface) was used. Since the above-mentioned API is targeted for the creation of multiplayer games, several functionalities had to be adjusted so as to align with the requirements of the framework. UNet can be realized through different layers. In Figure 5, the functionalities per layer of the UNet are presented.

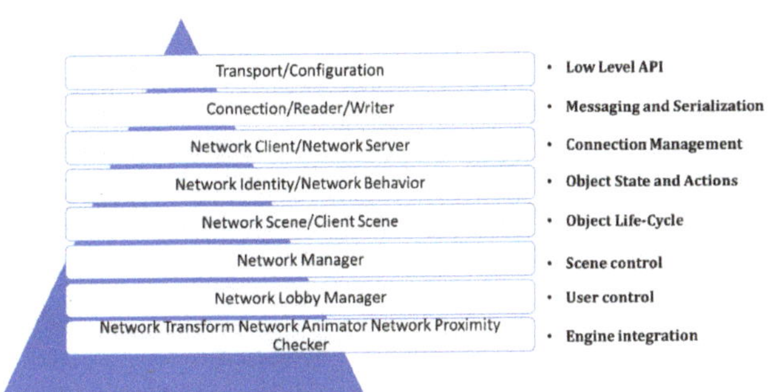

Figure 5. Functionalities per Layer of Unity 3D UNet API.

As far as the communication module is concerned and, more specifically, the teleconference functionality, a variety of solutions has been examined, including Skype, Agora.io [34], GIGA Video Streamer [35], and the development of a custom teleconference tool. Among the available solutions, the latter was selected, i.e., the development of a custom teleconference interface. To begin with, the integration of Skype, although it is an appealing solution, and the quality of the service itself is sufficient. It required that it was run as an external application on the Microsoft HoloLens HMD, which had a negative effect on the performance of the app, as well as being unpleasant for the end-users to switch over to two different applications including one for the communication and one for the AR remote support. On the other hand, the Agora.io and the GIGA Video Streamer are very capable SDKs (Software Development Kit) that could be integrated in the proposed framework. However, these two SDKs are not free. Therefore, the development of a custom module based on the UNet API, discussed above, was the only choice satisfying the framework's requirements. Concretely, from the expert engineer side, the video stream from the camera and the microphone of the PC was captured and transmitted to the Microsoft Hololens HMD. Form the technician side, a pop-up window appears where the technician can view the expert engineer and, through the embedded speakers of the HMD, to hear the expert's voice.

From a hardware point of view, for the on-site technician, a Microsoft HoloLens [36] device has been utilized in order to take advantage of its four environment-understanding cameras and the mixed reality capture in terms of sensors, whereas, for the development of the framework and the validation experiments, a laptop PC has been utilized. More specifically, the laptop PC is equipped with an Intel core i7 processor clocked at 2.20 GHz, 8 GB DDR4 RAM, and a NVIDIA GeForce 1060 GPU with 6 GB dedicated memory.

5. Case Study and Results

The applicability of the proposed framework is tested and validated in-vitro in a laboratory-based machine shop as well as in a real-life industrial scenario. The industrial scenario involves the machine tools used in an existing machine-shop. Taking into consideration the technological level of the machines, this use case is ideal since the machine shop consists of machines coming from different technological eras. Thus, it is of great importance to examine the actual impact of the proposed framework in varying machine tools. In Figure 6, the validation of the framework in the real-life scenario are presented. More specifically, in Figure 6a, the malfunctioning machine is visible along with the shop-floor technician who is wearing the HoloLens HMD in order to perform the maintenance operations, as instructed by the expert engineer. Similarly, in Figure 6b, another engineer is testing the developed framework. It is stressed out that, in the background, the expert engineers using the laptop PC in order to communicate with the technician are visible.

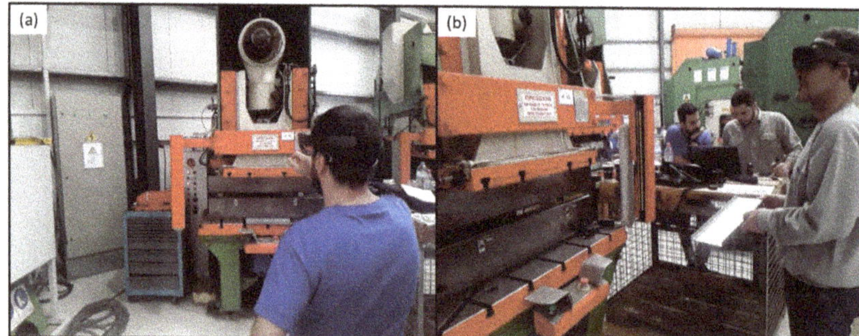

Figure 6. (**a**) Validation of the proposed framework by shop-floor technician; (**b**) Another technician tests the framework.

Similarly, in Figure 7, the practical implementation and validation of the framework in the lab-based scenario is presented. Concretely, in Figure 7a, the expert engineer, from his desktop, can view the user's FoV, which contains the malfunctioning machine tool. As can be seen, the spatial recognition algorithm of the Microsoft HoloLens device has recognized the technician's physical environment. Thus, the engineer based on that feature can place the 3D augmentations on the user's physical environment. On the other hand, in Figure 7b, the shop-floor technician is presented. What is worth noticing is that, in the top right corner of that figure, a snapshot from the HoloLens head up display is presented. The shop floor technician, after requesting assistance from the engineer, the connection has been established and a video teleconference has begun. Therefore, a communication window has opened in the technician's FoV, where the audio and video stream from the engineer are received and displayed.

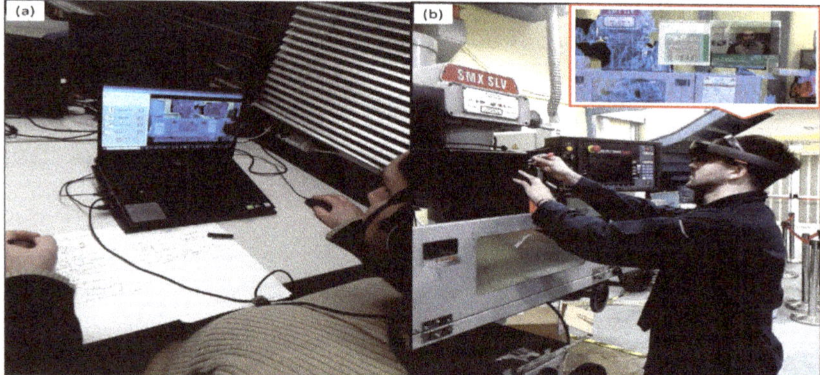

Figure 7. (**a**) Expert engineer visualizes video stream from technician; (**b**) Technician is performing maintenance tasks based on expert's instructions.

Validation Strategy—Design of Experiments

In the following paragraphs, the steps followed for the validation of the proposed framework in the real-life industrial scenario will be presented. Since maintenance is a very important aspect regarding the flawless and continuous operation of manufacturing systems, the validation strategy should be carefully designed and also should provide a quantified result of the impact. Therefore, for each of the machine-tools examined in this process, two basic metrics will be used. According to the guidelines given by Chryssolouris in Reference [21] regarding the mathematical programming of manufacturing systems, the operating and maintenance costs for a machine can be calculated by the equation below.

$$Operating\ and\ Maintenance\ Costs = \sum_{t=1}^{T}(P/F, I, t)\sum_{i=1}^{N} m_{it}x_{it} \tag{3}$$

where $(P/F, I, t)$ ≡ Present a worthy interest factor for discrete compounding when the discount rate is *I*% per period and the discount interval is *t* periods. m_{it} ≡ The cost of operating and maintaining a machine of type *i* during period *t*. Is the number of machines, on one hand, in the work center *i* for period *t*.

In order to quantify the outcome of the validation process for the proposed framework, each of the above-mentioned values will be calculated in order to conclude on the efficiency of the real-time remote maintenance.

Therefore, the cost for operation and maintenance for a CNC milling machine has been calculated, based on data derived from a real-life machine-shop. The cost of operation and maintenance has been calculated on an annual basis for time horizon of five years, and the results are depicted in Figure 8. Additionally, in Figure 8, a prediction of the cost is made for the oncoming five years based on statistical regression, presented by the red dot trendline. It is stressed out that, due to the non-linear relation between the individual costs of operation and maintenance of machine tools during their lifecycle [37], a 2nd degree polynomial has been calculated. The same calculations were made for the cost estimation based on the proposed methodology and, similarly, a prediction is made for future maintenance costs. As a result, with the proposed methodology, a reduction of approximately 100 k euros is feasible in a time span of 10 years, which indicates that the framework could eventually be implemented in manufacturing systems and improve their operation.

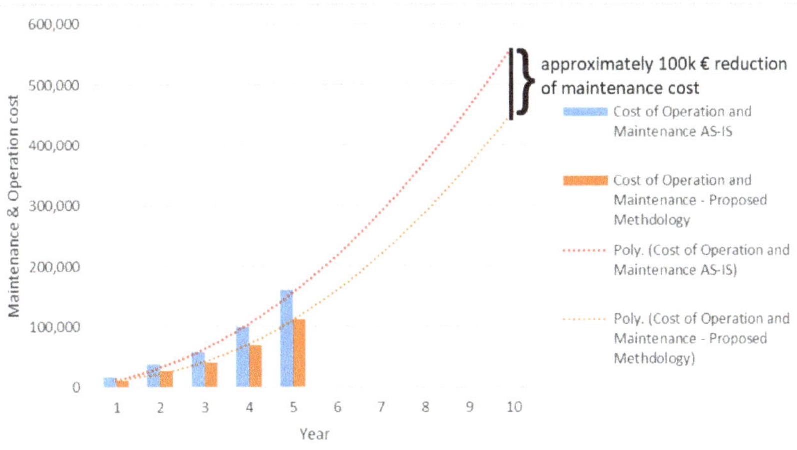

Figure 8. Cost of operation and maintenance analysis.

Besides the cost for operating and maintenance, another metric has been calculated and compared regarding the downtime of the above-mentioned machine. In manufacturing systems, time resources are equally important for the financial resources. Therefore, a maintenance scenario has been investigated. For this repair scenario, the machine tool breaks down at an unpredicted time. At that time, the production process for this machine stops and the expert engineer is called to come on-site, inspect the machine, and repair it. The same scenario has also been investigated, but instead of requesting an expert engineer to visit the machine shop, a remote support session is established, and the repair is conducted by the technicians available in the machine shop. In Figure 8, the distribution of time per maintenance task is displayed for each of the scenarios examined. From Figure 6, it is undeniable that a significant amount of time can be saved with the adoption of the proposed framework, as the time for travelling and inspection are considerably longer, which affects the overall time the machine tool has been off the production schedule. Through the use of Figure 9, it is intended to reflect the difference of time distribution between the two scenarios examined during the validation process of the proposed framework. Therefore, from Figure 9, it can be concluded that, in the first scenario, i.e., the "AS-IS-SITUATION", the most time spent was on inspection and travelling, as the expert engineer had to travel on site and physically inspect the malfunctioning machine. On the contrary, in the second scenario, i.e., "PROPOSED METHODOLOGY", a totally different time distribution has been observed with the most time spent on activities relative to the acquisition of spare parts for the machine repair. In addition to that, it can be concluded that the actual repair time, which is the black portion of the pie graph, in the second scenario is bigger. This observation can be considered normal as, in the second scenario, the maintenance operations have been carried out by a field technician, whereas, in the first scenario, the same operations have been carried out by an expert engineer with solid practical experience on such machines.

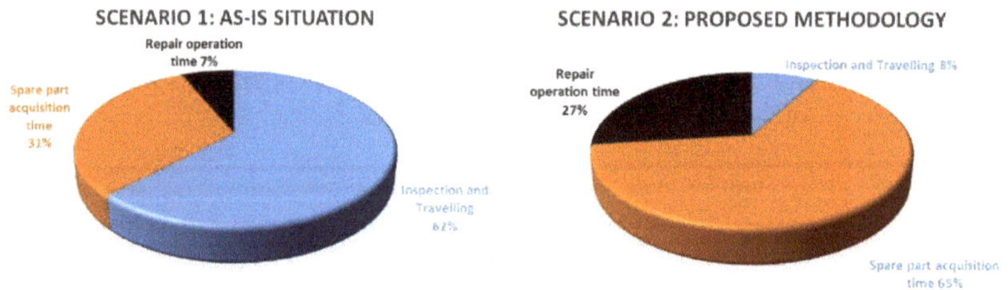

Figure 9. Distribution of time AS-IS situation vs. Proposed Methodology.

In Figure 10, a side-by-side comparison of the specific times for the two maintenance scenarios is presented. It is stressed that, with the adoption of the proposed framework, the time for inspection of the machine can be shorter as no travelling time is involved in order to reduce *MTTR*. However, due to the inexperience of the technician carrying out the maintenance activities, the actual repair operation time is longer, as the expert engineer had to perform additional checks in order to ensure that the repair was successful.

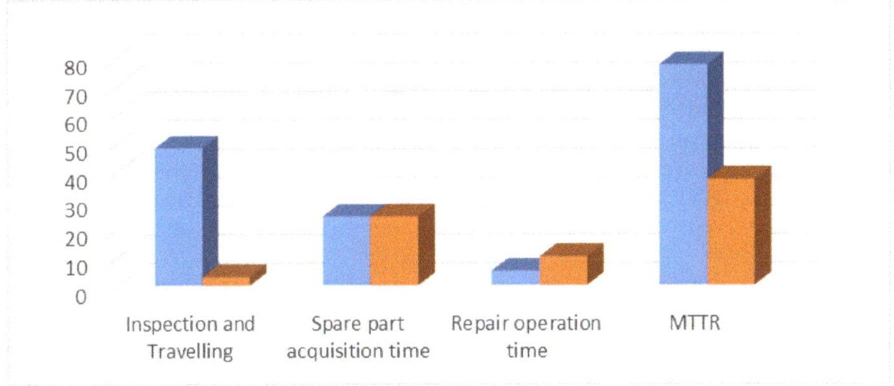

Figure 10. Time per maintenance action AS-IS situation vs. Proposed Methodology.

6. Discussion

The outcome of the current research work, is a multi-sided, multi-device application, enabling the remote maintenance of complex machine tools under the framework of Industry 4.0. The term multi-sided is used in order to describe that multiple operators can connect to the platform simultaneously and discuss on a malfunction that has occurred unexpectedly. On the other hand, the term multi-device, is used in order to describe the compatibility of the developed framework in a variety of devices. It is stressed out that the most common devices used for AR implementations are HDMs and handheld devices, including smart-phones and tablets. However, since the scope of the current research was mainly focused on the provision of a totally marker-less AR solution, the use of Microsoft's HoloLens HMD is preferred, since the device supports spatial recognition, which, in turn, has facilitated the development of the framework. It is stressed out that the use of the HMD is initially intended only for the shop-floor technician as a means to capture their surrounding and, based on that, the expert engineer is capable of placing/registering the virtual content on the user's physical environment. The development of the framework was performed in a set of consecutive stages. The first stage was based on the conceptualization of the system's architecture, the discussion of the modules comprising the architecture, and the tools to be used for the final development. Then, in stage two, the basic modules were developed, utilizing the software described in Section 4. As soon as the alpha version of the framework was released, the framework has been internally tested in vitro, in the lab-based machine shop described in Section 5. During this stage, colleague engineers have been asked to test the functionalities of the developed application either from the side of the expert engineer or from the side of the shop-floor technician aimed at gathering insightful feedback. Then, based on the feedback gathered during the previous stage, the framework has been improved in certain areas, regarding the layout of the GUIs and the functioning of the communication services. Then the beta version of the developed framework has been prepared and tested with the industrial partner in their premises.

Throughout the development and validation stages, meaningful feedback has been gathered. The main strengths of the proposed framework can be summarized to the minimum development effort required by the expert engineer. More specifically, the remote maintenance frameworks are based on the development of AR scenes, which are then implemented on an AR-ready device. Therefore, this requires advanced development skills and a certain amount of preparation time. On the contrary with the proposed framework, an expert engineer is capable of creating AR content on the user's FoV with the use of simple Drag-and-Drop actions. As such, no special skills are required and the time for AR content preparation and implementation is eliminated. Another strength of this framework is the ability to create online sessions. The online sessions are based on the Unity 3D multiplayer API offering the remote connection of multiple users. As discussed in previous paragraphs, the developed application is compatible with most devices. Thus, there is no practical limitation on the use of equipment. In comparison with other approaches, presented in References [3–6], the functionality of remote connectivity combined with the broad compatibility of the app further extends mobility of the user, i.e., the shop-floor technician in an industrial environment. More specifically, in Reference [4], the technicians have to carry a laptop computer along with a USB web camera in order to establish communication with the expert engineer and proceed with the maintenance operations. The frameworks presented in the above-mentioned publications are using frame markers for the registration of the content on the physical environment. With that said, the performance of the system relies heavily on the quality of the camera used for the detection of the marker. The user's mobility is limited, as the camera has to maintain visibility of the marker at all times. Otherwise, the augmentations cannot be registered and updated as the user moves around. In addition to that, the proposed framework, since it utilizes newer equipment with increased computational power, the use of more vivid/realistic virtual objects is feasible. Not only that, but in previous approaches, the augmentations are limited to the display of simple schematics, e.g., arrows, text boxes, or indicate the component of interest via a wireframe model. On the contrary, the current approach can recognize the objects by the utilization of Microsoft HoloLens.

7. Conclusions

In this research work, the most recent and relevant publications regarding remote maintenance support have been investigated. Through the literature review process, it became apparent that, although there are numerous approaches contributing in the field of remote maintenance support, there is a small amount of research work done on the provision of real-time support. Therefore, an opportunity for a new research work has been identified. Concretely, the design and development of a framework for the real-time remote maintenance and repair operations based on AR has been presented in detail. The framework aimed at creating interactive, intuitive, and collaborative communication channels between the shop-floor technicians and the expert engineers. Furthermore, the developed application was tested in a real-life industrial scenario in order to verify that all the functional requirements are met. The results indicate that the present research work has managed to successfully address the functional requirements discussed during the modelling phase of the problem and has succeeded in delivering a zero-time content authoring AR tool, which further minimizes the *MTTR*. However, there is fertile ground for future improvement. As discussed in the literature review section, one of the key elements consisting of an AR application is the content authorization methodology. Towards that end, future work could focus on integrating the developed framework to an automated methodology for content authoring, which could considerably reduce the effort and the skills needed from engineers in creating AR content.

Author Contributions: All authors participated in the modelling of the research project. More specifically, D.M. Supervisor, V.S. Development and methodology, J.A. Research, conceptualization, and writing. All authors have read and agreed to the published version of the manuscript.

Funding: This research received no external funding.

Acknowledgments: The H2020 EC funded project "Balancing Human and automation levels for the manufacturing workplaces of the future—MANUWORK" (GA No: 723711) partially supported this work.

Conflicts of Interest: The authors declare no conflict of interest.

References

1. Mourtzis, D.; Vlachou, E.; Milas, N.; Xanthopoulos, N. a Cloud-based Approach for Maintenance of Machine Tools and Equipment Based on Shop-floor Monitoring. In Proceedings of the 48th CIRP Conference on Manufacturing Systems, Ischia, Italy, 24–26 June 2015; pp. 655–660. [CrossRef]
2. Predictive Maintenance: Taking Pro-Active Measures Based on Advanced Data Analytics to Predict and Avoid Machine Failure. Available online: https://www2.deloitte.com/content/dam/Deloitte/de/Documents/deloitte-analytics/Deloitte_Predictive-Maintenance_PositionPaper.pdf (accessed on 2 January 2020).
3. Wang, W.; Peter, W.T.; Lee, J. Remote machine maintenance system through Internet and mobile communication. *Int. J. Adv. Manuf. Technol.* **2007**, *31*, 783–789. [CrossRef]
4. Wang, J.; Feng, Y.; Zeng, C.; Li, S. An augmented reality based system for remote collaborative maintenance instruction of complex products. In Proceedings of the 2014 IEEE International Conference on Automation Science and Engineering (CASE), Taipei, Taiwan, 18–22 August 2014; pp. 309–314. [CrossRef]
5. Masoni, R.; Ferrise, F.; Bordegoni, M.; Gattullo, M.; Uva, A.E.; Fiorentino, M.; Carrabba, E.; Di Donato, M. Supporting remote maintenance in industry 4.0 through augmented reality. *Procedia Manuf.* **2017**, *11*, 1296–1302. [CrossRef]
6. Ong, S.K.; Zhu, J. a novel maintenance system for equipment serviceability improvement. *CIRP Ann.* **2013**, *62*, 39–42. [CrossRef]
7. Mourtzis, D.; Vlachou, A.; Zogopoulos, V. Cloud-based augmented reality remote maintenance through shop-floor monitoring: a product-service system approach. *J. Manuf. Sci. Eng.* **2017**, *139*, 061011. [CrossRef]
8. Mourtzis, D.; Zogopoulos, V.; Katagis, I.; Lagios, P. Augmented Reality based Visualization of CAM Instructions towards Industry 4.0 paradigm: a CNC Bending Machine case study. In Proceedings of the 28th CIRP Design Conference 2018, Nantes, France, 23–25 May 2018; pp. 368–373. [CrossRef]
9. Cardoso, S.F.L.; Mariano, Q.M.C.F.; Zorzal, R.E. a Survey of Industrial Augmented Reality. *Comput. Ind. Eng.* **2020**, *139*, 106159. [CrossRef]
10. Palmarini, R.; Erkoyuncu, A.J.; Roy, R.; Torabmostaedi, H. a systematic review of augmented reality applications in maintenance. *Robot. Comput. Integr. Manuf.* **2018**, *49*, 215–228. [CrossRef]
11. Fernández del Amo, I.; Erkoyuncu, A.J.; Roy, R.; Palmarini, R.; Onoufriou, D. a systematic review of Augmented Reality content-related techniques for knowledge transfer in maintenance applications. *Comput. Ind.* **2018**, *103*, 47–71. [CrossRef]
12. Berglund, F.Å.; Gong, L.; Li, D. Testing and validating Extended Reality (xR) technologies in manufacturing. In Proceedings of the 8th Swedish Production Symposium (SPS 2018), Stokholm, Sweeden, 16–18 May 2018; pp. 31–38. [CrossRef]
13. What Is Extended Reality Technology? a Simple Explanation for Anyone. Available online: https://www.forbes.com/sites/bernardmarr/2019/08/12/what-is-extended-reality-technology-a-simple-explanation-for-anyone/#69b6421f7249 (accessed on 16 February 2020).
14. Mourtzis, D. Simulation in the design and operation of manufacturing systems: State of the art and new trends. *Int. J. Prod. Res.* **2019**, 1–23. [CrossRef]
15. 4 Types of Maintenance Strategy, Which one to Choose? Available online: https://new.abb.com/medium-voltage/service/maintenance/feature-articles/4-types-of-maintenance-strategy-which-one-to-choose (accessed on 2 January 2020).
16. Mourtzis, D.; Doukas, M. Design and Planning of Manufacturing Networks for Mass Customisation and Personalisation: Challenges and Outlook. In Proceedings of the 2nd CIRP Robust Manufacturing Conference (RoMac 2014), Bremen, Germany, 7–9 July 2014; pp. 1–13. [CrossRef]
17. Xun, X. From cloud computing to cloud manufacturing. *Robot. Comput. Integr. Manuf.* **2012**, *28*, 75–86. [CrossRef]
18. Mourtzis, D.; Vlachou, A. a cloud-based cyber-physical system for adaptive shop-floor scheduling and condition-based maintenance. *J. Manuf. Syst.* **2018**, *47*, 179–198. [CrossRef]
19. Azuma, T.R. a Survey of Augmented Reality. *Presence Teleoperators Virtual Environ.* **1997**, *6*, 355–385. [CrossRef]
20. Azuma, T.R. The Most Important Challenge Facing Augmenting reality. *Presence Teleoperators Virtual Environ.* **2017**, *25*, 234–238. [CrossRef]

21. Mourtzis, D.; Zogopoulos, B.; Vlachou, E. Augmented Reality Application to Support Remote Maintenance as a Service in the Robotics Industry. In Proceedings of the 50th CIRP conference on Manufacturing Systems, Taichung City, Taiwan, 3–5 May 2017; pp. 46–51. [CrossRef]
22. Henderson, S.; Feiner, S. Exploring the Benefits of Augmented Reality Documentation for Maintenance and Repair. *IEEE Trans. Vis. Comput. Graph.* **2011**, *17*, 1355–1368. [CrossRef] [PubMed]
23. Mourtzis, D.; Angelopoulos, J.; Boli, N. Maintenance assistance application of Engineering to Order manufacturing equipment: a Product Service System (PSS) approach. In Proceedings of the 16th IFAC Symposium on Information Control Problems in Manufacturing INCOM 2018, Bergamo, Italy, 11–13 June 2018; pp. 217–222. [CrossRef]
24. Mourtzis, D.; Xanthi, F.; Zogopoulos, V. An Adaptive Framework for Augmented Reality Instructions Considering Workforce Skill. In Proceedings of the 52nd CIRP Conference on Manufacturing Systems (CMS), Ljubljana, Slovenia, 12–14 June 2019; pp. 363–368. [CrossRef]
25. Siew, Y.C.; Ong, K.S.; Nee, C.Y.A. a practical augmented reality-assisted maintenance system framework for adaptive user support. *Robot. Comput. Integr. Manuf.* **2019**, *59*, 115–129. [CrossRef]
26. Chryssolouris, G. *Manufacturing Systems: Theory and Practice*, 2nd ed.; Springer: New York, NY, USA, 2006.
27. Chen, J.; Gusikhin, O.; Finkenstaedt, W.; Liu, Y. Maintenance, Repair, and Operations Parts Inventory Management in the Era of Industry 4.0. In Proceedings of the 9th IFAC Conference on Manufacturing Modelling, Management and Control MIM 2019, Berlin, Germany, 28–30 August 2019; pp. 171–176. [CrossRef]
28. Bokrantz, J.; Skoogh, A.; Berlin, C.; Wuest, T.; Stahre, J. Smart Maintenance: An empirically grounded conceptualization. *Int. J. Prod. Econ.* **2019**, 107534. [CrossRef]
29. Sahal, R.; Breslin, G.J.; Ali, I.M. Big data and stream processing platforms for Industry 4.0 requirements mapping for a predictive maintenance use case. *J. Manuf. Syst.* **2020**, *54*, 138–151. [CrossRef]
30. Ruiz-Sarmiento, J.R.; Monroy, J.; Moreno, F.A.; Galindo, C.; Bonelo, J.M.; Gonzalez-Jimenez, J. a predictive model for the maintenance of industrial machinery in the context of industry 4.0. *Eng. Appl. Artif. Intell.* **2020**, *87*, 103289. [CrossRef]
31. Bousdekis, A.; Lepenioti, K.; Apostolou, D.; Mentzas, G. Decision Making in Predictive Maintenance: Literature Review and Research Agenda for Industry 4.0. *IFAC-PapersOnLine* **2019**, *52*, 607–612. [CrossRef]
32. Efthymiou, K.; Papakostas, N.; Mourtzis, D.; Chryssolouris, G. On a Predictive Maintenance Platform for Production Systems. *Procedia CIRP* **2012**, *3*, 221–226. [CrossRef]
33. Nikolakis, N.; Papavasileiou, A.; Dimoulas, K.; Bourmpouchakis, K.; Makris, S. On a versatile scheduling concept of maintenance activities for increased availability of production resources. *Procedia CIRP* **2018**, *78*, 172–177. [CrossRef]
34. Agora.io. Available online: https://www.agora.io/en/ (accessed on 26 February 2020).
35. GIGA Video Streamer. Available online: https://assetstore.unity.com/packages/tools/video/giga-video-streamer-125456?_ga=2.75845011.740357347.1582886552-1421986738.1578039987 (accessed on 26 February 2020).
36. Microsoft HoloLens Specification. Available online: https://www.windowscentral.com/hololens-hardware-specs (accessed on 26 February 2020).
37. Bengtsson, M.; Kurdve, M. Machining Equipment Life Cycle Costing Model with Dynamic Maintenance Cost. In Proceedings of the 23rd CIRP Conference on Life Cycle Engineering, Berlin, Germany, 22–24 May 2016; pp. 102–107. [CrossRef]

Article

Cutting Path Planning Technology of Shearer Based on Virtual Reality

Juanli Li [1,2,*], Yang Liu [1,2], Jiacheng Xie [1,2,3], Xuewen Wang [1,2] and Xing Ge [1,2]

1 College of Mechanical and Vehicle Engineering, Taiyuan University of Technology, Taiyuan 030024, China; 18435167533@163.com (Y.L.); xiejiacheng@tyut.edu.cn (J.X.); wxuew@163.com (X.W.); Gexing9511@163.com (X.G.)

2 Shanxi Key Laboratory of Fully Mechanized Coal Mining Equipment, Taiyuan 030024, China

3 Post—Doctoral Scientific Research Station, Taiyuan Heavy Industry Co., Taiyuan 030024, China

* Correspondence: lijuanli@tyut.edu.cn

Received: 13 December 2019; Accepted: 20 January 2020; Published: 22 January 2020

Abstract: With regards to the low degree of digitization, lack of real geological terrain, and low degree of automation in the cutting process of the traditional virtual fully mechanized mining face, we studied the key technologies of virtual operation and cutting path planning of the shearer on the three-dimensional (3D) roof and floor based on the virtual reality engine (Unity3D). Firstly, the virtual 3D coal seam was constructed through the 3D geological coordinate data of the mine. On this basis, the shape function of the scraper conveyor with the adaptive configuration on the floor was constructed to obtain the combined operation of the virtual shearer and the scraper conveyor. The movement of the shearer's walking and height-adjustment was then, analyzed. A strategy for automatic height-adjustment based on the adjustment of the direction of the drum movement is hence, proposed to control the cutting path of the shearer. Finally, different experimental schemes were simulated in the developed prototype system after which each of the schemes was evaluated using the fuzzy comprehensive evaluation method. The results show that the proposed strategy for trajectory control can improve the accuracy and stability of the shearer's motion trajectory. In Unity3D, the pre-selected schemes and digital and visual planning of coal production are previewed ahead of time, the whole production process can be mapped synchronously in the production process. It is also obtained that the virtual preview and evaluation of the production process can provide some guidance for actual production.

Keywords: shearer; virtual reality; path planning; automatic height-adjusting; Unity3D technology

---

1. Introduction

With the advent of Industry 4.0 and Internet Plus, new concepts such as smart mine and digital mine have been continuously proposed. Cloud computing, mixed reality, big data, Internet of Things, and information physics systems have gradually merged with modern industries [1]. Consequently, coal-mining equipment has been continuously transformed into automated, intelligent, and unmanned equipment [2]. Virtual reality is a technology, developed in the 20th century, which involves a computer, electronic information, and simulation technology. Its basic implementation method is to create a simulated environment by computer technologies so as to give people a sense of environmental immersion [3,4]. Due to the unique working conditions of the coal production process, its complexity, and uncertainty, most of the traditional simulation tools and simulation methods are used to simulate a certain process or a single device in the production process, and it is difficult to achieve an overall and real effect, the integration of virtual reality technology into the traditional coal-mining field also presents new demands and broad prospects. Combined with the information from the actual production equipment, assembly conditions, geological environment, and operation data, a realistic three-dimensional (3D) scene was constructed to show and simulate the production process of the fully mechanized mining face [5,6].

The shearer is one of the key equipment for the fully mechanized mining face. It is an important technological equipment to achieve high-efficiency intensive coal mining, reduce major malignant accidents in coal mines, and improve the working conditions of the working face. It is, therefore, of great significance to the virtual simulation of the shearer's cutting operations. Several studies on the problems of virtual assembly [7], operation demonstration, and attitude analysis [8–10] of the shearer in the virtual environment have been carried out. However, most of these studies were carried out under the condition of a horizontal floor. In the real mining process, there are often complex geological conditions such as coal seam dip and undulations in the roof and floor, which leads to the disparity between the equipment operation state presented in the virtual simulation and that in the actual situation.

The planning of the cutting path of the shearer is an important means to achieve intelligent coal mining. At present, the researches on the cutting path planning of the shearer are mainly based on the accurate positioning of the shearer to obtain memory cutting [11,12], and automatic adjustment of drum based on intelligent decision-making or coal rock recognition [13]. Li Wei [14] proposed a hidden Markov model (HMM) memory cutting method for the shearer, which prevented large residual errors and frequent adjustments of the shearer cutting drum. Gospodarczyk Piotr [15] focused on the coal mining process for different variants of the shearer construction and kinematic parameters. Xu Jing [16] proposed an online cutting-pattern recognition method with high accuracy and speed. The cutting sound was collected as the recognition criterion. Although there are many reported studies on the cutting path planning of shearer, those on the demonstration and simulation using virtual reality technology are limited.

The 3D coal seam model is an important aspect to study the virtual operation of the shearer under complex geological conditions. With the help of a 3D visualization platform, the spatial geometry of the geological body in the mining area can be described more accurately and comprehensively by the real 3D coal seam model. Alan M. Lemon [17] developed a 3D geological model based on the drilling data using the horizons method. The user-defined profile method was further used to dynamically modify the geological model. Wu Lixin [18] proposed a generalized triangular prism (GTP) data model for coal seam modeling to solve the problem of geometric cracks. Most of the existing coal mine virtual environment systems are developed by independent GIS Engine [19,20], which makes it difficult for the virtual coal seam to carry out simulation analysis jointly with the shearer and other equipment. In the Unity3D, the 3D coal seam and the fully mechanized mining equipment model are built for joint analysis; all elements of the production process are digitized, and a more comprehensive and real virtual scene is restored for simulation, so as to enhance the actual production.

In an attempt to solve the above-stated challenges, this study was carried out to improve the authenticity and accuracy of simulation. Based on the analysis of the shearer walking and heightening motion model, the equipment modeling and 3D coal seam modeling, combined operation of the shearer and the scraper conveyor were further obtained in the virtual reality engine Unity3D. A cutting path control strategy of the shearer based on adjusting the direction of the drum movement is proposed. Finally, the simulation and the fuzzy comprehensive evaluation [21,22] of the simulation results were carried out.

2. An Idea of Cutting Path Planning Technology of Shearer Based on Virtual Reality

The main technical equipment in the fully mechanized mining face includes shearer, scraper conveyor and hydraulic support. The hydraulic support is mainly used for face advance and supporting the roof of the working face. The shearer and scraper conveyor jointly complete the circular coal cutting. Among them, the scraper conveyor is arranged on the coal seam floor. When the coal seam floor is undulating, the scraper conveyor section by section adapts to the terrain to produce slight bending. The shearer is meshed in the chute of the scraper conveyor through the traveling wheel, which is connected with the scraper conveyor to move forward smoothly, and the shearer sends the drum to the designated position by adjusting the rocker arm to complete the cutting work.

In the selection and design stage of comprehensive mining projects, geological conditions, coal storage, equipment selection, production capacity, and other factors are often considered. Therefore, the selection and design cycle of comprehensive mining face is long and difficult, and coal mining enterprises need to select quickly, and previewed the pre-selected schemes as much as possible, so as to reduce the problems that may occur in the future.

Using the virtual reality engine Unity3D, we studied the cutting path planning method of the shearer drum and obtained the operation simulation of the virtual shearer on the 3D terrain. The overall research framework is shown in Figure 1, which was realized through the following steps.

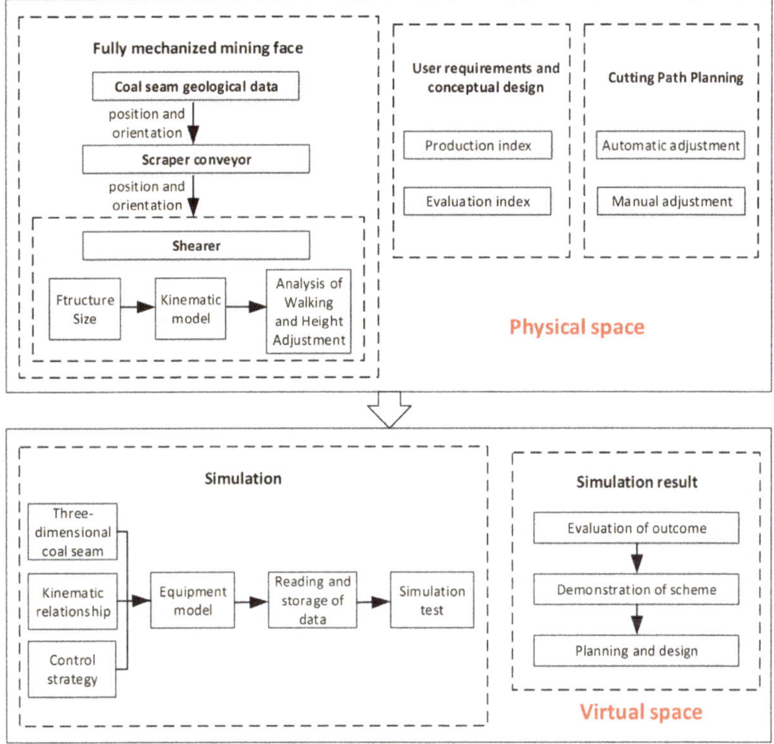

Figure 1. Overall framework.

- Based on the geometry model, movement model, and rule model of real equipment, the virtual model of the shearer operation was developed, which is completely consistent with reality.
- The Unity3D engine was used to build the model for the coal seam roof and floor (3D coal seam), and to obtain the orientation of the shearer and scraper conveyor; based on the geometry model, movement model, and rule model of real equipment, the virtual model of the shearer operation was developed, which is completely consistent with reality.
- The virtual Shearer was obtained to move and adjust the drum automatically on the 3D terrain;
- The prototype system was developed, different schemes simulated, and the simulation results processed and evaluated.

3. Kinematics Analysis and Cutting Path Planning Method of Real Shearer

3.1. Kinematics Model of Shearer Height-Adjusting

The simplified model for the height-adjusting system of the drum shearer is shown in Figure 2. A MG/TY 250/600 double-drum shearer (TZ coal mine whole set equipment Company Ltd, Taiyuan, Shanxi, China) was used in this research. The hydro-cylinder and the cylinder rods are articulated to the shearer fuselage and the rocker arm, respectively. When the cylinder rod moves, the short arm of the rocker arm rotates hence, drives the long arm to obtain the change in the drum position. In the process of lifting and descending the shearer drum, the drum has the same trajectory and law of movement, that is, the movement parameters of the drum in lifting and descending are only distinguished by positive and negative signs. Therefore, the kinematic analysis and height-adjusting strategy of shearer in this paper only focus on the process of drum lifting.

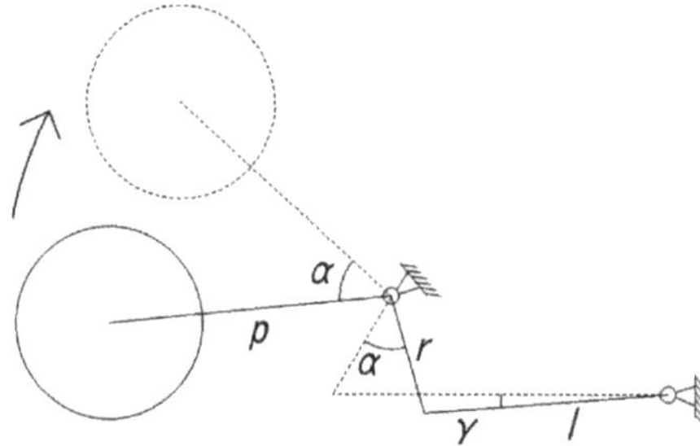

Figure 2. Height-adjusting system of the drum shearer.

3.2. Kinematics Analysis of Shearer Height-Adjusting and Walking

While the shearer is working, it performs two main functions; traction speed regulation and drum-height regulation. The analysis of the movement of the shearer is mainly the analysis of the above two functions. The traction motion and the drum-height adjustment motion of coal mining on the horizontal and uneven ground are analyzed as follows:

- Stationary shearer and the drum lifts

Setting the expansion speed of the height-adjusting cylinder rod to v_c and the running time to t, the rotation angle of the rocker arm can be calculated as follows:

$$\alpha = \arccos\left[\frac{r^2 + a^2 - (l + v_c \cdot t)^2}{2ra}\right] - \arccos\left[\frac{r^2 + a^2 - l^2}{2ra}\right]. \quad (1)$$

The rotation angle of the height-adjusting cylinder is given by:

$$\gamma = \arccos\left[\frac{l^2 + a^2 - r^2}{2ra}\right] - \arccos\left[\frac{(l + v_c \cdot t)^2 + a^2 - r^2}{2(l + v_c \cdot t)a}\right]. \quad (2)$$

The tangential velocity of the drum at any time is:

$$v_\tau = v_c \cdot \sin\left[arc\sin\left(\frac{r^2 + (l + v_c \cdot t)^2 - a^2}{2 \cdot r(l + v_c \cdot t)}\right)\right], \quad (3)$$

where r and p are the lengths of the short and the long arms (values are 543 and 1962 mm), respectively; a is the distance between the joint of the cylinder body and the fuselage and the joint of rocker arm and fuselage, its value is 1303 mm; l is the total initial length of the cylinder rod and the cylinder body, its value is 1701 mm. It can be observed that the expansion speed of the cylinder and the rotation angle of the rocker arm have a non-linear relationship.

- Shearer walks on the flat floor and the drum lifts

Under this condition, the rotation angle of the shearer rocker arm and that of the cylinder remain unchanged. The horizontal velocity of the shearer drum is expressed thus:

$$v_{\text{h1}} = -v_{\text{c}} \times \sin\left[\cos^{-1}\left(\frac{r^2 + (l + v_{\text{c}} \cdot t)^2 - a^2}{2r(l + v_{\text{c}} \cdot t)}\right)\right] \times \frac{p}{r} \times \sin(0.0872 + \alpha) + v_{\text{s}}. \quad (4)$$

The vertical velocity of the shearer drum is:

$$v_{\text{v1}} = v_{\text{c}} \times \sin\left[\cos^{-1}\left(\frac{r^2 + (l + v_{\text{c}} \cdot t)^2 - a^2}{2r(l + v_{\text{c}} \cdot t)}\right)\right] \times \frac{p}{r} \times \cos(0.0872 + \alpha). \quad (5)$$

The angle between the direction of the drum movement and the horizontal direction is:

$$\theta_1 = \tan^{-1}\left(\frac{v_{\text{v1}}}{v_{\text{h1}}}\right), \quad (6)$$

where the angle (in rad) between rocker arm and fuselage of shearer at initial state is 0.0872, v_{s} is the traction speed (in mm/s) of the shearer.

- Shearer walks on the unflat floor and the drum lifts

The horizontal velocity of the shearer drum is given by:

$$v_{\text{h}} = v_{\text{h1}} \times \cos\beta - v_{\text{v1}} \times \sin\beta. \quad (7)$$

The vertical velocity of the shearer drum is:

$$v_{\text{v}} = v_{\text{h1}} \times \sin\beta + v_{\text{v1}} \times \cos\beta. \quad (8)$$

The angle between the direction of the drum movement and the horizontal direction is:

$$\theta = \tan^{-1}\left(\frac{v_{\text{v}}}{v_{\text{h}}}\right), \quad (9)$$

where β is the inclination angle (in rad) of shearer fuselage.

- Joint analysis of shearer traction speed, cylinder expansion speed, and direction of drum movement

Set the simplified expansion speed of the height-adjusting cylinder to $v_{\text{c}} = 5\ \text{mm/s}$ and the range of traction speed of the shearer to 0–12.9 m/min (0–215 mm/s). In the process of lifting the drum from the horizontal floor, the image of the angle between the direction of the drum movement and the horizontal is as shown in Figure 3 (the upward direction is the positive direction).

As can be seen from the figure, although there is an obvious non-linear relationship between the expansion speed of the height-adjusting cylinder and the direction of the drum movement, the direction of the drum movement does not fluctuate much with increase in the running time when the traction speed of the shearer and the expansion speed of the cylinder are fixed.

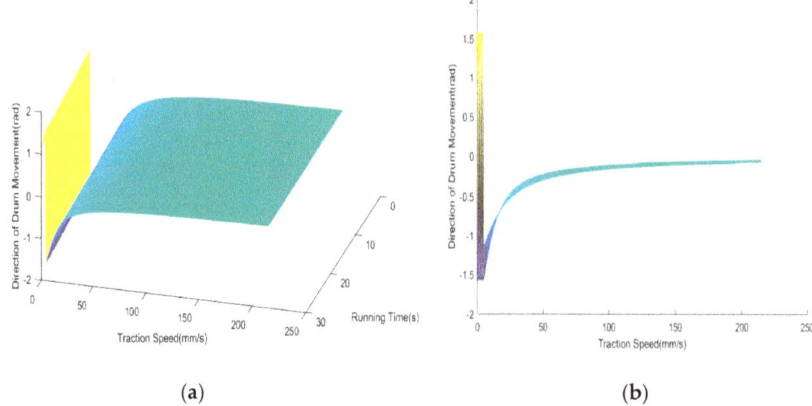

(a) (b)

Figure 3. Relationship between the direction of drum movement, running time and traction speed of the shearer: (**a**) oblique view; (**b**) Y–Z view.

3.3. Automatic Height-Adjusting Strategy

In the traditional memory cutting, the driver controls the shearer to cut a knife along the working face of the coal seam first, and the information such as the travel position of the sampling point, the height position of the drum, the working attitude of the shearer and the traction speed is stored into the computer. The later cutting process of the shearer drum is automatically controlled by the computer according to the stored parameters. When the geological structure of coal seam changes greatly, if the shearer continues to work in the original cutting direction using the parameters recorded by the memory cutting, the shearer drum may interfere with the roof and floor of the coal seam, thus causing damage to the comprehensive mining equipment or requiring manual adjustment of the Shearer to reduce the mining efficiency. The most fundamental way to solve this problem is to achieve self-adaptive cutting in the process of each cutting operation. When the geological structure of coal seam changes, the shearer drum can adjust the cutting posture timely and accurately, instead of keeping the same as the traditional memory cutting.

During the cutting process of the shearer, the inclination angle of the fuselage changes with the dip angle of the coal seam while the cutting height of the drum changes accordingly. Therefore, the lack of suitable positioning reference standards for the shearer while running on real terrain is one of the difficulties in achieving accurate and smooth control of the cutting path. In this study, a method of automatic height-adjusting of shearer based on the fine modeling of the coal seam is proposed. The mine itself is a real 3D static geological environment in which all mine activities are carried out. The position of the drum, any position on the coal seam, and the roof and floor are represented by absolute 3D coordinates. The direction of the drum movement, which integrates the information from the shearer traction speed, expansion speed of the height-adjusting cylinder, and inclination angle of the fuselage, are calculated. According to the analysis in 3.2, in the process of drum height-adjusting, the direction of the drum movement mainly changes with the change of haulage speed of shearer. When the roof and floor of the coal seam fluctuate, the posture of the shearer and the position of the coal to be cut will change accordingly. By adjusting the haulage speed of the shearer, the direction of the drum movement can be adjusted to move to the predetermined target position, so as to achieve self-adaptive cutting.

Based on the geological survey data and 3D coal seam modeling, the roof and floor curves of the current working face are obtained (as shown by the curve in Figure 4). The roof and floor curves are composed of a series of key cutting points with known coordinates. In the actual coal mining process, the key cutting point generally refers to the point where the roof and floor curves of the coal seam have significant fluctuations. In this study, the key cutting points are obtained by building a fine three-dimensional coal model based on geological exploration data of a coal mine [23]. As shown in Figure 4, P_i and P_{i+1} are selected as the two key cutting points. When the right drum of the shearer runs to the horizontal position of P_i, the direction of the drum movement can be changed by increasing the traction speed of the shearer and by lifting the drum, so that it moves to the next key cutting point P_{i+1} (In the figure, β is the inclination angle of the shearer fuselage, and K is the angle between the direction of the drum movement and the horizontal at the key cutting point (K is the specific value of θ in formula 9 in 3.2 at the key cutting point). Q is the angle between the line connecting the next key cutting point and the instant apex of the drum and the horizontal).

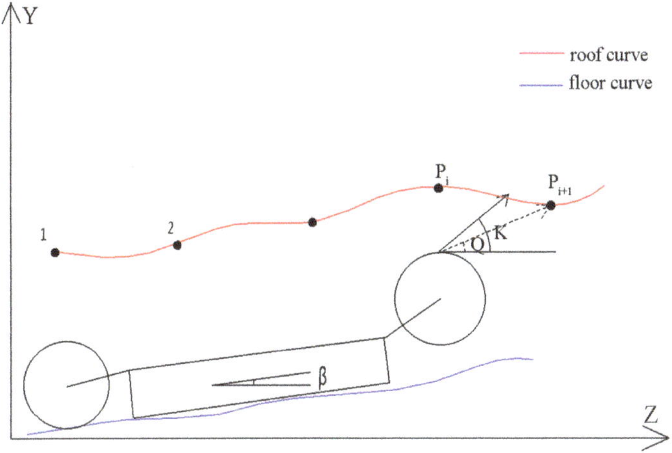

Figure 4. Automatic height-adjusting diagram.

The specific process of automatic height-adjusting is as follows:

- Reading data and calculating parameters

If the right drum of the shearer runs to the horizontal position of P_i, the coordinates of the center of the shearer drum at this time (X_{a1}, Y_{a1}, Z_{a1}) are recorded. The coordinates of the next key cutting point $P_{i(j+1)}$ (X_{b1}, Y_{b1}, Z_{b1}) are then read, after K (the angle of the direction of drum movement) is recorded. This is followed by the calculation of Q (the angle of the target direction):

$$Q = \tan^{-1}\left(\frac{Y_{b1} - (Y_{a1} + 9)}{Z_{b1} - Z_{a1}}\right), \tag{10}$$

where 9 is the radius of the drum, (in decimeter). The height compensation value of the inclination angle of the shearer fuselage given $M = L \times \sin\beta$ (where L is the length of a scraper conveyor, in dm; β is the inclination angle of the shearer fuselage, in rad) is calculated. When the shearer fuselage is inclined at an angle, the absolute height of the drum changes if not adjusted.

- Judgment of action

After reaching a key cutting point, the drum can experience three actions; up, down, and no adjustment. The concrete action is determined by the height difference between the apex of the drum and the next key cutting point. The thresholds of the height difference are determined by the inclination angle of the fuselage and the feasible region of the shearer traction speed. These thresholds include:

- the maximum height difference for lifting given by $\triangle H_{max} = L \times (\sin\beta + \sin\theta_{max})$,
- the minimum height difference for lifting $\triangle H_{up} = L \times \sin\beta + \frac{1}{2}\varepsilon$,

- the maximum height difference for lowering $\Delta H_{down} = L \times \sin\beta - \frac{1}{2}\varepsilon$,
- the range of height difference for no adjustment $\Delta H_{up} > \Delta H > \Delta H_{down}$,
- the minimum height difference for descending $\Delta H_{min} = L \times (\sin\beta - \sin\theta_{min})$.

In the above equations, ΔH is the height difference between adjacent key cutting points, β is the inclination angle of the shearer fuselage, θ_{max} and θ_{min} are the maximum and minimum angle, respectively, between the direction of the drum movement and the horizontal direction during the lifting process, ε indicates the range of the relative height difference between the adjacent key cutting points within which the drum may not be adjusted. In this study, we set ε to 60 mm.

When the adjustment direction is determined, the K is compared with the Q. The haulage speed of the shearer is then adjusted to redirect the drum toward the target point. Specific execution actions at the key points include acceleration-up, deceleration-up, acceleration-down, deceleration-down, constant speed-up, constant speed-down, and no adjustment. When matching the moving direction and the target direction of the drum, considering that there is still a slight change in the process of drum lifting, the analysis law of the shearer drum movement obtained in 3.2 is taken as the judgment knowledge and included in the strategy. The specific action judgment is shown in Table 1.

Table 1. Strategy of action judgment.

Judgment of Action		Before Adjustment	Specific Action	After Adjustment
$\Delta H > \Delta H_{max}$ or $\Delta H < \Delta H_{min}$	Too high or too low		Traction speed is reduced to a minimum, drum lifts	
$\Delta H < \Delta H_{max}$ and $\Delta H > \Delta H_{up}$	Lifting	K-Q > 0.02 Q-K > 0.02 |Q-K| ≤ 0.02	Traction speed is increased, drum lifts Traction speed is decreased, drum lifts Traction speed is decreased, drum lifts	|Q-K| ≤ 0.02
$\Delta H < \Delta H_{up}$ and $\Delta H > \Delta H_{down}$	No adjustment		Traction speed is constant, drum is not adjusted	
$\Delta H < \Delta H_{down}$ and $\Delta H > \Delta H_{min}$	Descending	K-Q > 0.02 Q-K > 0.02 |Q-K| ≤ 0.02	Traction speed is increased, drum descends Traction speed is decreased, drum descends Traction speed is constant, drum descends	|Q-K| ≤ 0.02

3.4. Selection of Operation Parameters of the Shearer

In the coal production process, the selection of the operating parameters of the shearer is key to achieving high-efficiency and high-quality cutting under various working conditions, and as well, meeting the production capacity of the working face. When the model of the fully mechanized mining equipment is selected, the performance of the shearer is mainly affected by the traction speed and the drum cutting speed; this study focuses on the traction speed of shearer. Under the condition of guaranteed normal coal falling and conveying, the higher the traction speed of the shearer, the better the coal quality, the lower the specific energy of cutting, and the higher the coal production efficiency. However, the traction speed is restricted by the traction and the cutting powers of the shearer motor. On the other hand, the higher the traction speed, the lower the recovery rate of the coal seam with larger fluctuation. In actual production, the mining replacement arrangement, mining technology, and mechanization degree vary in each mining operation. In most cases, higher production capacity of the working face and recovery rate of the coal seam is required.

With reference to the mine production capacity verification standard, the coal mining face capacity is calculated as follows:

$$A_C = 10^{-4} l \cdot h \cdot r \cdot b \cdot n \cdot N \cdot c \cdot a, \tag{11}$$

where A_C is the annual production capacity of the coal mining face, 10,000 tons/year; l is the average length of coal mining face, m; h is the average mining height of coal seam in coal mining face, m; r is the raw coal density, m^3; b is the average daily advancement of coal mining face, m/d; n is the annual working days, d; N is the normal cycle operation coefficient, %, according to the technical performance of coal mining equipment, the production organization, staff quality, and other factors, generally taking 0.8; c is the recovery rate of coal mining face, %; a is the average number of coal mining face. It can be seen that for annual production capacity, the speed of coal-mining equipment needs to be adjusted when other conditions are fixed. With reference to the actual production experience of a coal mine, the minimum average traction speed of the shearer may be up to 3 m/min for higher production capacity.

4. Virtual Cutting Path Planning Method of Shearer and Design of Prototype System

Comparing the characteristics of the traditional CAD software and that of the Unity3D engine, the following main processes are needed to simulate the working process of shearer under the 3D coal seam (Figure 5):

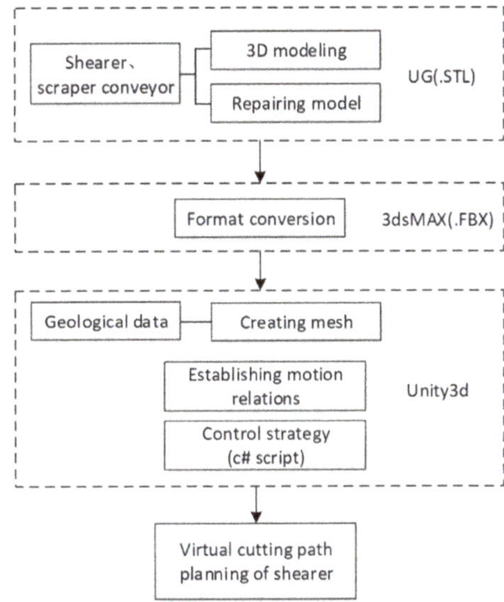

Figure 5. Algorithm of the main process of the virtual cutting of shearer.

4.1. Key Technology for Modeling Coal Seam Roof and Floor

Developing a 3D model that accurately expresses the coal seams is the key to achieving digital, visual, and intelligent production management in the fully mechanized mining face in coal mines. In this study, the discrete detection points are transformed into a series of continuous data using the geological exploration data of a coal mine, through the Kriging [20] interpolation. After processing, the 3D coordinates of $16 \times 40 \times 2$ points are obtained, so as to build the coal seam roof and floor model in Unity3D.

The mesh grid component in Unity provides a way to form a surface closely through a large number of small triangles. The MeshFilter records the grid data whereas the MeshRenderer components set the grid material. Since only one side of the triangle generated by the mesh grid component is visible (visible on one side of the clockwise connection), the opposite vertex order is required to make the coal seam top and bottom plates visible under normal viewing angles. During the implementation of the mesh grid of the coal seam roof, the vertices of the triangle are selected according to the specific rules with reference to Figure 6. As shown in Figure 6, assuming that there are altogether 640 3D geological coordinates of the coal seam roof (No. 0–639), the three vertices of the first triangle are 0, 40, 1 in turn, the three vertices of the second triangle are 1, 41, 2 in turn. Similarly, the three vertices of the last triangle were 639, 598, 638 in turn.

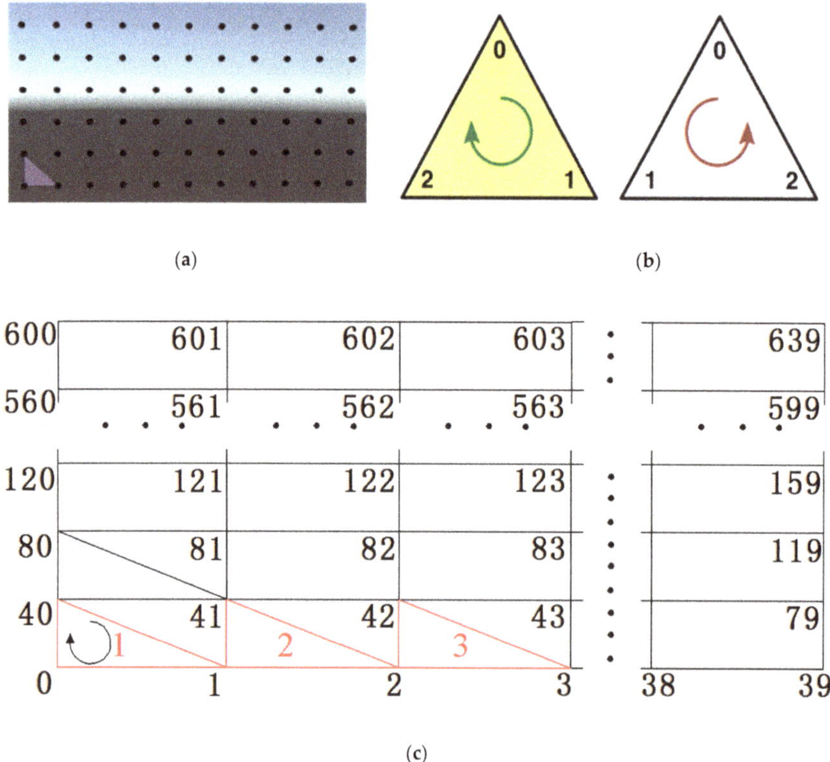

Figure 6. Rules for sorting triangular vertices in roof and floor mesh: (**a**,**c**) connection order of triangle vertices; (**b**) clockwise and counterclockwise connections.

The processed data of the coal seam roof and floor are stored as an XML file, and the mesh is created by a c# script. According to the above method, the final effect in Unity3D is as shown in Figure 7:

Figure 7. Mesh of virtual coal seam roof and floor.

4.2. Key Technology for Combined Operation of Shearer and Scraper Conveyor

This paper presents a method of combined operation of the shearer and scraper conveyor on the real terrain of Unity3D.

The collider component in Unity3D provides a physical collision between the colliding bodies under gravity. Based on the above-mentioned modeling of coal seam, the mesh collider component is added to the GameObject of the coal seam floor and scraper conveyor to measure the position and the inclination of each scraper conveyor on the floor. Using the measured parameters of each scraper conveyor, they are arranged according to the actual arrangement rules. In the course of the shearer walking on the scraper conveyor, the attitude of its fuselage is determined by the left and right supporting slippers. Therefore, it is necessary to obtain the shape function of the scraper conveyor so that the slippers can move strictly along the obtained shape while the virtual walking of the shearer is controlled by coordinates. The shape function of the scraper conveyor can be calculated as shown in Figure 8.

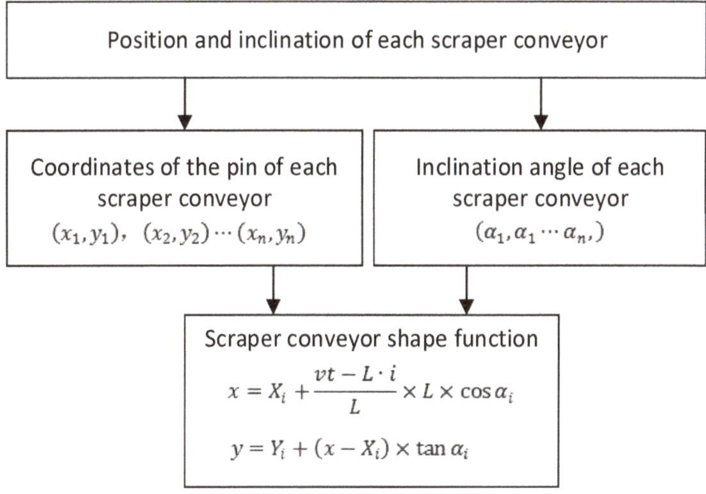

Figure 8. Shape function of the scraper conveyor.

Where (X_i, Y_i) is the coordinate of the pin of the ith scraper conveyor, mm; α_i is the inclination angle of the ith scraper conveyor, rad; i is the serial number of the scraper conveyor where the slipper of the shearer is located; L is the length of a scraper conveyor, mm; v is the traction speed of the shearer, mm/s; t is the walking time of the shearer, s; x and y are the corresponding coordinate values on the shape function.

The ultimate effect in Unity3D is as shown in Figure 9:

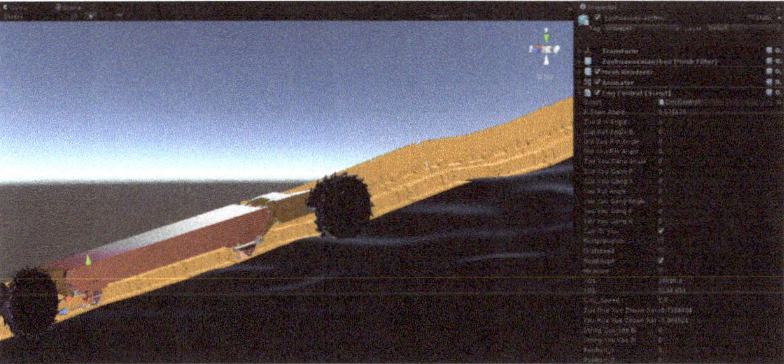

Figure 9. Effect diagram of combined operation of shearer and scraper conveyor.

4.3. Key Technology of Virtual Cutting of Shearer

In the Hierarchy interface of Unity3D, the hierarchical relationship (parent-child relationship) of each component is established. By defining the Transform property of each component through C# script, the parameters such as position, rotation, and size of each component can be obtained and varied through the script. The virtual shearer can be controlled by a programmed C# script based on the collaborative mathematical model, obtaining the start and stop, variable speed walking, positioning of the shearer, expansion, and rotation of the cylinder, adjustment of the drum height, and rotation of the drum. The components are moved using the 3D coordinate, whereas, they are rotated by the quaternion. The parent-child relationship of shearer components is shown in Figure 10.

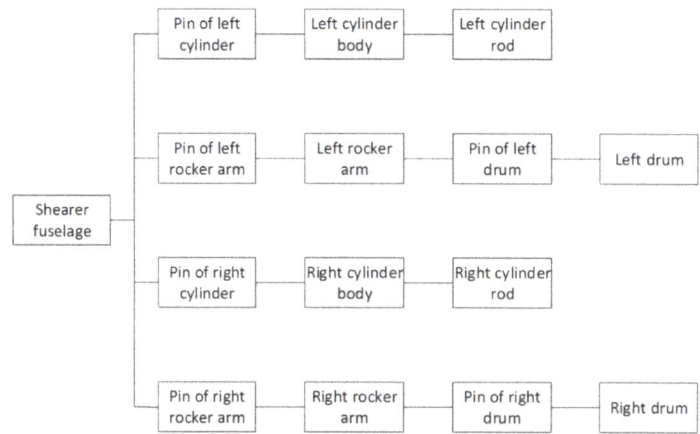

Figure 10. Parent–child relationship of shearer components.

In the process of creating a coal seam roof and floor, the cutting point data was stored as an XML file. As the shearer runs, the current number of the cutting feed is i, and the cutting point number within the range of one knife is j. The key cut point of the path in the entire coal seam is therefore, the set $P_{ij}(i = 1,2,3\cdots, j = 1,2,3\cdots)$. When the system starts to run, the current cutting point of the working face can be read and the LineRender component in Unity3D is used to draw the curve of the coal roof and the cutting curve from time to time. The cut point data on the cut curve is also stored in the XML file for subsequent analysis of the simulation results. The schematic diagram of the cut point and the effect obtained in Unity3D are shown in Figure 11a,b, respectively.

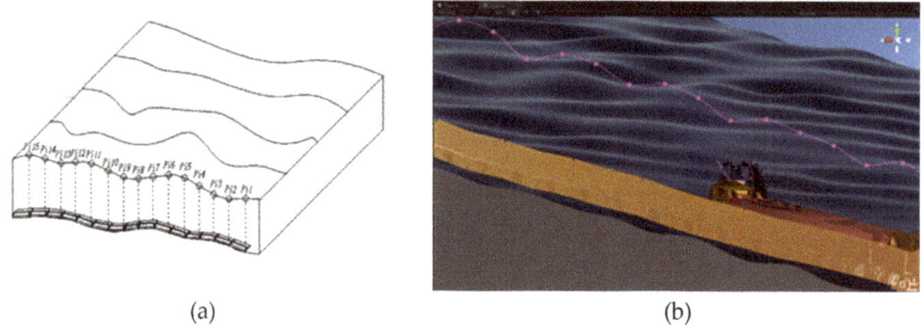

Figure 11. Schematic diagram: (**a**) the map of cutting points (**b**) the effect of cutting points in Unity3D.

5. Simulation Experiment

5.1. Setting of Simulation Conditions

Two different height-adjusting modes were designed in the experiment, namely; the automatic height-adjusting mode and the manual operation mode. In the planning stage of the fully mechanized mining face, the production capacity of the working face is often verified. In this study, the influence of shearer traction speed on production capacity was considered most. The lowest traction speed of the cutting process was set to ensure that the production capacity meets the requirements of different working faces. The feasible range of the haulage speed of the shearer was 2.5–12.9 m/min, 3.5–12.9 m/min, 4.5–12.9 m/min, and 0–12.9 m/min (operated by experienced workers).

5.2. Analysis of Simulation Results

In scheme 1, the actual cutting curve coincides with the preset roof curve, and the maximum error was obtained to be 12 cm. However, in schemes 2 and 3, an increase in the minimum traction speed of shearer decreased the coincidence, and the maximum error was obtained to be 48 cm. This could be attributed to the fact that the direction of the drum movement cannot meet the changing trend of the preset roof curve when the shearer is at the minimum traction speed. During the height-adjusting of the shearer while walking, if the shearer is running at a relatively large traction speed, K (the angle of the direction of drum movement) is relatively small, and the drum height-adjusting cannot be performed in a large scale. At the position where the roof and floor of the coal seam fluctuates greatly, the cutting curve will produce a large error. The effect in Unity3D is shown in Figure 12a and the cutting errors of the four schemes are shown in Figure 12b.

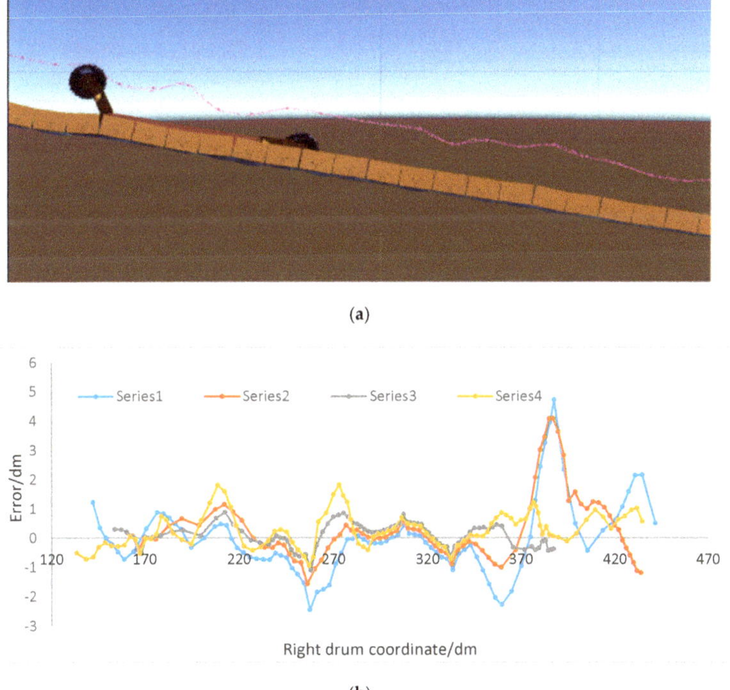

Figure 12. Simulation results: (**a**) three-dimensional graph of cutting curve (**b**) error of the actual cutting points and the preset points.

In production, scientific research and our daily life, we need to evaluate the quality of goods, the reasonable degree of some engineering design parameters, etc. to determine the best way to deal with them. In the cutting process of the fully mechanized mining face, there are many indexes to measure the quality of the cutting results, including the coal quality, recovery rate, production efficiency, economic benefit, and safety degree. Evaluating coal production is not an easy task, as it involves dimensionless index, human decision making, which are imprecise, fuzzy, and uncertain. Hence, using a scientific method to evaluate coal production comprehensively and effectively plays a crucial role in the analysis of simulation results.

The evaluation that needs to be made according to many factors is called comprehensive evaluation. When some specific evaluation factors are fuzzy, the comprehensive evaluation is called a fuzzy comprehensive evaluation. The fuzzy theory bridges the gap between the precise classical mathematics and the fuzzy real world. Fuzzy comprehensive evaluation method is a method which transforms qualitative evaluations into quantitative evaluations through the membership degree theory of fuzzy mathematics. It, therefore, has a good theoretical basis for the overall evaluation of things or objects having many constraints. Hence, the fuzzy comprehensive evaluation method is suitable for evaluating the cutting results. The general steps of fuzzy comprehensive evaluation include the construction of evaluation indexes, the determination of the weight of each index, the construction of the evaluation matrix, the synthesis of evaluation matrix and weight.

The specific steps for evaluating the cutting results using the fuzzy comprehensive evaluation are as follows:

- Determine the evaluation object

$$X = \{x_1, x_2, x_3, x_4\} = \{\text{scheme1}, \text{scheme2}, \text{scheme3}, \text{scheme4}\}. \tag{12}$$

- Determine the index set

$$\begin{aligned} U &= \{u_1, u_2, u_3\} \\ &= \{\text{recovery rate, production efficiency, others(safety degree, labor cost, etc)}\}. \end{aligned} \tag{13}$$

When the index set is determined, each of the indicators of each scheme is analyzed quantitatively. The recovery rate is expressed by the sum of the squares due to the error of the cutting points thus: $SSE = \sum_{i=1}^{n} w_i(y_i - y)^2$, where y_i and y are the heights of the actual cutting point and the preset roof point, respectively. The efficiency of production is represented by the time used for cutting one knife, however, other factors are scored according to the production experience of the general enterprises (0–9 point). The scores of each index are shown in Table 2.

Table 2. Evaluation index.

Index	Recovery Rate (SSE)	Production Efficiency (min)	Others (Safety Degree, Labor Cost, etc.)
Scheme1	13.89	8	8
Scheme2	92.5	7	8
Scheme3	125.73	5.5	8
Scheme4	40.96	8.5	2

- Determine the weight vector of the evaluation index

Set the weight vector to $A = (W_1, W_2 \ldots W_n)$, where W_i indicates the i-th index weight coefficient. The matrix gives different weights to each index according to the different functions of each index in the overall evaluation. Under the condition that the safety level and other production costs are guaranteed, coal production pays more attention to the production efficiency directly linked to economic benefits and the recovery rate that can maximize the use of resources. According to the general enterprise production experience [24], $A = (0.4,\ 0.4,\ 0.2)$.

- Determine the fuzzy evaluation matrix

After determining the weights of the indicators, the evaluation index needs to be normalized and quantified one by one. A row in the fuzzy evaluation matrix R reflects the degree of membership of different schemes on this indicator. For the evaluation index with a specific value, the value of the optimal evaluation object can be set to 1 whereas, it is set according to the scoring level (0 ≤ E ≤ 1) for other objects. For the index whose value is bigger and whose evaluation is better, the calculation method of evaluation value is: Evaluation value = specific value/optimal value. For the index whose value is smaller and whose evaluation is better, the calculation method of evaluation value is: Evaluation value = optimal value/specific value. When calculating the evaluation value of the recovery rate, a higher SSE indicates a lower recovery rate, and the evaluation value should be smaller. Among the four schemes, the one with the highest recovery rate is scheme 1, and the corresponding SSE value is 13.89. Then, the evaluation value of scheme 1 is set to 1, the evaluation value of scheme 2 is set to $0.15 = \frac{13.89}{92.5}$, the evaluation value of scheme 3 is set to $0.11 = \frac{13.89}{125.73}$, the evaluation value of scheme 4 is set to $0.339 = \frac{13.89}{40.96}$.

Finally, the fuzzy evaluation matrix is obtained thus:

$$R = \begin{bmatrix} 1 & 0.15 & 0.11 & 0.339 \\ 0.7 & 0.808 & 1 & 0.646 \\ 1 & 1 & 1 & 0.25 \end{bmatrix}. \tag{14}$$

- Comprehensive evaluation

Comprehensive fuzzy evaluation vector B is calculated by the combination of the fuzzy multiplication ⊙ and fuzzy evaluation matrix R; $B = A \odot R$. The common fuzzy multiplications mainly include single-factor prominent type and weighted average type, in order to be able to objectively evaluate the impact of each index, the fuzzy multiplication chosen in this paper is the weighted average method [25]; $M(\cdot,\ \mathrm{v}),\ \ b_j = \sum \left(a_i{\cdot}r_{ij}\right)(j = 1, 2 \cdots m)$. Hence, the final evaluation was calculated as follows:

$$\begin{aligned} B &= A \odot R \\ &= \begin{bmatrix} 0.88 & 0.5832 & 0.644 & 0.444 \end{bmatrix} \end{aligned} \tag{15}$$

It can be observed that the four schemes above are sorted as schemes 1, 3, 4, and 2. The higher value of the test has a better comprehensive evaluation. When the minimum traction speed of the shearer is small, more accurate trajectory tracking can be achieved. With an increase in the minimum traction speed, the production efficiency of the shearer improves, whereas the accuracy of trajectory tracking decreases. On this basis, according to the requirements of different enterprises, the production process of different working parameters can be previewed and evaluated in a virtual environment.

6. Conclusions

Based on the motion analysis of the shearer and drum-height adjustment, and with the drum movement law as the judgment knowledge, an automatic height-adjusting strategy of shearer was designed in the Unity3D engine, which integrates the information of the roof and floor fluctuation, coal seam inclination, and shearer fuselage inclination.

The modeling of the roof and floor was accomplished in the Unity3D engine using the mesh and LineRender components. The key cutting points could be read and the actual cutting curve drawn.

A real shearer height-adjusting model was developed, which can be controlled by the c# script. By measuring the position and posture of the scraper conveyor on real terrain, and constructing the shape function of the scraper conveyor, the combined operation of the shearer and scraper conveyor was established. This provides theoretical support for shearer simulation on real terrain.

A conceptual idea is put forward: on the basis of geological exploration of coal seam, a virtual fine 3D coal seam is established. According to the production requirements of different enterprises, the pre-selected schemes and the decision-making are previewed in Unity3D, hence, digitized and visualized planning is carried out ahead of time. The entire production process can, therefore, be synchronously mapped in the production process, so as to obtain safe and efficient transparent mining.

Author Contributions: All authors (J.L.;Y.L.;J.X.;X.W.;X.G.) designed the project and participated in the experiments and the interpretation of the results. Conceptualization, J.L. and Y.L.; methodology, J.L.; software, J.X.; writing—original draft preparation, J.L.; writing—review and editing, J.L. and Y.L. All authors have read and agreed to the published version of the manuscript.

Funding: This work was supported by the Shanxi Province Nature Science Fund (201901D111056), the Shanxi province graduate education reform research topic (2019JG047), China Postdoctoral Science Foundation (No.2019M651081), and the Shanxi Youth Science and Technology Research Fund Project (201901D211022).

Conflicts of Interest: The authors declare no conflict of interest.

References

1. Ralston, J.C.; Hargrave, C.O.; Dunn, M.T. Longwall automation: Trend, challenges and opportunities. *Int. J. Min. Sci. Technol.* **2017**, *27*, 15–21. [CrossRef]
2. Chad, O.H.; James, C.A.; Ralston, J.C. Infrastructure-based localization of automated coal mining equipment. *Int. J. Coal Sci. Technol.* **2017**, *4*, 252–264.
3. Ottogalli, K.; Rosquete, D.; Amundarain, A.; Aguinaga, I. Flexible Framework to Model Industry 4.0 Processes for Virtual Simulators. *Appl. Sci.* **2019**, *9*, 4983. [CrossRef]
4. Ralston, J.C.; Reid, D.C.; Dunn, M.T. Longwall automation: Delivering enabling technology to achieve safer and more productive underground mining. *Int. J. Min. Sci. Technol.* **2015**, *25*, 865–876. [CrossRef]
5. Tibbett, A.E.; Greenberg, S.; Brazil, E. The use of virtual reality scientific visualization for investigation and exploration of block cave mining system data. In Proceedings of the Virtual Reality and Spatial Information Applications in the Mining Industry Conference, University of Pretoria, Pretoria, South Africa, 15–17 July 2015.
6. Xie, J.C.; Yang, Z.J.; Wang, X.W. A Virtual Reality Collaborative Planning Simulator and its Method for Three Machines in a Fully Mechanized coal Mining Face. *Arab. J. Sci. Eng.* **2018**, *43*, 4835–4854. [CrossRef]
7. Alam, K.M.; Saddik, A.E. C2PS: A Digital Twin Architecture Reference Model for the Cloud Based Cyber Physical System. *IEEE Acess.* **2017**, *5*, 2050–2062. [CrossRef]
8. Stothard, P.; Squelch, A.; Stone, R.; Van Wyk, E.; Kizil, M. Taxonomy of interactive computer-based visualization systems and content for the mining industry—Part 2. *Min. Technol.* **2015**, *124*, 83–96. [CrossRef]
9. Xie, J.C.; Wang, X.W.; Li, X. Research status and prospect of virtual reality technology in field of coal mine. *Coal Sci. Technol.* **2019**, *47*, 53–59.
10. Tichon, J.; Burgesslimerick, R. A review of Virtual Reality as a medium for safety related training in Mining. *J. Health Saf. Res. Pract.* **2011**, *3*, 33–40.
11. Zhang, L.L.; Tan, C.; Wang, Z.B.; Yang, X.F. Memory Cutting Path Optimization of the Shearer Based on Genetic Algorithm. *Coal Eng.* **2011**, *2*, 111–113.
12. Liu, C.S.; Yang, Q. Simulation of shearer drum cutting with memory program controlling by fuzzy control. *J. China Coal Soc.* **2008**, *7*, 822–825.
13. Si, L.; Wang, Z.B.; Liu, Y.W. Online Identification of Shearer Cutting State Using Infrared Thermal Images of Cutting Unit. *Appl. Sci.* **2018**, *8*, 1772. [CrossRef]
14. Li, W.; Luo, C.M.; Yang, H.; Fan, Q.G. Memory cutting of adjacent coal seams based on a hidden Markov model. *Arab. J. Geosci.* **2014**, *4*, 5051–5060. [CrossRef]
15. Gospodarczyk, P. Modeling and simulation of coal loading by cutting drum in flat seams. *Arch. Min. Sci.* **2016**, *61*, 365–379.

16. Xu, J.; Wang, Z.B.; Tan, C. A Cutting Pattern Recognition Method for Shearers Based on Improved Ensemble Empirical Mode Decomposition and a Probabilistic Neural Network. *Sensors* **2015**, *15*, 27721–27737. [CrossRef] [PubMed]
17. Lemon, A.M.; Jones, N.L. Building solid models from boreholes and user-defined cross-section. *Comput. Geo-Inf. Sci.* **2003**, *29*, 547–555. [CrossRef]
18. Che, D.F.; Wu, L.X.; Chen, X.X. Modeling and visualizing methods for real 3D geosciences model based on amended generalized tri-prism(GTP). *J. China Coal Soc.* **2006**, *31*, 576–580.
19. Michel, P.; Zhu, B.T. Knowledge-driven applications for geological modeling. *J. Petrol. Sci. Eng.* **2005**, *47*, 89–104.
20. Tacher, L.; Pomian-srzednicki, I.; Parriaux, A. Geological uncertainties associated with 3D subsurface model. *Comput. Geosci.* **2006**, *2*, 212–221. [CrossRef]
21. Chen, J.F.; Hsieh, H.N.; Do, Q.H. Evaluating teaching performance based on fuzzy AHP and comprehensive evaluation approach. *Appl. Soft Comput.* **2015**, *28*, 100–108. [CrossRef]
22. Ko, Y.C. Application of Fuzzy Theory to the Evaluation Model of Product Assembly Design and Usability Operation Complexity. *Appl. Sci.* **2019**, *9*, 4055. [CrossRef]
23. Wang, T.J. *Research on Generalized Memory Cutting Technology of Thin Seam Shearer Based on Dynamic Fine Modeling;* China University of mining and Technology (Beijing Campus): Beijing, China, 2003.
24. Gao, C.Z.; Li, Y.Z.; Wang, T.L. Safety production evaluation of coal mine based on fuzzy comprehensive evaluation and AHP. *Opencast Min. Technol.* **2013**, *11*, 72–77.
25. Mu, Y.Z.; Lu, Z.X.; Qiao, Y. A Comprehensive Evaluation Index System of Power Grid Security and Benefit Based on Multi-Operator Fuzzy Hierarchy Evaluation Method. *Power System Technol.* **2015**, *1*, 23–28.

Article

Classification of Small- and Medium-Sized Enterprises Based on the Level of Industry 4.0 Implementation

Martin Pech and Jaroslav Vrchota *

Department of Management, Faculty of Economics, University of South Bohemia in Ceske Budejovice, Studentska 13, 370 05 Ceske Budejovice, Czech Republic; mpechac@ef.jcu.cz
* Correspondence: vrchota@ef.jcu.cz

Received: 22 June 2020; Accepted: 24 July 2020; Published: 27 July 2020

Abstract: Due to Industry 4.0 technologies, small- and medium-sized enterprises have a great opportunity to increase their competitiveness. However, the question remains as to whether they are truly able to implement such modern technologies faster and carry out digital transformation. The main aim of the paper is to classify small- and medium-sized enterprises into various groups, according to the level of implementation of Industry 4.0, using the Index of Industry 4.0. Based on the results of the cluster analysis, the small and medium enterprises are categorized into four different groups, according to the level of implementation of Industry 4.0. There are top Industry 4.0 technological enterprises, I4 start enterprises, noobs enterprises, and I4 advances enterprises. So far, the largest group consists of the small- and medium-sized enterprises that are just starting out with the introduction of Industry 4.0 technologies, such as IT infrastructure, digitalization (data, cloud, data analysis, and information systems), and sensors. On the other hand, the top I4 technological enterprises group is the least numerous. The analysis carried out comparing the small- and medium-sized enterprises with the large enterprises shows that the SMEs still have a lower level of Industry 4.0 implementation. This confirms the assumption that the large enterprises have greater opportunities to use new technologies and transform them into smart factories. However, this situation may change in the future if new technologies become more accessible, and SMEs are worth investing in Industry 4.0 in terms of the return on investment.

Keywords: Industry 4.0; SMEs; technologies; cluster analysis; maturity model; implementation

---

1. Introduction

Industry 4.0 (I4) is understood to be a revolutionary method of production. It brings a new perspective on the synergy of production with modern technologies, the maximum possible output, and the minimum resources used (German Standardization Roadmap Industrie 4.0—Version 3, b.r.). The new industrial revolution is changing the rules of competition, as the business models are completely changing as a result of the digitization of factories and the adoption of the Internet of Things concept [1]. The modern Industry 4.0 concept was originally designed as a national initiative for the development of the German economy in 2011.

Initially, Industry 4.0 technology was only available to large enterprises, due to high capital requirements [2]. However, in many countries, the small- and medium-sized enterprises (SMEs) represent more than 90% of all the enterprises [3]. Even such enterprises are beginning to introduce modern technologies into production gradually [4]. For this reason, it is necessary to map their situation and the degree of implementation of Industry 4.0 so that the enterprises are able to compare each other and get a closer overview of their position on the market. In relation, it is also necessary to better evaluate future steps in implementing Industry 4.0, so that they do not face unnecessary implementation difficulties, such as, by Ingaldi [5] and Turkes [6], the narrow product portfolio of SMEs, which does not guarantee full use of the efficiency of automated and autonomous production systems, the cost of obtaining money, the turbulence of the environment from the micro and macro aspect, lack of knowledge about Industry 4.0, more focus on operation at the costs of developing the company, lack of understanding of the strategic importance of industry 4.0, too few human resources, need for continued education of employees, and lack of standards.

The paper is structured as follows: Section 1 is the introduction; Section 2 offers the theoretical background with a current review of the literature; Section 3 covers the materials and methods, describing the main objectives, methods (cluster analysis), and hypotheses; Section 4 contains the results of the analysis and includes SMEs categorization, comparison with large enterprises, and validation of results; Section 5 offers a discussion focused on the main contrubutions and limitations of the research; and Section 6 is the conclusion, which contains the main results and further research challenges.

2. Theoretical Background

Industry 4.0, or the fourth industrial revolution, means the use of digital technology in the manufacturing process to produce higher-quality goods at reduced costs. This process has different names throughout the world, e.g., Industry 4.0 in Germany, New Industrial France in France, the National Network for Manufacturing Innovation in the USA, Society 5.0 in Japan, and Made in China 2025 in China, etc.

In general, Industry 4.0 is consists of several elements for small- and medium-sized enterprises: autonomous robots, horizontal and vertical integration, the aforementioned Internet of Things (IoT), cloud computing, high-volume data, and additive manufacturing [7]. Taking into account the general paradigm shift from mass production to mass customization, there is also a need for configurable automatic technologies such as the robots [8]. Although the role of humans in the vision of a new industrial revolution is still considered irreplaceable, with the advances in data visualization, new technical interactions seem to make robots suitable for integration into the industrial environment [9].

This article understands Industry 4.0 as a revolutionary industrial concept of the production process in manufacturing, focused on new technologies that interconnect machines and equipment with digital data into automatic, intelligent systems. This definition is based on a number of authors [10–17], mainly Schumacher [18,19], who analyzed more than 70 works focused on Industry 4.0; Rainer [20], who examined Industry 4.0 on more than 1000 questionnaires; and Kaltenbach [21], who researched Industry 4.0 on 30 German enterprises.

Advanced manufacturing systems, together with Information and Communication Technologies (ICT) analysis tools in small- and medium-sized enterprises, transform production into a modern form of ICT known as the Internet of Things [22]. ICT is a prerequisite for efficient supply chain management, and it plays a vital role in the ability to integrate suppliers and customers to improve supply chain performance [23].

Internet of Things is a new era of computing that is completely outside the traditional desktops. In the new industrial revolution, RFID (Radio Frequency Identification) technologies should meet the requirements of identifiable objects that are located in a computer network in one form or another and in which ICT is invisibly built into the environment around us [24]. Madakam, Ramaswamy, and Tripathi [25] define Internet of Things as an open and comprehensive network of intelligent objects capable of organizing and sharing information, data, and resources and responding to sudden changes in the environment.

One of the other indispensable elements of Industry 4.0 is the largescale data, Big Data, which is gradually coming from the Smart cities, through large enterprises to SMEs, as mentioned by Dwevedi [26]. Big data refers to large data files with a more diverse and complex structure and related processes of visualization, analysis, and storage [27]. Babiceanu and Seker [28] noticed that data analysis alone is the most important aspect of Big Data, without which sub-aspects such as storage or collection would not be of great value, and appropriate data analysis can reveal new information and facilitate timely response to emerging opportunities and threats [29].

The basic building block is Cloud Computing, which is more and more often used by SMEs [30]. Assuming an enterprise has the resources to deal with information on the network, Cloud Computing is an alternative for the enterprises that do not intend to invest in their own Information and Communication Technologies resources [31]. The combination of Big Data and Cloud Computing enables process participants to collaborate from different locations, in real time, to improve productivity and security, and ensure project feasibility [32]. Another element of the Industry 4.0 concept, which is increasingly being introduced by SMEs, is additive manufacturing [33], a technology that is rapidly evolving and integrated into production processes and into our daily lives [34]. Additive manufacturing is defined as a group of processes that create objects arising from the deposition of material on layers [35]). Urhal, Weightman, Diver, and Bartolo [36] also noticed that additive production disrupts the traditional supply chain, as the products are produced closer to the intended place of use at the time of the need.

The modern production concept is complemented by other elements of the holistic system, such as modern enterprise resource planning (ERP) cores, simulation, cyber security, augmented/virtual reality, and 3D printing [37,38]. Together, these elements are embedded systems with decentralized control and advanced connectivity that collect and exchange real-time information to identify, track, monitor, and optimize the production process [39]. The importance of connectivity and the continuous flow of information created the new machine-to-machine (M2M) interactions wherin products, machines, and factories are connected and they communicate through the industrial Internet.

Another type of interaction is human-to-machine (H2M), which is necessary, as complex and unstructured assignments to robots and production lines are too complex to be fully automated, while these systems are still predominantly tested [40], and SMEs are afraid of it. With increasing digitization and the availability of industry-based data, many new jobs are emerging [41]. Compared to the SMEs, the large enterprises are able to quickly identify new opportunities [42] brought by new technologies, for example, in the areas of ecology [43], monitoring [44], and project management [45]. These and other factors are a critical success factor for SMEs in the current extremely strong and competitive environment [32]. It is assumed that Industry 4.0 will result in a comprehensive communications network that will bring together factories, suppliers, logistics, resources, customers, etc. Each participant in the process will be able to optimize the configuration in real time, depending on the current requirements and status of other participants [46].

SMEs are not yet convinced of the benefits of the Industry 4.0 concepts. The high investment costs of technology and the question of its return rate are factors that hinder the persuasion of enterprises [47]. The most important technologies for SMEs in Romania are, according to Reference [6], robots, vertical and horizontal system integration, Big Data, the Internet of Things, and Cyber-Security. They investigated that the main barriers of implementation are lack of knowledge about Industry 4.0, operation and cost focus, other strategic priorities, lack of skilled employees, etc. The SMEs have a competitive disadvantage compared to large enterprises in up-to-date information technology and systems, lack of advance manufacturing technologies, lack of strategic management, and lack of standards [48]. A method for deciding on the best factors affecting the implementation of Industry 4.0 in Peruvian SMEs [49] was introduced. Their conclusion is that lack of capital and investment is the most important factor, followed by technology, management vision, and workers' skills. The commitment of management in Pakistan in SMEs plays an important role in accessing and exploiting innovative capabilities [50]. The implementation process, research, development, and innovation management systems in industrial SME are analyzed in a case study [51].

The level of Industry 4.0 implementation and readiness in SMEs can be evaluated by various assessment tools. Hamidi, Aziz, Shuhidan, and Mokhsin [52] created the IMPULS maturity assessment model for SMEs in Malaysia. Twenty Italian SMEs examined Pirola, Chimini, and Pinto [53] to find out digital readiness of SMEs, with respect to Industry 4.0. Ganzarain and Errasti [54] presented an Industry 4.0 maturity model for SMEs which starts with vision, followed by roadmap and appropriate projects to achieve transformation of an enterprise business model. Kolla, Minufekr, and Plapper [55] linked lean with Industry 4.0 in an assessment model for SMEs. The strategy, manufacturing and operations, technology, digitalization, and people capability are dimensions of the Industry 4.0 maturity model by Reference [56]. However, most of these models do not categorize enterprises. Another maturity model for SMEs was developed by Reference [57], who used a three-dimensional axis composed of organizational dimensions, toolboxes, and maturity levels based on critical review.

The reviewed literature of various approaches, technologies, barriers, and assessment models (tools) illustrates that implementation of Industry 4.0 technologies in SMEs is a contemporary challenge. For evaluating the levels of implementation of Industry 4.0, it is appropriate to use some tools, e.g., the Vrchota–Pech Industry 4.0 index (VPi4 index) [58]. However, there are still possibilities to improve. For the comparison of SMEs, it is preferable to create special groups of mutually comparable enterprises (categorize SMEs). In our paper, we also analyze the level of implementation of Industry 4.0 in SMEs and, based on these results, create their categorization.

3. Materials and Methods

The main aim of the paper was the classifation of small- and medium-sized enterprises into various groups, according to the level of implementation of Industry 4.0. The first partial objective of the paper was to analyze the readiness of the small and medium enterprises for the implementation of Industry 4.0. The second objective (a sub-objective) was to compare the level of Industry 4.0 of small- and medium-sized enterprises to that of large enterprises, through the VPi4 index.

3.1. Data

The research sample was compiled on the basis data from the Czech Statistical Office, which states that, in the Czech Republic, there were 175,894 enterprises in the manufacturing industry in 2005 [59].

In February 2018 to May 2019, approximately 2500 randomly selected enterprises were asked through the managers to ensure that the 95% confidence requirement was met at 5% error margin and 15% questionnaire return. For equal representation of these 2500 enterprises, we used stratified random sampling based on the size (about 50% micro and small enterprises; 50% medium-sized enterprises) and technological intensity (about 50% of high-tech and low-tech enterprises) of enterprises. The composition of the sample corresponds to this distribution (Table 1).

Table 1. Characteristics of research sample for cluster analysis.

Group	Category of Group	Number (%)
Size	Micro Enterprises (0–9 Employees)	16.13
	Small Enterprise (10–49 Employees)	40.32
	Medium Enterprise (50–249 Employees)	43.55
Technological Intensity	High-Tech and Medium High-Tech Intensity (HTI)	48.92
	of which High-Tech Sector (HTS)	5.91
	and Medium High-Tech Sector (MHTS)	43.01
	Low-Tech and Medium Low-Tech Intensity (LTI)	51.08
	of which Low-Tech Sector (LTS)	39.25
	and Medium Low-Tech Sector (MLTS)	11.83

The actual rate of return of the questionnaires was 12.5%, and 314 questionnaires were completed (18 questionnaires were further excluded due to the existence of one year on the market and completeness of the survey). A total of 90 questionnaires were filled in by the large enterprises, not included in the cluster analysis for the purpose of the paper; however, they were used in comparison with the small- and medium-sized enterprises.

The sample of small- and medium-sized enterprises therefore comprises a total of 186 enterprises. The questionnaire focused on the main characteristics of Industry 4.0, and its questions were defined in cooperation with the managers in the framework of quality research. The main areas of the questionnaire were 13 variables characterizing the various technologies of Industry 4.0 used by enterprises. Table 1, below, summarizes the characteristics of SMEs in terms of technological demands. Classification of the enterprises by size was based on the number of employees of the enterprise, as defined by the methodology of the European Commission [60].

3.2. Industry 4.0 Index Methodology

An Industry 4.0 index (VPi4) was used to determine the level of implementation of Industry 4.0 in the small- and medium-sized enterprises. The VPi4 index was developed based on a survey which asked managers of enterprises if they use different Industry 4.0 technologies and processes. The index is based on exploratory factor analysis which identified 3 factors consisting of 13 variables (technologies) described in Reference [58]. These factors and variables were chosen for classification of small- and medium-sized enterprises into various groups, according to the level of implementation of Industry 4.0.

The answers from the questionnaire survey of a sample of small- and medium-sized enterprises were transformed into VPi4 variables, using factor score (as weights), and the levels of the VPi4 index were further calculated. The connection of the VPi4 and questionnaire is described in a previous study [58], which also explains the procedures for creating VPi4. The index consists of three follow-up levels (the factor scores of variables are reported in brackets):

1. The First Level of VPi4 (Digitization and Human Resources Infrastructure)—skilled people (0.61), collecting data (0.82), data storage in cloud (0.63), and data analysis (0.86).
2. The Second Level of VPi4 (Automation and information system/information technology (IS/IT) Infrastructure)—information technology (IT) infrastructure (0.53), MES and ERP (0.75), linked data M2M (0.58), robots in production (0.54), mobile terminals (0.54), and sensors (0.58).
3. The Third Level of VPi4 (Learning and AI Infrastructure)—learning software (0.44), sharing data with suppliers (0.67), virtual reality (VR), and simulations (0.68).

As the factor analysis provides the factors that do not correlate with each other, three levels of VPi4 were appropriate for the cluster analysis. Principal Component Analysis (PCA) was performed to check the level independence, which is also part of the results of the paper. In general, its aim is to reduce high-dimensional data to a few so-called "principal" dimensions, to reveal structure in the data, and so to facilitate their interpretation (for more, see Reference [61]). Consequently, VPi4 levels are to be used for the division and categorization of the SMEs through cluster analysis. Specifically, factor scores of the enterprises, standardized by *z*-score, are used. The standardization process consists of transforming the variables such that they have mean zero and standard deviation one.

3.3. K-Means Clustering

Cluster analysis employs a measure of similarity or dissimilarity for assigning points in space to a cluster [62]. K-means clustering is the most popular unsupervised clustering technique for partitioning a given dataset into a set of *k*-groups (clusters). It is an iterative optimization method based on the initial division of objects into k-clusters. The distribution is based on the determination of the k-centroids that form the center of the clusters. The distance of each object to the centroids is then examined based on the Euclidean distance. The object is then assigned to the nearest centroid. A new centroid is then calculated for each cluster as an m-dimensional vector of the average values of each variable. The cycle is then repeated by calculating distances and assigning objects to these clusters. The process is performed as long as the objects are moved between the clusters [63].

The purpose of cluster analysis was to discover a system for categorizing and grouping the enterprises based on correlation obtained by evaluating factor variables. The enterprises with high positive correlation were grouped and separated by negative correlators. Cluster analysis calculations were performed by software R, using the following packages: ggplot2 [64], factoextra [65] and cluster [66], rgl [67], scatterplot3d [68], gridExtra [69], and dplyr [70].

K-means algorithm can be summarized as follows [71]:

1. Specify the number of clusters (k) based on the Elbow method (graphical form) and Average Silhouette method (Equation (1)). Silhouette coefficient (SC) finds the average distance to the best-fitting cluster, compared to the average distance between a data point, $x \in C_k$, and other points of C_k, for determining cluster system appropriateness [62]:

$$SC = \frac{1}{N} \sum_{i=1}^{N} \frac{b(x) - a(x)}{max\{a(x), b(x)\}}, \tag{1}$$

where x_i is a data object belonging to the cluster C_k, *a(x)* is cohesion (average distance of x to all other vectors in the same cluster), and *b(x)* is separation (average distance of x to the vectors in other clusters). A value of +1 indicates a perfect clustering choice, and a value below 0 indicates a bad clustering choice. We try to find the minimum among the clusters.

2. Initialize Cluster Centroids by randomly selecting k-objects from the dataset as the initial cluster centers or means.
3. Form k-clusters by assigning each object (x_i) to its closest centroid, based on the Euclidean distance (Equation (2)) between the object and the centroid [72]:

$$W(C_k) = \sum_{x_i \in C_k} (x_i - \mu_k)^2, \tag{2}$$

where x_i is a data object belonging to the cluster C_k, and μ_k is the mean value of the objects assigned to the cluster C_k.

4. Re-compute the centroid of each cluster. For each of the k-clusters, update the cluster centroid by calculating the new mean values of all the data points in the cluster.
5. Repeat Steps 3 and 4, until the cluster assignments stop changing or the maximum number of iterations (usually 10) is reached. Iteratively minimize the total within sum of square (Equation (3)) of the objects to their assigned cluster centers (μ_k). Total within sum of square is defined as follows [72]:

$$\sum_{k=1}^{k} W(C_k) = \prod_{k=1}^{k} \sum_{x_i \in C_k} (x_i - \mu_k)^2, \tag{3}$$

The smaller the value $W(C_k)$, the better the clustering (C_k, x_i). Although, finding an optimal pair (C_k, x_i) is quite a computationally intensive task; finding either optimal S or optimal c is fairly easy.

6. Clusters validation consists of measuring the goodness of clustering results, using one-way Analysis Of Variance (ANOVA test). The ANOVA F-test evaluates if there are any differences between group means of clusters. Further, the Tukey method of "Honest Significant Difference" is performed for multiple pairwise comparisons of the means of clusters in the analysis of variance (for more, see Reference [61]).

3.4. Statistical Analysis and Hypotheses

The research results were also tested by a statistical analysis. The aim of this analysis was to compare the results of the Industry 4.0 VPi4 index of the small- and medium-sized enterprises with those of the large enterprises. A comparison was performed at three VPi4 levels, including the overall index:

$H1_0$: The VPi4 index of small- and medium-sized enterprises and VPi4 index of large enterprises are identical populations.

$H1_A$: The VPi4 index of small- and medium-sized enterprises and VPi4 index of large enterprises are different populations.

Furthermore, the dependence between subjective perception of Industry 4.0 level and VPi4 Index was investigated by using Pearson and Spearman correlation coefficients. The index was expected to correlate, to some extent, with the subjective perception of the situation in the enterprise. The working hypotheses, verified at the 5% significance level, are as follows:

$H2_0$: There is no dependency between the perception of Industry 4.0 in enterprises and theVPi4 index of small- and medium-sized enterprises.
$H2_A$: There is dependency between the perception of Industry 4.0 in enterprises and the VPi4 index of small- and medium-sized enterprises.

Statistical evaluation of tests was performed by using Statistica 12 and R software.

4. Results

This is a summary of the results obtained at various stages of the research. The results are divided into two parts: cluster of SMEs analysis and comparison SMEs index of VPi4 with large enterprises.

4.1. Cluster Analysis of Small- and Medium-Sized Enterprises

The aim of the cluster analysis was to categorize and categorize SMEs based on three levels of the Industry 4.0 index (VPi4). First, the procedure of selecting variables (factors) for clustering is summarized, and then the optimal number of clusters is found, using Elbow a Silhouette methods. The next part presents the results of k-means cluster analysis, including the description of the clusters. Finally, the results of the cluster analysis are validated.

4.1.1. Variables Selection and Principal Component Analysis

Before performing the cluster analysis, it was necessary to choose the variables to be used in clustering. Based on the results of the questionnaire survey, the VPi4 index was calculated for the SMEs. The index consists of three factors (levels): VPi4 Level 1, VPi4 Level 2, and VPi4 Level 3. These factors should not correlate with each other, so it was advisable to use three levels of the VPi4 index. Principal Component Analysis (PCA) was performed to check the independence of these factors. The results are summarized in Table 2. All three dimensions have very similar % of total variance explained.

Table 2. Principal Component Analysis.

Variable *	Eigenvalue	Total Variance %	Cumulative Eigenvalue	Cumulative %
VPi4 Level 1	1.0954	36.41	1.0954	36.51
VPi4 Level 2	1.0097	33.66	2.1051	70.17
VPi4 Level 3	0.8949	29.83	3.0000	100.00

* Note: variables are Industry 4.0 index values (VPi4) at different levels.

4.1.2. Optimal Number of Clusters Determination

In this step, the goal is to determine the optimal number of clusters suitable for k-means cluster analysis. It is advisable to use more methods for this purpose, i.e., the Elbow and Silhouette methods in particular (Figure 1). The procedure consists in performing decompositions for different numbers of k-clusters. A value of total within sum of square is calculated for each decomposition (in Elbow method Equation (3)) and Silhouette coefficient (Equation (1)). Using the k-means algorithm, we divide the data sequentially into two to ten clusters and construct the corresponding function of the number of the clusters. In this case, there is a steep bending in four clusters (Figure 1a). In the case of the Silhouette method, the analysis is performed similarly, obtaining the optimal number of clusters based on the maximum value of the coefficient (Figure 1b). Both methods choose four clusters as the best solution for k-means clustering.

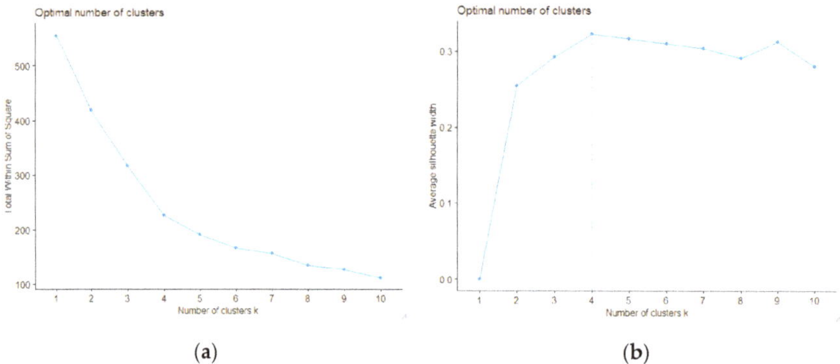

(a) (b)

Figure 1. Optimal number of clusters methods results: (**a**) Elbow method and (**b**) Silhouette method.

4.1.3. Results of Cluster Analysis

Based on the performed PCA to select the clustering variables and to determine the optimal number of analysis clusters, a k-means cluster analysis was performed. The calculations were performed in software R, using factoextra and cluster packages. The results are summarized in Table 3, which shows that four rather consistent clusters with 32 to 65 elements were created. Table 3 reports the values within sum of squares (WSS), maximum within cluster distances as diameter, average distance within clusters (which should be as small as possible), and separation with minimum distances of a point in the cluster to a point to another cluster reported for each cluster.

Table 3. Results of k-means clustering.

Variable	Cluster 1	Cluster 2	Cluster 3	Cluster 4
Within Sum of Squares	57.6031	61.8557	73.8122	33.4195
Diameter	3.9006	3.7375	3.5976	2.7862
Average Distance	1.7915	1.4810	1.4024	1.1872
Separation	0.2584	0.3370	0.2584	0.3583
Cluster Size	32	49	65	40

The graphical depiction of the cluster analysis is shown in Figure 2, which shows the results of the cluster analysis in relation to factor variables (VPi4 Level 1, Level 2, and Level 3) in 3D and then in 2D space. It is also evident from Figure 2b that the dimensions of the variables have different directions, confirming their independence. However, it is also evident that there is some overlap in the clusters. This overlap is also explained by the fact that the factor variables are partly complementary and reflect the levels of Industry 4.0. Achieving the second level is conditional by the first level; the achievement of the third level is partly conditional by reaching the first and second levels.

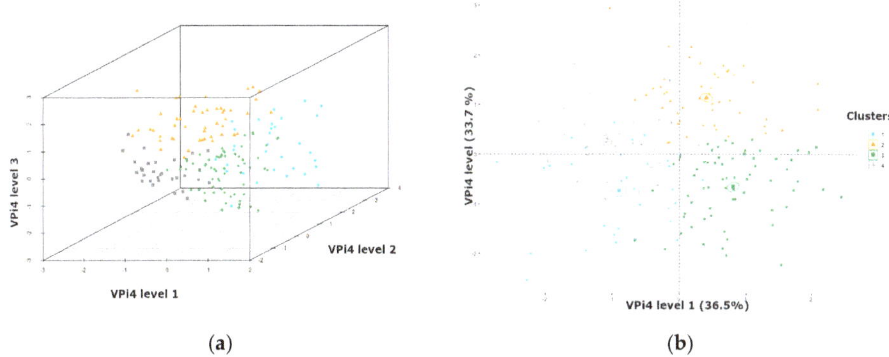

(a) (b)

Figure 2. K-means and PCA analysis results: (**a**) clusters based on three factors (i.e., F1, F2, and F3) in 3D space; (**b**) biplot—PCA combined with cluster analysis (relations of the clusters and factors).

Table 4 shows results of k-means clustering according to the centroids average characteristics of the Industry 4.0 levels in individual clusters. Cluster 1 is strongly represented by the first and second levels of the VPi4 index. This means that Cluster 1 contains the enterprises with the highest industry implementation rate of 4.0 on average. Cluster 2 only shows a high rate for Factor 3, which is a certain extension of the implementation of Industry 4.0 technologies. Cluster 3 achieves positive values only for the first level of the VPi4 index. So far, its average enterprise achieves only a certain initial industry-leading implementation level of 4.0. The latest cluster, Cluster 4, contains the lowest average values at almost all levels. On average, it is an enterprise with a very low implementation rate of Industry 4.0.

Table 4. Results of k-means clustering—cluster means (centroids).

Factor	Cluster 1	Cluster 2	Cluster 3	Cluster 4
VPi4 Level 1	0.6123	−0.2678	0.7175	−1.3276
VPi4 Level 2	1.3700	−0.0685	−0.7312	0.0931
VPi4 Level 3	−0.4001	1.5000	−0.2943	−0.7022

4.1.4. Comprehensive Description of the Clusters

The next part describes the characteristics of clusters in details, based on the characteristics of the enterprises that make them up. It is mainly the relation of the clusters to VPi4 index, its levels, the technologies used by the enterprises, the existence of Industry 4.0 strategies, subjective perception of Industry 4.0 levels by the managers, composition in terms of size, and technological demands of the enterprises in the clusters and others. Based on the prevailing tendencies, the clusters are named as follows:

Cluster 1—top I4 technological enterprises;
Cluster 2—advances I4 enterprises;
Cluster 3—I4 starting enterprises;
Cluster 4—I4 noobs enterprises.

First, the characteristics related to the VPi4 index are defined: It is clear from Figure 3a that the clusters differ in the level (0–100%) achieved by the VPi4 index values. Cluster 4 enterprises ("noobs enterprises") account for more than 80% of the very low 0–25% index. On the other hand, Cluster 1 ("top technological enterprises") and Cluster 2 ("advances enterprises") have the largest share of the 50–100% index. Both of these categories of the enterprises thus achieve a relatively significant index value. Cluster 3 ("start enterprises") contains mainly the enterprises with an index of 25–50%.

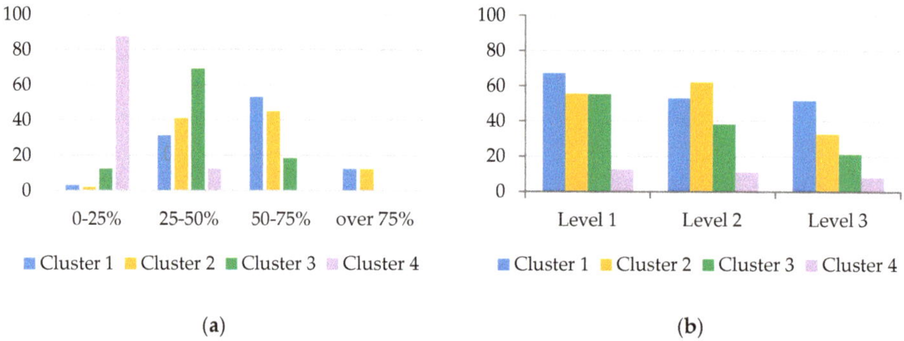

Figure 3. Characteristics of clusters based on the VPi4 index: (**a**) VPi4 index results of enterprises (0–25% minimum, over 75% maximum of VPi4 index level) and (**b**) VPi4 index levels 1–3 (Factors 1–3) results.

Figure 3b shows a more detailed resolution of each index level. Cluster 4 has low VPi4 index value at all levels; Cluster 3 has low VPi4 index value at the first two levels, in particular; and Clusters 1 and 2 have a decent rating at all the levels of the index. The distinction between Cluster 1 and Cluster 2 is subjected to further analysis with regard to the similarity of the VPi4 index level.

Other criteria in describing the characteristics of the clusters are related to a subjective evaluation of VPi4 level by the managers and the existence of Industry 4 strategy. Figure 4a summarizes the results of the subjective evaluation of Industry 4 level by the managers on a scale of 1 (low) to 5 (high). From the results, it is clear that, even in the case of Cluster 4, the managers perceived the level of Industry 4.0 at most companies at a low level. A relatively large proportion of managers reported their ranking to be low, similarly, in Cluster 3. The enterprises in Clusters 3 and 4 (mostly 80% of them) have no Industry 4.0 strategy (Figure 4b). The best subjective perception of the implementation of Industry 4.0 was in Clusters 1 and 2. The enterprises in these clusters (approximately 50% of them) also have an Industry 4.0 strategy. These characteristics divide the clusters into two groups (Figure 4b).

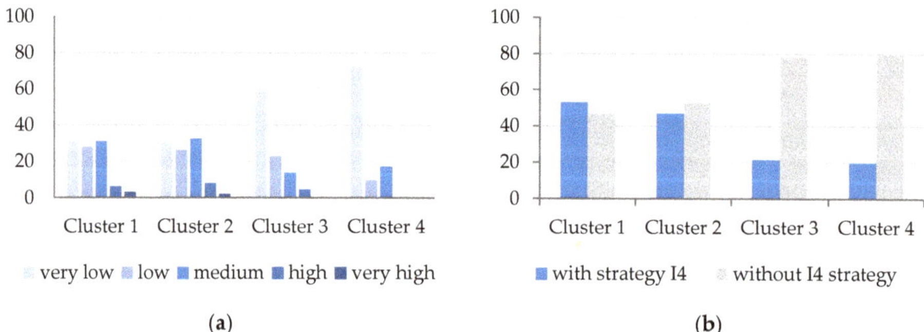

Figure 4. Characteristics of clusters, based on subjective assessment of Industry 4.0 (I4): (**a**) subjective assessment of Industry 4.0 implementation by enterprises (very low, low, medium, high, or very high implementation of I4) and (**b**) clusters according to the existence of I4 strategy in enterprises.

4.1.5. Description of Each Cluster

Further, the clusters are described separately for better clarity of the differences in the use of the technologies and characteristics of the enterprises. Following four clusters are described:

1. Top Industry 4.0 Technological Enterprises (Cluster 1)

   There are 32 enterprises in the cluster. In terms of size, these are medium-sized enterprises (40.65%). Cluster 1 is characterized by 53.13% enterprises that implement the strategy of Industry 4.0. Mostly, there are the enterprises (56.25%) with high-tech and medium high-tech intensity (HTI), mostly electro and engineering (53.13%). These enterprises use a large variety of I4 technologies (Figure 5a). Technological enterprises have a very high value of variables at the first level of the VPi4, such as people, data, cloud, and analysis. These enterprises have a high value of variables as mobile platforms and IT and an average value of sensors and M2M at the second level of the VPi4. At the third level of the VPi4, these enterprises have an average value of VR, sharing data, and learning software. Some enterprises already use nanotechnologies and 3D printers. In general, the technology is the largest in these enterprises, especially at the first and second levels of the index. Due to this, they are called "I4 top technological enterprises".

2. Industry 4.0 Advances Enterprises (Cluster 2)

   The second cluster comprises 49 enterprises, of which 65.31% are the medium-sized enterprises. A total of 46.94% of the enterprises in this cluster have an Industry 4.0 strategy. Similar to the first cluster, there are predominantly (65.31%) high-tech and medium-high intensity (HTI) enterprises in the electro and engineering sector (63.27%). The cluster is characterized by the enterprises with a high rating for some technologies (Figure 5b). These enterprises have average values of variables at first VPi4 level, such as analysis, data, and people. However, most of them still have not using cloud. At the second VPi4 level, they use a high level of IT and IS (ERP and MES). They already introduced some technologies as sharing data and learning software from the third VPi4 level. Some companies use, to a lesser extent, other Industry 4.0 applications, such as nanotechnologies, 3D printers, or auto-vehicles. Overall, enterprises in this cluster use technologies that are supported mainly by IT and infrastructure at the second level of the index. However, some higher-level technologies are used. Due to this, they are called "Advances I4 enterprises".

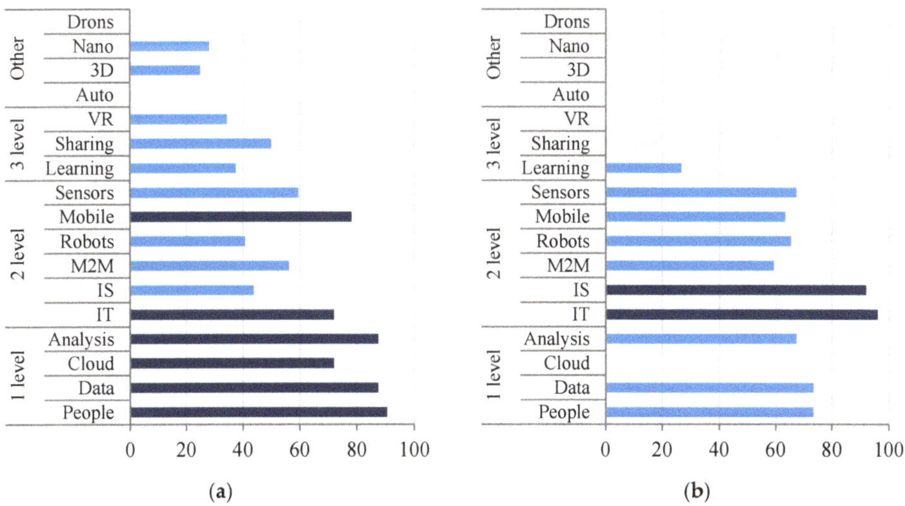

Figure 5. Clusters' characteristics: (**a**) Cluster 1—I4 technological enterprises; (**b**) Cluster 2—I4 advances enterprises.

3. Iindustry 4.0 Starting Enterprises (Cluster 3)

The third cluster consists of a total of 65 enterprises. The vast majority of these enterprises is small (47.69%). However, only 21.54% of them developed an Industry 4.0 strategy. In terms of technological intensity, there are mostly the (61.54%) enterprises with low-tech and medium low-tech intensity (LTI), predominantly the electro and engineering enterprises (38.47%). This cluster includes enterprises which have started implementing Industry 4.0 technologies (Figure 6a). These enterprises have a very high value of variables at the first level of VPi4, such as in people, data, and analysis and average value of cloud. They already have IT infrastructure and, at the average level, have IS and sensors. These enterprises did not use robots, M2M, or mobile technologies. The third level value of VPi4 is very low, as compared to other applications of Industry 4.0. In general, these enterprises are characterized by the introduction of technology only at the first level of the index. Due to this, they are called "I4 starting enterprises".

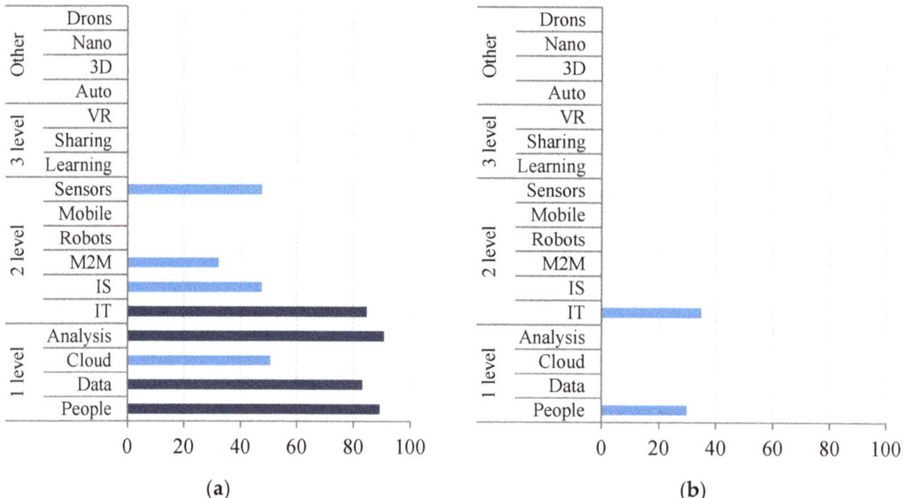

Figure 6. Clusters characteristics: (**a**) Cluster 3—I4 starting enterprises; (**b**) Cluster 4—I4 noobs enterprises.

4. Industry 4.0 Noobs Enterprises (Cluster 4)

The fourth cluster consists of 40 enterprises of a very low level of technology. Of the total, 40% are the small enterprises, and 30% are the micro-enterprises. Only 20% of them have an Industry 4.0 strategy. These are mostly (60%) the enterprises with low-tech and medium low-tech intensity (LTI). Almost 32.5% of them produce the products for domestic market, so that they probably do not require high technology use. These enterprises have very low values of almost all technologies of Industry 4.0 (Figure 6b). They are usually without new technologies or they are using only basic IT infrastructure. Second or third level of VPi4 and other applications of Industry 4.0 are not presented yet. Due to this, they are called "I4 noobs".

4.1.6. Validation of Cluster Analysis

The validity of the cluster analysis results was verified by one-way ANOVA F-tests for each clustered factor (VPi4 Level 1, Level 2, and Level 3). The results of this analysis should verify whether there are statistically significant differences between the clusters. In the one-way ANOVA test, we used a p-value significance level 0.05, but some of the cluster means are different. If the p-value for all four variables is greater than 0.05 (labeled by '*' in Table 5), excluding them from the analysis should be considered. The results of the factor ANOVA are shown in Table 5. The results show that, for all variables, there are some differences among the clusters.

Table 5. Results of ANOVA test.

Clusters	Sum Sq	Mean Sq	F-Value	Pr(>F) [1]
VPi4 Level 1	144.45	48.15	110.6	2×10^{-16} ***
VPi4 Level 2	86.98	28.994	79.33	2×10^{-16} ***
VPi4 Level 3	109.84	36.61	77.91	2×10^{-16} *** [1]

[1] Significance codes: '***' 0.001, '**' 0.01, and '*' 0.05.

However, the ANOVA results cannot be interpreted unambiguously, as it is not clear which pairs of clusters are different. It is advisable to use the Tukey Honest Significant Difference (HSD) test (for multiple pairwise comparison) to find out if there is a statistically significant difference between the means of certain cluster pairs. The results of this analysis are summarized in Table 6. In most cases, the differences between the clusters are statistically significant. However, there were no differences between Cluster 3 and Cluster 1 for VPi4 Level 1; there were no the differences between Clusters 4 and 2 for VPi4 level 2, and three differences for VPi4 level 3. This means that there are some overlaps between the clusters, especially for VPi4 Level 3 factor variables. Such lack of differences in the clusters may partly be due to the fact that all factors are considered follow-up levels of the VPi4 index. In the case of the small- and medium-sized enterprises, it is typical that most of these enterprises do not reach the high third level of the index. Therefore, their values are rather lower, and this causes the similarity of the clusters. Therefore, the differences in clusters of VPi4 Level 3 are not so significant. However, such conclusion should not diminish the significance of the cluster analysis and the fact that the clusters are different. The differences are also reported by Table 3 (see within sum of squares, diameter, average distance, separation, etc.).

Table 6. Results of the Tukey Honest Significant Difference (HSD) test.

Dependent Variable	Clusters Comparison	Mean Difference	Lower Bound [2]	Upper Bound [2]	*p*-Value [1]
VPi4 Level 1	2–1	−0.9677	−1.3565	−0.5788	2×10^{-5} ***
	3–1	0.1157	−0.2538	0.4851	0.848
	4–1	−2.1330	−2.5388	−1.7272	2×10^{-5} ***
	3–2	−1.0833	−1.4070	−0.7597	2×10^{-5} ***
	4–2	1.1654	0.8008	1.5299	2×10^{-5} ***
	4–3	−2.2487	−2.5925	−1.9049	2×10^{-5} ***
VPi4 Level 2	2–1	−1.4048	−1.7611	−1.0486	2×10^{-4} ***
	3–1	−2.0085	−2.3470	−1.6700	2×10^{-4} ***
	4–1	−1.2576	−1.6294	−0.8858	2×10^{-4} ***
	3–2	0.6036	0.3071	0.9002	2×10^{-4} ***
	4–2	−0.1473	−0.4813	0.1868	0.662
	4–3	0.7509	0.4359	1.0659	2×10^{-4} ***
VPi4 Level 3	2–1	1.6700	1.2660	2.0740	0.001 ***
	3–1	0.1087	−0.2751	0.4926	0.8824
	4–1	−0.3104	−0.7320	0.1111	0.2261
	3–2	1.5613	1.2250	1.8976	0.001 ***
	4–2	1.9805	1.6017	2.3592	0.001 ***
	4–3	−0.4191	−0.7763	−0.0619	0.0142 *

[1] Significant codes: '***' 0.001, '**' 0.01, and '*' 0.05. [2] The lower and the upper end point of the confidence interval at 95% (default).

Another method of cluster analysis validation is related to the abovementioned methods, to determining the optimal number of clusters, the Silhouette and Elbow method. The methods were used to determine the parameters of the cluster analysis in order to achieve the most accurate results. However, it should be noted that the average value of Silhouette coefficient is only 0.32. The reason for this is the overlap of the factor VPi4 Level 3. Another way to verify the validity of a cluster analysis is, for example, the Dunn index, calculated as the ratio of minimum separation/maximum diameter [73,74], 0.06623, in the paper. The higher the Dunn index value, the better the clustering is.

4.2. Comparison SMEs Index of VPi4 with Large Enterprises

Subsequently, the large enterprises (90 enterprises) and the SMEs were compared in terms of VPi4%. After the Mann–Whitney–Wilcox test was used to test the statistics, at a significance level of 0.05, the differences between the two groups of the enterprises were tested in the overall VPi4% index and also at the different levels (Levels 1–3). In all cases, the null hypothesis ($H1_0$) on the agreement of both groups) was rejected in favor of the alternative hypothesis ($H1_A$, claiming that the large enterprises reach a higher VPi4% level). The results in Table 7 show that the VPi4 index of small- and medium-sized enterprises (SMEs) and VPi4 index of large enterprises (LEs) is different populations.

Table 7. Industry 4.0 index (VPi4) distribution, using Mann–Whitney–Wilcox test.

Variable	Median Les [1]	Median SMEs [1]	Z	*p*-Value
VPi4% Total	57.212	39.070	6.061	0.000
VPi4% Level 1	64.741	50.512	5.002	0.000
VPi4% Level 2	62.855	42.451	6.512	0.000
VPi4% Level 3	43.654	25.198	5.667	0.000

[1] Note: differences between VPi4 index median of large enterprises (LEs) and small- and medium-sized enterprises (SMEs).

The normalization of the data of both variables was verified by the Shapiro–Wilk tests, with the *p*-value of VPi4% = 0.002 and the perception of Industry 4.0 by the enterprises *p*-value = 0.000, assuming the normality of the data, as also shown by both histograms. After that, the VPi4% and the subjective perception rate of Industry 4.0 were compared by using both Pearson and Spearman correlation coefficients. The Table 8 below shows the values of the tested statistics for both samples. It is apparent from *p*-values close to zero that, in both large and small enterprises, VPi4% correlates with the subjective perception of enterprises. In both cases, the correlation rate is very similar. Results show that there is dependency between the perception of Industry 4.0 in the enterprises and the VPi4 index of the small- and medium-sized enterprises. The $H2_0$ hypothesis was rejected in favor of $H2_A$ (Hypothesis 2_A) at a significance level of 0.05.

Table 8. Relation to subjective perception of Industry 4.0, based on Pearson and Spearman coefficients.

-		Pearson			Sperman		
Research [1]	Variables	Perception	VPi4	*p*-Value	Perception	VPi4	*p*-Value
LEs	Perception	1.000	0.384	0.000	1.000	0.383	0.000
	Index VPi4	0.384	1.000		0.383	1.000	
SMEs	Perception	1.000	0.385	0.000	1.000	0.387	0.000
	Index VPi4	0.385	1.000		0.387	1.000	

[1] Note: Research was conducted between large enterprises (LEs) and small- and medium-sized enterprises (SMEs).

Overall, the results show that these small- and medium-sized enterprise groups differ statistically significantly from large enterprises in their implementation and use of Industry 4.0 technologies.

5. Discussion

Here we present a discussion of our major findings and results. This section offers contributions and identifies limitations of the research.

5.1. General Discussion of Results

The cluster approach has gained extensive acceptance among academics and policymakers as an effective development strategy both for an industry and society [75], as evidenced by a number of studies focusing on innovation in SMEs [76–79] or technology SMEs [80] or their performance [81–83]. In this paper, according to statistical models, SMEs were divided into four clusters. This number is very common for this type of enterprises, as evidenced by the studies by Refereces [84,85]; or a very similar number is used, as with References [86,87]. An appropriate study of Industry 4.0 clusters and levels is also supported by the Austrian study on the implementation of Industry 4.0 [88], where deployed technologies also determine the level of implementation.

As confirmed by the results, Cluster 1 enterprises have the highest level, and they are characterized by a high level of Industry 4.0 deployment and have all the technology levels that they use successfully. These conclusions are also supported by a Romanian study on barriers to deployment [6], when the importance of technology for Industry 4.0 is emphasized by Reference [89], an article by a Finnish author on production systems. Brozzi [90] used self-assessment tools for Industry 4.0 readiness in SMEs with current and expected levels of digitalization. This author distinguished an average score of the current four-level groups of enterprises: traditional craftsman (very low score), digital newcomer (low score), ambitious (high score), and digital champion (very high score). These categories are almost similar to the categories in our research.

The contributions were in agreement with the research of Woods [84], who confirmed differences in Industry 4.0 levels among SMEs and large enterprises, while supporting the claim of Dubrova [87] that small businesses often do not have high technology; this can also be due to their difficulty in accessing free capital and the associated IT spending, as reported by Statista [91], and they start to increase significantly year-on-year (5% increase between 2017 and 2018; 15% increase between 2018 and 2019). Data for the largest expenditure areas around the world from 2006 to 2021 [92] show that SMEs increased investment into business services (7.1%) and software (6.9%). By contrast, only 54% of small- and medium-sized businesses store data in the cloud, compared to 92% of large businesses [93], where the research from Cluster 2 is lagging behind most. According to Computer Economics [94], differences in robotic process automation adoption rate in 2019 among small- and medium-sized enterprises (9%) and large enterprises (24%) are very high. These differences are more pronounced for the rate of investment in the robotic process, where large enterprises covar about 49%, while small-sized ones have only 14%, and middle-sized enterprises about 17% investment.

As the results showed, technological enterprises have very high value of variables at the first level VPi4. The importance of people is also emphasized by Nickel et al. [95], data by Chumnumporn et al. [96], cloud by Erasmus et al. [97] and analysis by Choy et al. [98]. These enterprises have high value of variables as mobile platforms and IT; average value of sensors and M2M at second-level VPi4, which is not yet widely used by SMEs [99]. At the third level of VPi4, these enterprises have an average value of VR, sharing data, and learning software. Some enterprises already use nanotechnologies and 3D printers, which is contrary to the study of Pallas [100], which claims that these technologies attribute predominantly to large companies; Gaudin [101] makes the same claim, but stresses their importance for sustainable development.

5.2. Theoretical and Practional Implications

Based on the VPi4 index, SMEs can be divided into four groups. The practical benefits include the possibility of comparing the company with others, both based on the classic VPi4 index and in terms of classification into one of four groups of clusters. Managers of SMEs can compare their own enterprises with those of competitors and understand which technologies may be suitable for further deployment in relation to the competitive advantage. The enterprise can better analyze the current weaknesses and strengths of technical factors. In practice, it is very difficult to determine the level of technology in an enterprise, especially in SMEs, which have limited capacities and funding opportunities. Thanks to our model, managers will get a better overview of the current situation, as they can use our model in planning and strategy development for the future.

Developed classification of enterprises extends the possibilities of VPi4 index especially for SMEs. SMEs are specific in terms of I4 technology implementation. The proposed categorization of enterprises into four groups also offers easy comparison of enterprises. Due to the limited capital and the necessity to choose only one of many possible technologies, the implementation process of I4 in SMEs is not usually gradual by individual index levels (from first to second level and above). Small- and medium-sized enterprises often have different needs when introducing new technologies. Their preferences are evidenced by clustering.

The results point out which Industry 4.0 technologies are most often used by small- and medium-sized enterprises. The results confirmed that these are mainly technologies at the first level, i.e., analysis, collecting data, cloud storage (and for them having skilled people). With the exception of I4 technological enterprises group and partly I4 advances enterprises group, small- and medium-sized enterprises do not use a higher level of VPi4 index technologies (virtual reality, data sharing, and learning software) or special applications, such as 3D printing, drones, or nanotechnology. From this perspective, the paper extends knowledge about the use of technology in SMEs.

Another theoretical contribution is confirmation of the hypothesis of different approaches to technology between SMEs and large enterprises. The differences between SMEs and large enterprises show that large enterprises have a higher level of Industry 4.0 implementation. On the other hand, SMEs have the advantage of implementing new technologies due to greater flexibility; more involved, empowered employees; less bureaucracy and coordination due to the usually leaner organizational structure; and the ability to produce customized products and close relationship with customers. These benefits should be used by SMEs to increase value-added services and tailored manufacturing processes.

5.3. Limitations

However, some study limitations should be acknowledged. In the paper we used for cluster analysis the VPi4 index [58], which has some limitations. First, the VPi4 index does not include some Industry 4.0 technologies (block chain, laser technology, RFID, drones, 3D printers, track-n-trace, nanotechnology, autonomous vehicles, etc.). Second, the VPi4 index is based on the results of a questionnaire survey. Some limitations of this kind of method should therefore be considered. Third, the VPi4 index was not verified by confirmatory factor analysis and structural model equation method. This problem will be resolved in the future, after the third wave of questionnaire survey research on Industry 4.0. Currently, the results and structure of the VPi4 index are confirmed by relation to subjective perception of the level of Industry 4.0 by the enterprises and re-test in second wave of the research [58]. Fourth, the VPi4 index is based on the research in the Czech Republic and can be limited to the specifics of this Country. Our study was conducted in the same country, so it should not be a problem. Another limit may be the use of only factors in the field of technology, without taking into account, for example, financial data, which also prove to be very important in terms of sustainable development [102].

The questionnaire research on small- and medium-sized enterprises had a lower rate of return than we expected. Therefore, the margin of error (5.86%) was higher than 5% at 95% confidence level. The error, therefore, slightly exceeded the planned level. However, we assume that this will not affect the results. Moreover, the sample size was not representative in the case of proportionality of the Czech Republic enterprise population. Rather, the research sample was created in regard to the size of the enterprises and technological demands of the industry.

Other limitations include problems related to k-cluster analysis methods. The main issues of this method are handling empty clusters (special situation) and outliers (centroids may not be representative for all cases). K-means is not suitable for all types of data. It cannot handle non-globular clusters or data without a specific center [71]. Another issue could be selecting the wrong number of initial clusters. This can be solved by Elbow, Silhouette, or other methods.

The results of the factory analysis validation based on the Tukey HSD test shows similarities between clusters especially at the third level of the VPi4 index. Similarly, this overlap is also confirmed by the average Silhouette coefficient and the lower Dunn index. This clustering issue is based on the fact that most small- and medium-sized enterprises do not reach the highest level of the VPi4 index. Similarity is therefore in low index values. However, in spite of this problem, the different categories of clusters are different from each other. With the increasing level of implementation of Industry 4.0 in SMEs, this issue will be eliminated in the future, and the differences between the clusters will become more noticeable. Hovewer, the relation between Industry 4.0 implementation and the size of an enterprise is still significant [103].

6. Conclusions

To conclude this paper, we summarize the results of the research. The paper deals with the analysis of the level of Industry 4.0 implementation in SMEs. Based on cluster analysis, we developed a categorization of the enterprises which consists of four groups, namely top I4 technological enterprises, advances I4 enterprises, I4 starting enterprises, and I4 noobs enterprises. The largest group is formed by SMEs, which are just-introduced Industry 4.0 technologies. Depending on their classification, enterprises differ in the use of preferred technologies. Generally, SMEs are more used technologies that can be found at lower levels of the VPi4 index (data collection, analysis, Cloud, IS, IT, and mobile platforms). At the same time, SMEs with a higher level of implementation of Industry 4.0 reported enough skilled human resources.

The research examined hypotheses about the differences between SMEs and large enterprises in the implementation of Industry 4.0. Results of comparisons show that SMEs have, so far, a lower level of Industry 4.0 implementation. This confirms the assumption that large enterprises have greater opportunities to use new technologies and transform them into smart factories. However, this situation may change in the future if new technologies become more accessible (for example, using leasing financing) and more appropriate for investment of SMEs.

As part of future research, we plan to conduct a deeper analysis of the implementation of Industry 4.0 in SMEs, focusing on the main barriers. Case studies on specific enterprises will be appropriate. These case studies could explain why SMEs are not yet implementing Industry 4.0 to a greater extent. We plan to show the procedure of how to rank the implementation steps or technologies in a single enterprise. It may be interesting to compare the current and required (or expected) level of Industry 4.0 in SMEs. Other challenges for future research include developing a methodology (step-by-step roadmap) to implement Industry 4.0 in enterprises. SMEs often do not have much information on how they can turn their business into a digital form; however, in the current situation (pandemic of coronavirus), they are forced to use home office and reduced working hours. New Industry 4.0 technologies can help them with this current challenge.

Author Contributions: Conceptualization, M.P. and J.V.; data curation, M.P. and J.V.; formal analysis, M.P. and J.V.; funding acquisition, M.P. and J.V.; investigation, M.P. and J.V.; methodology, M.P. and J.V.; project administration, M.P. and J.V.; resources, M.P. and J.V.; software M.P. and J.V.; supervision, M.P.; validation, M.P. and J.V.; visualization, M.P.; writing—original draft, M.P.; writing—review and editing, M.P. and J.V. All authors have read and agreed to the published version of the manuscript.

Funding: This research was funded by "EF-150-GAJU 047/2019/S".

Acknowledgments: The authors thank the enterprises taking part in the research.

Conflicts of Interest: The authors declare no conflict of interest.

References

1. Dalenogare, L.S.; Benitez, G.B.; Ayala, N.F.; Frank, A.G. The expected contribution of Industry 4.0 technologies for industrial performance. *Int. J. Prod. Econ.* **2018**, *204*, 383–394. [CrossRef]
2. Novotná, M.; Volek, T. Efficiency of production factors in the EU. *Cent. Eur. J. Reg. Dev. Tour.* **2018**, *10*, 147–168.
3. Rolínek, L.; Řehoř, P. Strategic management and measurement of competitiveness of regions on example of countries EU. *J. Cent. Eur. Agric.* **2008**, *9*, 17–22.
4. Sahi, G.K.; Gupta, M.C.; Cheng, T.C.E. The effects of strategic orientation on operational ambidexterity: A study of indian SMEs in the industry 4.0 era. *Int. J. Prod. Econ.* **2020**, *220*, 107395. [CrossRef]
5. Ingaldi, M.; Ulewicz, R. Problems with the implementation of industry 4.0 in enterprises from the SME sector. *Sustainability* **2020**, *12*, 217. [CrossRef]
6. Turkes, M.C.; Oncioiu, I.; Aslam, H.D.; Marin-Pantelescu, A.; Topor, D.I.; Capusneanu, S. Drivers and barriers in using industry 4.0: A perspective of SMEs in Romania. *Processes* **2019**, *7*, 153. [CrossRef]
7. Ahmi, A.; Elbardan, H.; Raja Mohd Ali, R.H. Bibliometric analysis of published literature on industry 4.0. In Proceedings of the 2019 International Conference on Electronics, Information, and Communication (ICEIC), Auckland, New Zealand, 22–25 January 2019; IEEE: Auckland, New Zealand, 2019; pp. 1–6.
8. Pedersen, M.R.; Nalpantidis, L.; Andersen, R.S.; Schou, C.; Bøgh, S.; Krüger, V.; Madsen, O. Robot skills for manufacturing: From concept to industrial deployment. *Robot. Comput. Integr. Manuf.* **2016**, *37*, 282–291. [CrossRef]
9. Maly, I.; Sedlacek, D.; Leitao, P. Augmented reality experiments with industrial robot in industry 4.0 environment. In Proceedings of the 2016 IEEE 14th International Conference on Industrial Informatics (INDIN), Poitiers, France, 19–21 July 2016; IEEE: Poitiers, France, 2016; pp. 176–181.
10. Zambon, I.; Cecchini, M.; Egidi, G.; Saporito, M.G.; Colantoni, A. Revolution 4.0: Industry vs. agriculture in a future development for SMEs. *Processes* **2019**, *7*, 36. [CrossRef]
11. Lin, B.; Wu, W.; Song, M. Industry 4.0: Driving factors and impacts on firm's performance: An empirical study on China's manufacturing industry. *Ann. Oper. Res.* **2019**. [CrossRef]
12. Kabugo, J.C.; Jämsä-Jounela, S.-L.; Schiemann, R.; Binder, C. Industry 4.0 based process data analytics platform: A waste-to-energy plant case study. *Int. J. Electr. Power Energy Syst.* **2020**, *115*, 105508. [CrossRef]
13. Madsen, D.Ø. The emergence and rise of industry 4.0 viewed through the Lens of management fashion theory. *Adm. Sci.* **2019**, *9*, 71. [CrossRef]
14. Hahn, G.J. Industry 4.0: A supply chain innovation perspective. *Int. J. Prod. Res.* **2020**, *58*, 1425–1441. [CrossRef]
15. Galati, F.; Bigliardi, B. Industry 4.0: Emerging themes and future research avenues using a text mining approach. *Comput. Ind.* **2019**, *109*, 100–113. [CrossRef]
16. Jena, M.C.; Mishra, S.K.; Moharana, H.S. Application of Industry 4.0 to enhance sustainable manufacturing. *Environ. Prog. Sustain. Energy* **2020**, *39*, 13360. [CrossRef]
17. Shukla, A.K.; Nath, R.; Muhuri, P.K.; Lohani, Q.M.D. Energy efficient multi-objective scheduling of tasks with interval type-2 fuzzy timing constraints in an Industry 4.0 ecosystem. *Eng. Appl. Artif. Intell.* **2020**, *87*, 103257. [CrossRef]
18. Schumacher, A.; Erol, S.; Sihn, W. A Maturity model for assessing Industry 4.0 readiness and maturity of manufacturing enterprises. *Procedia Cirp* **2016**, *52*, 161–166. [CrossRef]
19. Schumacher, A.; Nemeth, T.; Sihn, W. Roadmapping towards industrial digitalization based on an Industry 4.0 maturity model for manufacturing enterprises. *Procedia Cirp* **2019**, *79*, 409–414. [CrossRef]
20. Rainer, A.; Hall, T. Key success factors for implementing software process improvement: A maturity-based analysis. *J. Syst. Softw.* **2002**, *62*, 71–84. [CrossRef]
21. Kaltenbach, F.; Marber, P.; Gosemann, C.; Bolts, T.; Kuhn, A. Smart services maturity level in Germany. In Proceedings of the 2018 IEEE International Conference on Engineering, Technology and Innovation (ICE/ITMC), Stuttgart, Germany, 17–20 June 2018; IEEE: Stuttgart, Germany, 2018; pp. 1–7.
22. Hamzeh, R.; Zhong, R.; Xu, X.W. A survey study on Industry 4.0 for New Zealand manufacturing. *Procedia Manuf.* **2018**, *26*, 49–57. [CrossRef]
23. Ben-Daya, M.; Hassini, E.; Bahroun, Z. Internet of things and supply chain management: A literature review. *Int. J. Prod. Res.* **2019**, *57*, 4719–4742. [CrossRef]

24. Gubbi, J.; Buyya, R.; Marusic, S.; Palaniswami, M. Internet of Things (IoT): A vision, architectural elements, and future directions. *Future Gener. Comput. Syst.* **2017**, *29*, 1645–1660. [CrossRef]
25. Madakam, S.; Ramaswamy, R.; Tripathi, S. Internet of Things (IoT): A literature review. *J. Comput. Commun.* **2015**, *3*, 164–173. [CrossRef]
26. Dwevedi, R.; Krishna, V.; Kumar, A. Environment and big data: Role in smart cities of India. *Resources* **2018**, *7*, 64. [CrossRef]
27. Sagiroglu, S.; Sinanc, D. Big data: A review. In Proceedings of the 2013 International Conference on Collaboration Technologies and Systems (CTS), San Diego, CA, USA, 20–24 May 2013; IEEE: San Diego, CA, USA, 2013; pp. 42–47.
28. Babiceanu, R.F.; Seker, R. Big Data and virtualization for manufacturing cyber-physical systems: A survey of the current status and future outlook. *Comput. Ind.* **2016**, *81*, 128–137. [CrossRef]
29. Sivarajah, U.; Kamal, M.M.; Irani, Z.; Weerakkody, V. Critical analysis of big data challenges and analytical methods. *J. Bus. Res.* **2017**, *70*, 263–286. [CrossRef]
30. Abubakr, M.; Abbas, A.T.; Tomaz, I.; Soliman, M.S.; Luqman, M.; Hegab, H. Sustainable and smart manufacturing: An integrated approach. *Sustainability* **2020**, *12*, 2280. [CrossRef]
31. Branco, T.; de Sá-Soares, F.; Rivero, A.L. Key issues for the successful adoption of cloud computing. *Procedia Comput. Sci.* **2017**, *121*, 115–122. [CrossRef]
32. Maskuriy, R.; Selamat, A.; Ali, K.N.; Maresova, P.; Krejcar, O. Industry 4.0 for the construction industry-how ready is the industry? *Appl. Sci.* **2019**, *9*, 2819. [CrossRef]
33. Bambach, M.; Sviridov, A.; Weisheit, A.; Schleifenbaum, J.H. Case studies on local reinforcement of sheet metal components by laser additive manufacturing. *Metals* **2017**, *7*, 113. [CrossRef]
34. Bandyopadhyay, A.; Susmita Bose, S. (Eds.) *Additive Manufacturing, 2nd ed*; CRC Press: Boca Raton, FL, USA, 2019; ISBN 978-1-138-60925-9.
35. Gibson, I.; Rosen, D.; Stucker, B. *Additive Manufacturing Technologies*; Springer: New York, NY, USA, 2015; ISBN 978-1-4939-2112-6.
36. Urhal, P.; Weightman, A.; Diver, C.; Bartolo, P. Robot assisted additive manufacturing: A review. *Robot. Comput. Integr. Manuf.* **2019**, *59*, 335–345. [CrossRef]
37. Venkatraman, S.; Fahd, K. Challenges and success factors of ERP systems in Australian SMEs. *Systems* **2016**, *4*, 20. [CrossRef]
38. Alcácer, V.; Cruz-Machado, V. Scanning the Industry 4.0: A literature review on technologies for manufacturing systems. *Eng. Sci. Technol. Int. J.* **2019**, *22*, 899–919.
39. Rojko, A. Industry 4.0 concept: Background and overview. *Int. J. Interact. Mob. Technol.* **2017**, *11*, 77–90. [CrossRef]
40. Elfekey, H.; Bastawrous, H.A.; Okamoto, S. A Touch sensing technique using the effects of extremely low frequency fields on the human body. *Sensors* **2016**, *16*, 2049. [CrossRef]
41. Stojanova, H.; Lietavcova, B.; Vrdoljak Raguž, I. The Dependence of unemployment of the senior workforce upon explanatory variables in the European Union in the context of Industry 4.0. *Soc. Sci.* **2019**, *8*, 29. [CrossRef]
42. Koilo, V. Evidence of the environmental kuznets curve: Unleashing the opportunity of Industry 4.0 in emerging economies. *J. Risk Financ. Manag.* **2019**, *12*, 122. [CrossRef]
43. Tsai, W.-H. Green Production planning and control for the textile industry by using mathematical programming and Industry 4.0 techniques. *Energies* **2018**, *11*, 2072. [CrossRef]
44. Ageyeva, T.; Horváth, S.; Kovács, J.G. In-mold sensors for injection molding: On the way to Industry 4.0. *Sensors* **2019**, *19*, 3551. [CrossRef]
45. Vrchota, J.; Řehoř, P. Project management and innovation in the manufacturing industry in Czech Republic. *Procedia Comput. Sci.* **2019**, *164*, 457–462. [CrossRef]
46. Maresova, P.; Soukal, I.; Svobodova, L.; Hedvicakova, M.; Javanmardi, E.; Selamat, A.; Krejcar, O. Consequences of Industry 4.0 in business and economics. *Economies* **2018**, *6*, 46. [CrossRef]
47. Sevinc, A.; Gur, S.; Eren, T. Analysis of the difficulties of SMEs in industry 4.0 applications by analytical hierarchy process and analytical network process. *Processes* **2018**, *6*, 264. [CrossRef]
48. Kleindienst, M.; Ramsauer, C. SMEs and industry 4.0-introducing a KPI based procedure model to identify focus areas in manufacturing industry. *Athens J. Bus. Econ.* **2016**, *2*, 109–122. [CrossRef]

49. Huang, C.J.; Chicoma, E.D.T.; Huang, Y.H. Evaluating the factors that are affecting the implementation of industry 4.0 technologies in manufacturing MSMEs, the case of Peru. *Processes* **2019**, *7*, 161. [CrossRef]
50. Imran, M.; Salisu, I.; Aslam, H.D.; Iqbal, J.; Hameed, I. Resource and information access for SME sustainability in the era of IR 4.0: The mediating and moderating roles of innovation capability and management commitment. *Processes* **2019**, *7*, 211. [CrossRef]
51. Cerezo-Narviez, A.; Garcia-Jurado, D.; Gonzalez-Cruz, M.C.; Pastor-Fernandez, A.; Otero-Mateo, M.; Ballesteros-Perez, P. Standardizing innovation management: An opportunity for SMEs in the aerospace industry. *Processes* **2019**, *7*, 282. [CrossRef]
52. Hamidi, S.R.; Aziz, A.A.; Shuhidan, S.M.; Mokhsin, M. *SMEs Maturity Model Assessment of IR4.0 Digital Transformation*; Springer: Singapore, 2018; Volume 739, ISBN 9789811086113, ISSN 21945357.
53. Pirola, F.; Cimini, C.; Pinto, R. Digital readiness assessment of Italian SMEs: A case-study research. *J. Manuf. Technol. Manag.* **2019**. [CrossRef]
54. Ganzarain, J.; Errasti, N. Three stage maturity model in SME's towards industry 4.0. *J. Ind. Eng. Manag.* **2016**, *9*, 1119–1128. [CrossRef]
55. Kolla, S.; Minufekr, M.; Plapper, P. *Deriving Essential Components of Lean and Industry 4.0 Assessment Model for Manufacturing SMEs.*; Butala, P., Govekar, E., Vrabic, R., Eds.; Elsevier Bv: Amsterdam, The Netherlands, 2019; Volume 81, pp. 753–758.
56. Chonsawat, N.; Sopadang, A. *The Development of the Maturity Model to Evaluate the Smart SMEs 4.0 Readiness*; IEOM Society: Southfield, MI, USA, 2019; Volume 2019, pp. 354–363.
57. Mittal, S.; Khan, M.A.; Romero, D.; Wuest, T. A critical review of smart manufacturing & Industry 4.0 maturity models: Implications for small and medium-sized enterprises (SMEs). *J. Manuf. Syst.* **2018**, *49*, 194–214.
58. Vrchota, J.; Pech, M. Readiness of enterprises in Czech Republic to implement industry 4.0: Index of industry 4.0. *Appl. Sci.* **2019**, *9*, 5405. [CrossRef]
59. Czech Statistical Office. Available online: https://www.czso.cz/documents/10180/23169600/ht_odvetvi.pdf/cb4dc782-a3e0-43a4-8d96-99b8d1f14cca (accessed on 20 February 2020).
60. European Commission. *Recommendation of 6 May 2003 Concerning the Definition of Micro, Small and Medium-Sized Enterprises (Text with EEA Relevance) (Notified under Document Number C (2003) 1422)*; The Publications Office of the European Union: Luxembourg, 2003.
61. Little, T. *The Oxford Handbook of Quantitative Methods, Vol.2: Statistical Analysis*; Oxford University Press: New York, NY, USA, 2013.
62. King, R.S. *Cluster Analysis and Data Mining: An Introduction*; Mercury Learning and Information: Boston, MA, USA, 2015.
63. Rezankova, H. *Shluková Analýza Dat (Cluster Data Analysis)*; Professional Publishing: Praha, Czech Republic, 2007.
64. Wickham, H. *Ggplot2: Elegant Graphics for Data Analysis*; Springer: New York, NY, USA, 2016.
65. Lemenkova, P. K-Means clustering in R libraries {cluster} and {factoextra} for Grouping Oceanographic Data. *Int. J. Inform. Appl. Math.* **2019**, *2*, 1–26.
66. Maechler, M.; Rousseeuw, P.; Struyf, A.; Hubert, M.; Hornik, K. Cluster: Cluster Analysis Basics and Extensions. R Package Version 2.1.0. Available online: https://cran.r-project.org/web/packages/cluster/ (accessed on 20 February 2020).
67. Adler, D.; Murdoch, D. Rgl: 3D Visualization Using OpenGL. Available online: https://www.researchgate.net/publication/318392813_Rgl_3D_Visualization_Using_OpenGL (accessed on 1 February 2020).
68. Ligges, U.; Mächler, M. Scatterplot3d—An R package for visualizing multivariate data. *J. Stat. Softw.* **2003**, *8*. [CrossRef]
69. Babtiste, A. GridExtra: Miscellaneous Functions for "Grid" Graphics. R Package Version 2.3.2017. Available online: https://CRAN.R-project.org/package=gridExtra (accessed on 20 February 2020).
70. Wickham, H.; François, R.; Henry, L.; Müller, K. Dplyr: A Grammar of Data Manipulation. R Package Version 0.7.6. Available online: https://CRAN.R-project.org/package=dplyr (accessed on 20 February 2020).
71. Tan, P.N.; Steinbach, M.; Karpatne, A.; Kumar, V. *Introduction to Data Mining*, 2nd ed.; Pearson: Harlow, UK, 2019.
72. Hennig, C.; Meila, M.; Murtagh, F.; Rocci, R. *Handbook of Cluster Analysis*; CRC Press: New York, NY, USA, 2016.
73. Halkidi, M.; Batistakis, Y.; Vazirgiannis, M. Cluster validity methods: Part I. *ACM Sigmod Rec.* **2002**, *31*, 40–45. [CrossRef]

74. Halkidi, M.; Batistakis, Y.; Vazirgiannis, M. Cluster validity methods: Part II. *ACM Sigmod Rec.* **2002**, *31*, 19–27. [CrossRef]
75. Rahman, S.M.T.; Kabir, A. Potential cluster regions for manufacturing small and medium enterprises in khulna city of bangladesh: A spatial examination. *Int. J. Recent Technol. Eng.* **2019**, *8*, 980–986.
76. Rodríguez, A.J.G.; Barón, N.J.; Martínez, J.M.G. Validity of dynamic capabilities in the operation based on new sustainability narratives on nature tourism SMEs and clusters. *Sustainability* **2020**, *12*, 1004. [CrossRef]
77. de Oliveira, L.S.; Soares Echeveste, M.E.; Cortimiglia, M.N.; Gularte, A.C. Open innovation in regional innovation systems: Assessment of critical success factors for implementation in SMEs. *J. Knowl. Econ.* **2019**, *10*, 1597–1619. [CrossRef]
78. Nestle, V.; Täube, F.A.; Heidenreich, S.; Bogers, M. Establishing open innovation culture in cluster initiatives: The role of trust and information asymmetry. *Technol. Forecast. Soc. Chang.* **2019**, *146*, 563–572. [CrossRef]
79. Bahena-Álvarez, I.L.; Cordón-Pozo, E.; Delgado-Cruz, A. Social entrepreneurship in the conduct of responsible innovation: Analysis cluster in Mexican SMEs. *Sustainability* **2019**, *11*, 3714. [CrossRef]
80. Pelletier, C.; Cloutier, L.M. Conceptualising digital transformation in SMEs: An ecosystemic perspective. *J. Small Bus. Enterp. Dev.* **2019**, *26*, 855–876. [CrossRef]
81. Jami Pour, M.; Asarian, M. Strategic orientations, knowledge management (KM) and business performance: An exploratory study in SMEs using clustering analysis. *Kybernetes* **2019**, *48*, 1942–1964. [CrossRef]
82. Cicea, C.; Popa, I.; Marinescu, C.; Ștefan, S.C. Determinants of SMEs' performance: Evidence from European countries. *Econ. Res. Ekon. Istraz.* **2019**, *32*, 1602–1620. [CrossRef]
83. Dubravska, M.; Sira, E. The analysis of the factors influencing the international trade of the slovak republic. In Proceedings of the 2nd Global Conference on Business, Economics, Management and Tourism, Prague, Czech Republic, 30–31 October 2014; Iacob, A.I., Ed.; Elsevier: Amsterdam, The Netherlands, 2015; Volume 23, pp. 1210–1216.
84. Woods, J.; Galbraith, B.; Hewitt-Dundas, N. Network centrality and open innovation: A social network analysis of an SME manufacturing cluster. *IEEE Trans. Eng. Manag.* **2019**, 1–14. [CrossRef]
85. Viloria, A.; Lezama, O.B.P. Improvements for determining the number of clusters in k-means for innovation databases in SMEs. *Procedia Comput. Sci.* **2019**, *151*, 1201–1206. [CrossRef]
86. Kiefer, C.P.; Carrillo-Hermosilla, J.; Del Río, P. Building a taxonomy of eco-innovation types in firms. A quantitative perspective. *Resour. Conserv. Recycl.* **2019**, *145*, 339–348. [CrossRef]
87. Dubrova, T.A.; Ermolina, A.A.; Esenin, M.A. Innovative activities of SMEs in Russia: Constraints and growth factors. *Int. J. Econ. Bus. Adm.* **2019**, *7*, 26–40.
88. Pessl, E.; Sorko, S.R.; Mayer, B. Roadmap industry 4.0—Implementation guideline for enterprises. *Int. J. Sci. Technol. Soc.* **2020**, *5*, 1728–1743. [CrossRef]
89. Fox, S. Moveable production systems for sustainable development and trade: Limitations, opportunities and barriers. *Sustainability* **2019**, *11*, 5154. [CrossRef]
90. Brozzi, R.; D'Amico, R.D.; Pasetti Monizza, G.; Marcher, C.; Riedl, M.; Matt, D. *Design of Self-Assessment Tools to Measure Industry 4.0 Readiness. A Methodological Approach for Craftsmanship SMEs*; Springer: New York, NY, USA, 2018; Volume 540, ISBN 9783030016135, ISSN 18684238.
91. Statista Worldwide Semiannual Small and Medium Business Spending Guide. Available online: https://www.statista.com/statistics/760799/worldwide-small-medium-business-it-spending/ (accessed on 15 December 2019).
92. IDC Worldwide Semiannual Small and Medium Business Spending Guide. Available online: https://www.statista.com/statistics/800684/worldwide-small-medium-business-it-spending-growth/ (accessed on 2 January 2020).
93. Statista In-depth: Industry 4.0. Available online: https://www.statista.com/study/66974/in-depth-industry-40/ (accessed on 12 December 2019).
94. Computer Economics Robotic Process Automation Adoption Trends and Customer Experience. Available online: https://www.statista.com/statistics/1017027/worldwide-robotic-process-automation-adoption-investment-rates-organization-size/ (accessed on 19 December 2019).
95. Nickel, P.; Bärenz, P.; Radandt, S.; Wichtl, M.; Kaufmann, U.; Monica, L.; Bischoff, H.-J.; Nellutla, M. Human-system interaction design requirements to improve machinery and systems safety. *Adv. Intell. Syst. Comput.* **2020**, *969*, 3–13.

96. Chumnumporn, K.; Jeenanunta, C.; Komolavanij, S.; Saenluang, N.; Onsri, K.; Fairat, K.; Itthidechakhachon, K. The Impact of IT knowledge capability and big data and analytics on firm's industry 4.0 capability. *Proceedings* **2020**, *39*, 22. [CrossRef]
97. Erasmus, J.; Grefen, P.; Vanderfeesten, I.; Traganos, K. Smart hybrid manufacturing control using cloud computing and the internet-of-things. *Machines* **2018**, *6*, 62. [CrossRef]
98. Choy, M.; Park, G. Sustaining innovative success: A case study on consumer-centric innovation in the ICT industry. *Sustainability* **2016**, *8*, 986. [CrossRef]
99. Lara, E.; Aguilar, L.; Sanchez, M.A.; García, J.A. Lightweight authentication protocol for M2M communications of resource-constrained devices in industrial internet of things. *Sensors* **2020**, *20*, 501. [CrossRef] [PubMed]
100. Pallas, G.; Peijnenburg, W.J.G.M.; Guinée, J.B.; Heijungs, R.; Vijver, M.G. Green and clean: Reviewing the justification of claims for nanomaterials from a sustainability point of view. *Sustainability* **2018**, *10*, 689. [CrossRef]
101. Gaudin, V. The growing interest in development of innovative optical aptasensors for the detection of antimicrobial residues in food products. *Biosensors* **2020**, *10*, 21. [CrossRef] [PubMed]
102. Kliestik, T.; Misankova, M.; Valaskova, K.; Svabova, L. Bankruptcy prevention: New effort to reflect on legal and social changes. *Sci. Eng. Ethics* **2018**, *24*, 791–803. [CrossRef] [PubMed]
103. Vrchota, J.; Volek, T.; Novotná, M. Factors Introducing Industry 4.0 to SMES. *Soc. Sci.* **2019**, *8*, 130. [CrossRef]

Article

Readiness of Enterprises in Czech Republic to Implement Industry 4.0: Index of Industry 4.0

Jaroslav Vrchota * and Martin Pech

Department of Management, Faculty of Economics, University of South Bohemia in Ceske Budejovice, Studentska 13, 370 05 Ceske Budejovice, Czech Republic; mpechac@ef.jcu.cz

* Correspondence: vrchota@ef.jcu.cz

Received: 14 November 2019; Accepted: 6 December 2019; Published: 10 December 2019

Abstract: Industry 4.0 includes digital process transformation, information technology (IT) development, mobile devices, learning software, automation, and robotics, as well as intelligent sensors to collect large datasets, store, analyze, and use them in business, including simulation, virtual reality, and digital twins. The aim of the paper is to characterize the readiness of the enterprise to use Industry 4.0. In the research, a questionnaire survey was carried out on a sample of 276 enterprises mainly from the manufacturing industry. Using explorative factor analysis, the index of Industry 4.0 (VPi4) was designed to determine the level of Industry 4.0 implementation in the enterprises. The results were further verified by a statistical analysis, using Mann–Whitney test and correlation coefficients. The results indicate that the VPi4 index was consistent in terms of distribution when comparing the results on the verification sample. Its results correlate with the subjective perception of the enterprises, and different levels of the index reflect the difference in technological intensity of the industry. The VPi4 index enables the enterprises to determine their own level of current state of readiness for Industry 4.0, to better prioritize business development. The proposed solution categorizes Industry 4.0 components into a useful theoretical framework. Further research offers the possibility of applying the index in other sectors, its relation to the size of enterprises, and updating with respect to new trends in information technology.

Keywords: Industry 4.0; index; smart; intensity of technology; manufacturing; implementation

---

1. Introduction

Industry 4.0 is a platform combining a variety of advanced modern technologies to meet today's challenges. Industry 4.0 elements are increasingly emerging as one of the main strategic management goals in recent years. The use of new technologies raises the need for long-term and strategic investments, intended to increase the competitiveness of the enterprises in the future. Most of the enterprises are already implementing smart technologies and smart processes. Some of the enterprises indicate their experience with such technologies practically. They use the new technologies at least partially. It means that they successfully completed the implementation, and they are now looking for the added value that these technologies offer. However, their use is still partial, used to deal with certain issues and probably without the overall interdependence of all the systems.

Finally, outside this area, there is a small group of innovators, looking for new developments and looking for ways to integrate them into their well-functioning organizations. They have in common mastering the basic and higher levels of Industry 4.0 brought by the Fourth Industrial Revolution, now preparing for further global changes brought about by advances in artificial intelligence, digitalization, computer science, robotics, complexity, and network theory. They include the enterprises that are at the heart of these changes, drawing their energy and position from these processes.

The current challenges of Industry 4.0 force managers to discuss whether they are prepared for such changes. They often wonder what the situation is in their own organizations and then which technology they should invest into in the near future.

Lower absorption capacity of the enterprises is related to the issue of resource constraints. Right now, the managements of a number of enterprises consider the future strategy and the steps to be taken to be competitive in the future. For this reason, this paper discusses how to assess and analyze the current state of business in the context of Industry 4.0. Based on the research, the authors suggest a methodology, with an index for easy evaluation of the preparedness of the enterprise for the future.

The structure of this paper is organized as follows: Section 2 defines the theory as used in the paper; Section 3 describes the methodology and methods used; Section 4 summarizes the most important results, including the evaluation of the hypotheses; Section 5 discusses the results with other authors; finally, Section 6 summarizes the most important conclusions of the research.

2. Theoretical Background

The cornerstone of Industry 4.0 is based on machines, equipment, logistics, and humans who are connected to each other to exchange data, process data, and make decisions, appropriately coordinating the ever-present machines [1]. Industry 4.0 is characterized primarily by digitization, robotics, and artificial intelligence. Kelkar [2] emphasized that 79% of the manufacturing enterprises (any size) perceive Industry 4.0 as very important for their development (research was conducted in 227 enterprises). Similarly, in Computer Science Corp (2015), 63% of United States (US) manufacturing companies (900 in the sample) identified Industry 4.0 as necessary for their further development. Consistent with these findings, there is a study of 235 German enterprises carried out by PricewaterhouseCoopers [3], reporting that the enterprises plan to increase digitization between 24% and 86% in the next five years. Dörfler [4], who stated that 94% of companies perceive digitization as important for their development, reported the highest percentage. This research was carried out regardless of the size and area of business using a sample of 1849 German enterprises [5].

The subsequent sub-sections describe theoretic background of the main technologies and processes which are necessary to create future intelligent factories and enterprises based on the conception of Industry 4.0.

2.1. Use of Sensors

Sensors are sources of information for the control system (computer, brain) and technical devices, which measure certain physical and technical quantities and convert them into a signal that is remotely transmitted and further processed. These are various global positioning system (GPS) sensors, cameras, and microphones, forming the digital nervous system. These devices acquire information on position, distance, motion, speed, displacement, temperature, drought, humidity, sound, vibration, gases, chemicals, flow, strength, load, pressure, level, electricity, acceleration, tilt, light, etc. The use of sensors in smart factories has many facets, as discussed by many authors [6–8]. To make full use of the sensors, the availability of efficient and affordable sensor networks (such as radio frequency identification, RFID) is a prerequisite [9]. Based on this, intelligent objects and devices are created, enabling real-time communication between computers, work resources, and application systems. Together, this technological development provides the basis for the introduction of new production processes and business models in smart factories [10]. As they are able to retrieve and process data, they can check certain tasks and communicate with people through an interface [6]. The importance of sensors for Industry 4.0 is also illustrated by the fact that, as mentioned by Reference [7], by 2020, nearly 20.8 billion devices will be connected and RFID will be fully utilized. Such a shift will have an impact on most industrial sectors and, in particular, manufacturing industries. RFID technology is used to identify various objects in warehouses, production halls, logistics companies, distribution centers, retail outlets, and disposal/recycling stages [11]. Analysis of monitored activities will be used for fault detection and predictive maintenance [12]. Based on the information gathered and also using machine-to-machine (M2M) communication, the resources will be remotely controlled to improve industrial processes [13].

2.2. Data Collection and Analysis

The process of data collection process includes retrieving, searching, selecting, and generating. As more and more business activities are digitized, new data sources arise, and the equipment to process these data becomes increasingly cheaper; we are entering a new era [14]. The volume of stored data is growing four times faster than the world economy, and computing power is increasing nine times faster. Interestingly, in 2000, only one-quarter of the world's information volume was stored in digital form; today, the figure is close to 95% [15]. The creation of large volumes of data is supported by digitization, aiming to convert all possible information and media—text, sounds, photos, videos, and instrument and sensor data—to the natural language of computers. Big data is a versatile term for any collection of datasets that are very large and complex. Big data are quantified in petabytes (10^{15}), as it is not possible to receive, store, secure, process, and visualize them with common hardware and software in a reasonable time [15].

Big data are defined as a cultural, technological, and scientific phenomenon that rests on interplay [16] of technology and analysis. Laney [17,18] defined big data through three characteristics: volume, variety, and velocity. The Gartner company introduced the term big data. Gantz and Reinsel [19] complemented the big data characteristics with a fourth Vs value. Reference [20] classified big data through a data life cycle that includes data, process, and management activities. Ge, Bangui, and Buhn [21] classified big data in eight areas: healthcare, automation, transport, energy, smart cities, agriculture, industry, and military.

Data analysis can be expressed through different terms such as data mining, clustering, classification, analytics, aggregation, annotation, combining, extraction, evaluation, and filtering. Data analysis is performed either directly through a variety of cloud computing services (PaaS — platform as a Service; SaaS — Software as a Service), or in a conventional way on the user's end computers [19]. The main advantage of cloud services is effective integration with other applications, scalability, performance, multitasking, and configurability [22]. According to Tsai et al. [23], data analysis methods have the following limitations for big data usage: centrality and unscalability, dynamics (inability to analyze data on the fly), and data structure format (data inconsistency).

2.3. Information Technology (IT) Infrastructure and Mobile Terminals

In particular, Industry 4.0 includes a radical shift in how IT infrastructure works, defined as the overall transformation of the manufacturing industry through the introduction of digitization and the internet.

These transformations mean a revolutionary improvement in the design and manufacturing processes, operations, and services of manufacturing products and systems. Tjahjono [24] defines Industry 4.0 requirements for IT infrastructure in terms of device automation, auto-driving, increased need for reality, an extremely large number of monitored and managed devices, and process automation.

The enterprises using the Internet of things (IoT) cannot rely solely on wireless networks such as WiFi, ZigBee, and low-power wide-area network (LPWAN) for their future critical related systems [25]. They demand more and more functionality now unavailable according to Rao and Prasad [26], particularly including very low latency, very high reliability, and very high bandwidth and bit rate.

Many of the current network technologies (2G, 3G, 4G — 2nd, 3rd, 4th Generation of mobile telecommunications technology, NFC — Near Field Communication, ANT — Adaptive Network Topology, Bluetooth, GSM — Global System for Mobile communication, WMAX — Woldwide Interoperability for Microwave Access, etc.) are not really good for the future. Future flexibility is offered by the 5G (5th generation of mobile telecommunications technology) standardization as recommended by Sriganesh [26] for infrastructure. Future infrastructure will lead to vertical and horizontal network connectivity and the use of the industrial internet concept. For example, General

Electric's practice Leber [27] described it as connecting industrial sensors and drives to local processing and the internet. Furthermore, links with other important industrial networks can independently create value. The main difference between consumer/social internet and industrial internet lies in the value created. For consumer/social websites, most of the value is created from ads [28].

2.4. Cloud Storage

Data storage includes recording, transportation, replication, compression, cleaning, indexing, stream processing, integration, and transformation of data. Given the increase in data volume (big data), the main question is how to store all data and where. Data warehouses and centers are the most often used. A data warehouse (DW) is an integrated collection of subject-oriented decision support data [29]. Data warehousing (DW) is a specific type of information system and should enable the acquisition of business data, its transformation into appropriate strategic business information, and subsequent storage of data in a format that facilitates business analysis [30]. The cloud is currently the hardware and software solution of the data center providing the services [31]. The cloud is a parallel and distributed computing system consisting of a collection of interconnected and virtualized computers that are dynamically delivered and presented as one or more unified computing resources based on service level agreements negotiated between a service provider and a customer [32]. Such resources are dynamically transformed to adapt to variable load, enabling the optimal use of resources [33]. Clouds are hardware services offering computing, networking, and storage capacity [34]. Mostly, clouds are operated on a cloud deployment models basis [35]: public cloud, private cloud, hybrid cloud, and community cloud [36–38]. The cloud inherently includes the concept of cloud computing, based on the provision of services or programs stored on servers and the internet. Cloud computing distinguishes three types of distribution models [22]: IaaS (infrastructure as a service), PaaS (platform as a service), and SaaS (software as a service). For the purposes of data storage, there are IaaS services, i.e., the use of virtualization, providing only infrastructure and hardware. Block chain technology [39] is emerging significantly in the architecture of the internet and is pushing out the standard model of client–server architectures. The point is that individual transactions between different entities are transparent to everyone, but no one other than the two entities that took the action can influence and change this record. Block chains remove the third-party distribution of information flow [40]. Block chain is basically a data structure that is used to create a digital transaction ledger that is not stored by a single provider but is shared across a distributed network of computers. Block chain is, therefore, a special type of distributed decentralized database in which records are stored.

2.5. Information Systems and Learning Software

The implementation of Industry 4.0 uses the concept of an automation pyramid in connection with information technology. The pyramid is closely related to the vertical integration of information systems in an enterprise. Typical solutions and technologies in this vertical integration include data acquisition sensors: programmable logic controllers (PLC) that control production processes and take control levels, supervisory control and data acquisition (SCADA), which allows managing different levels of support processes and supervision, manufacturing execution systems (MES) controlling production processes, and intelligent enterprise resource planning (ERP) management for the enterprise level, the highest level in this hierarchical image [41–43]. ERP supports enterprise-wide planning such as business planning, supply chain management, sales and distribution, accounting, human resources management, and the like. These are usually commercially available solutions.

German SAP SE (Systems, Aplications & Product in Data Processing Service) is currently the leading SAP solution. In traditional ERP tools, the decision-making process is centralized at the highest level; most available ERP solutions do not support rapid adaptation in production planning due to unplanned events. MES supports reporting, scheduling, dispatching, product tracking, maintenance, performance analysis, workforce tracking, resource allocation, and more [44]. Most systems work with their own way of storing data and, often, with their own data format. The use of a production system

(MES) as a central database is a solution to such an issue [45]. In the future, a decentralized IT solution in smart factories might be used. In order to give the employees the right data in time, a support system is needed [46]. As Haddara [47] noted, it is the right time to check the readiness of ERP systems to meet the demands of the factories of the future. SAP developed its predictive maintenance module, based on firm integration of robots, machines (to be maintained), and ERP. Predictive maintenance is based both on the integration of data from ERP data sensors and the predictive algorithms.

Learning software includes pattern and machine learning (ML), which embodies some of the aspects of the human mind that allow us to deal with an extremely complex solution to the problem with the speed overcoming even the fastest computers [48]. Wen et al. [49] reported the most commonly used techniques: case-based reasoning (CBR) [50], artificial neural networks (ANN) decision trees (DT) [51], Bayesian networks (BN) [52], support vector regression (SVR) [53], genetic algorithms (GA) [54], genetic programming (GP) [55,56], association rules (AR) [57], rule induction (RI) [58], and fuzzy algorithms [59].

2.6. Robots

Production process automation began in the 1960s with the introduction of industrial robots into the automotive manufacturing process. The automation of production systems by the introduction of industrial robots is an ongoing process and is now in line with the evolution of information technology [7]. Industrial robots, ranked in Industry 4.0, are divided into the following two types [60]:

- The machines that help and facilitate the handling of physical objects by reducing human effort without deducting sensitivity and accuracy;
- The machines that learn from their errors and are, therefore, able to gradually function better and manage their own development.

The area of collaborative robots was extensively explored, but it is necessary to define precisely what type of robot can be specified as cooperative. Even with many products available [61] and after the completion of many research projects [62], the definition of a collaborative robot remains unclear. Based on SICK AG (sensor intelligence), there are three types of human–robot interaction [63]: coexistence, cooperation, and collaboration. Robots play an important role in the modern manufacturing industry. Since 2004, the number of multipurpose industrial robots developed by enterprises in the 4.0 sector in Europe almost doubled [15]. The number of installed industrial robots is calculated per 10,000 employees in the manufacturing industry. The highest robot densities in 2017 according to the International Federation of Robotics [62] were found in the Republic of Korea (710), Singapore (658), and Germany (322). The world average was 85 robots per 10,000 employees; however, during the period 2013–2017, global sales of industrial robots increased by 114%. The use of robots is expanding to include a variety of functions: production, logistics, office management (for document distribution), maintenance, and repair of manufacturing defects [64]. An autonomous robot is a robotic device that works independently (it is not controlled in real time by a human, but by a program). In the future, they will be based on artificial intelligence and they will be capable of learning [65].

2.7. M2M Communication

Digital production includes a wide range of applied sciences. Studies in these fields attract a lot of effort both in academia and in industry, especially in connection with machine connectivity and communication (M2M), vitally important for machine collaboration and process optimization [66]. Computer-to-computer communication brings much greater efficiency and extraordinary security in production units, from factory halls to agriculture. Literally, machine-to-machine is synonymous with technology that communicates without human intervention. M2M communications change some processes by giving more data to the enterprises, and they will require companies to train employees for these purposes. In addition, the integration of M2M elements will require better integration capabilities and the creation of reliable complex networks with a higher level of security [67]. Rao [26] and his team described a farm of "no farmers" where cows can be detected by the feeding machines through sensors and M2M communication, and where the digital sensor capsules inside the cow send reports that the cow is fertile. Worldwide, the automotive, energy, transportation, logistics, consumer electronics, and ultimately retail industries are becoming the new view of new M2M applications [66]. M2M communication offers autonomous communication between intelligent encoders and drives and delivers greater value in the transport sector [17]. M2M communication systems implement automated data communication between machine-type communication (MTC) devices, creating a basic communication infrastructure for IoT and 5G networks [68,69]. M2M communication will be provided both between physical objects and between their cloud-based digital counterparts [70]. Depending on the location of the distant objects relative to the network, cloud computer technology is referred to [71]. In the future, cloud robotics will be used with real-time connectivity. A higher level of M2M communication is related to the Internet of things (IoT), which is a designation for a much more intelligent interconnection of various products, devices, etc. [72]. The key elements are miniature sensors, representing an almost ubiquitous image recognition technology capable of recognizing people, buildings, and other objects [73].

2.8. Sharing and Using Data with Suppliers and Customers

Enterprises face a precarious environment and strive to achieve greater cooperation in the supply chain to leverage the resources and knowledge of their suppliers and customers [74]. In such a chain, this cooperation takes place through electronic data interchange (EDI) [53]. Using and evaluating multidimensional process knowledge is considered an effective strategy to improve the competitiveness of the enterprises [75]. Sharing forecasting information helps supply chain parties better match demand and supply [76]. The information is used to update variations in seasonal product demand [77]. Information needs to be shared to achieve an efficient supply chain [78]. Optimum supply chain performance requires manufacturers to truly inform other partners of their original forecast [79]. By Croson and Donohue [80], it is useful for the enterprises to share sales data (POS — Point of Sale materials), especially to reduce the bullwhip effect. Christopher, in connection with data sharing and supplier and customer integration, discussed "demand chain management", linking supply chain management with marketing, bringing agile and lean properties to chains [81]. Demand chain is defined by (1) managing integration between demand and supply processes, (2) managing the structure between integrated processes and customer segments, and (3) managing the working relationship between the marketing and supply chain [82]. In addition, the enterprises are able to share product life-cycle information and focus on product design [83]. In practice, it is the co-design, visualization and production analysis, and joint research and design [84]. This creates a variety of systems for exchanging and sharing product information between users and platforms [85,86].

2.9. Use of Virtual Reality, Simulation, and Digital Twins

Simulation is defined as an imitation of a real thing, a state, or a process. Generally, it implies displaying or modeling some key features and behavior of some physical or abstract systems for testing, optimization, and education. Product and process simulations are used extensively in production, especially processes of visualization, representation, simulation, modeling, and interpretation. Enriching digital simulations with sensor data brings reality closer and improves the accuracy of simulation results [87]. Virtual reality (VR) is broadly defined as a computer-generated three-dimensional (3D) world [88], and an environment that simulates complex situations and contexts in real life and allows people to immerse, navigate, and communicate [89]. A key feature of virtual reality is real-time interactivity. VR systems generally track the movement of hand-held objects and the user's head and limbs, and the received data are used to determine the user's view, navigation, interaction with objects, and possible movement of the virtual body, known as an "avatar" [90]. Virtual reality by Steuer [91] is technological hardware that includes a computer, an imaging helmet, headphones, and motion-sensory gloves. The main areas of VR application include healthcare [92]. The concept of augmented reality must be distinguished from the concept of virtual reality. Augmented reality (AR) is a special application providing its users with a direct or indirect view of the real world, whose parts are complemented, expanded, and enriched with additional digital visual elements [93,94]. Examples include end-to-end applications, viewing glasses, and projection of information in a car directly onto the windshield. The use of simulation to control and optimize products and manufacturing systems in real time is a concept known as the digital twin [95], which is considered as another step in modeling, simulation, and optimization of technologies [96]. Digital twins are defined as a digital replication of both living and inanimate entities that enable seamless data transfer between the physical and virtual worlds [97]. Digital twins are a mirror image of a real-time physical process [98]. The concept of using "twins" comes from the Apollo NASA (National Aeronautics and Space Administration) program; later, it was used also in aviation, such as the "Iron Bird" [96]. Digital twin devices offer a platform for the development, testing, improvement, and upscale of the manufacturing environment [99]. Digital twin technology is considered a key technology for the realization of cyber physical systems [100]. The application of simulation techniques brings digital twins to life and makes them experimentable; the digital twins become known as experimentable digital twins (EDTs). Initially, these EDTs communicate with each other purely in the virtual world. In this way, complete digital representations of the respective real assets and their behavior are created. Real-estate EDT networking leads to hybrid application scenarios in which EDT is used in combination with real hardware, delivering complex control algorithms, innovative user interfaces, and smart models for smart systems [101].

3. Materials and Methods

The main aim of the paper was to analyze the readiness of enterprises to implement Industry 4.0 in the period 2018–2019. The first partial aim of the paper was to compile an index of evaluation of the level of Industry 4.0 in enterprises based on the results of a survey. The second partial aim was the statistical verification of the consistency of the index with further results from the questionnaire survey.

The preparation of the research sample firstly included the identification of the number of enterprises used for the questionnaire survey. Based on CSU (Czech Statistical Office) data, it was found that, in the Czech Republic, there were 175,894 enterprises in the manufacturing industry in 2017, of which 7.1% were small, medium, and large enterprises, i.e., a total of 12,470 subjects [102]. Approximately 2500 enterprises were approached to ensure that a 95% confidence level condition was met at a 5% margin of error and at a discarded 15% return on the questionnaires. The data were collected on the basis of interviews with business managers, firstly addressed electronically. Of the total number of respondents, 314 enterprises agreed to cooperate and participate in a questionnaire survey with a return rate of 12.5%. The authors of the paper and university students were present at the meetings with the enterprises and in the process of completing the questionnaires. The establishment of the research was approached in two stages (two research waves): first in February–March 2018 and then in January–May 2019.

As part of Industry 4.0 research, the research sample consisted of 276 enterprises reporting their data (38 out of 314 questionnaires were excluded based on these criteria: at least 10 employees, one year on the market, and completeness of survey). The amount of obtained data was further specified in terms of business characteristics, i.e., size and technology demands (Table 1). The first wave of the research was used to create the Industry 4.0 index (VPi4), whereas the second wave of the research was used to check and compare the results achieved. Characteristics of the research samples according to the research waves were as follows:

- First wave of research (year 2018)—164 enterprises (60% of data sample);
- Second wave of research (year 2019)—112 enterprises (40% of data sample).

Table 1. Research sample characteristics.

Group	Category of Group	1st Wave	2nd Wave
Size	Small enterprise (10–49 employees)	39.0%	36.6%
	Medium enterprise (50–249 employees)	28.7%	30.4%
	Large enterprise (over 250 employees)	32.3%	33.0%
Technological intensity	High-tech and medium high-tech intensity (HTI)	51.2%	58.0%
	Of which high-tech sector (HTS)	7.9%	8.9%
	and medium high-tech sector (MHTS)	43.3%	49.1%
	Low-tech and medium low-tech intensity (LTI)	48.8%	42.0%
	Of which low-tech sector (LTS)	12.2%	27.7%
	and medium low-tech sector (MLTS)	36.6%	14.3%

Table 1 describes the research sample in terms of the size of the enterprises and their technological intensity.

Classification of the enterprises by size was based on the number of employees of the enterprise, as defined by the methodology of the European Commission [103]. Table 1 shows that, in the first wave sample, there were 39.0% small enterprises, 28.7% medium-sized enterprises, and 32.3% large enterprises as the most common. The composition of enterprises in the second wave of research was very similar.

Table 1 shows the distribution of enterprises in terms of their technological intensity, with the enterprises with higher technological intensity (HTI) and the enterprises with lower technological intensity (LTI) according to the methodology of the Czech Statistical Office [102]. In the Czech Republic and in our research in both waves, the groups were comparable. The only difference was the representation of the enterprises from the low-tech sector (LTS) and medium low-tech sector (MLTS) in the area with lower technological intensity in the first and second waves of research.

- HTI: Engineering and electro-technical production (CZ-NACE groups 24–30), chemical, paper, and non-metallic production (CZ-NACE groups 17–23).
- LTI: Production of products for domestic use (CZ-NACE Groups 13–16, 31–32), food production industry (CZ-NACE Groups 10–12).

The questionnaire focused on main groups of Industry 4.0 characteristics (observed phenomena). The items of the questionnaire were defined with the support of 34 managers and their expert evaluation within the framework of the qualitative research. The main part of the questionnaire consisted of 17 variables characterizing different technologies of Industry 4.0 used by the enterprises (data collection, cloud storage, data analysis, people capability, IT infrastructure, information systems, M2M, robots, mobile terminals, using sensors, learning software, sharing data, virtual reality, additive manufacturing, i.e., 3D print, nanotechnology, drones, and autonomous vehicles). The areas are described in detail in Section 2. In addition, four identification characteristics were measured for the enterprises, i.e., size according to the number of employees, field of activity, technological intensity, and type of owner. The questionnaire also included questions about whether the enterprises had a formulated strategy, whether they planned on investing in technology, and a subjective assessment of the level of Industry 4.0 in their organization.

3.1. Exploratory Factor Analysis

The factor analysis was chosen to classify the most important variables affecting the level of enterprise readiness for Industry 4.0 into groups. The central aim of factor analysis is the orderly simplification [104] of several interrelated measures using mathematical procedures. The goal of the analysis is to reduce the number of variables through fewer common factors and to reveal the structure of relationships between the variables. Factor analysis in the broad sense comprises both a number of statistical models and a number of simplifying procedures for the approximate description of data [105]. The basis of factor analysis is the assumption that the observed covariance (relationships, i.e., correlations) between the variables is the result of the action of common factors and not the interrelationship between the variables. Gorsuch [106] pointed out that the aim of factor analysis is to summarize the interrelationships among the variables in a concise but accurate manner as an aid in conceptualization. Each factor represents an area of generalization that is qualitatively distinct from that represented by any other factor. A measure of the degree of generalizability found between each variable and each factor is calculated and referred to as a factor loading.

We used exploratory factor analysis (EFA) to explore the main dimensions and generate a new index of Industry 4.0. The scales of the items used in factor analysis were assessed on a scale of 1–4, using the same range as Veza [107] in the survey, to evaluate the Industry 4.0 maturity level of Croatian enterprises. This scale achieved better pilot research results than 1–5 used by Frank [108] or Schumacher [109] to determine the implementation of different technologies in manufacturing companies.

The factor analysis helped in particular to determine the internal structure of covariance of variable indexes and to differentiate different groups of the factors. The suitability of data structure for factor analysis was analyzed by Bartlett's test of sphericity [110] and the Kaiser–Meyer–Olkin (KMO) test [111]. Bartlett's test checked that the observed correlation matrix diverged significantly from the identity matrix at $\alpha = 0.05$ with a p-value of 3.021×10^{-15} ($\chi^2 = 96.243$, degrees of freedom (df) = 12). Subsequently, the Kaiser–Meyer–Olkin sample adequacy ratio was calculated, and the value was 0.8495. Such a value was deemed high (higher than 0.7), making factor analysis very appropriate [112]. Tabachnick and Fidell [113] recommended inspecting the correlation matrix for correlation coefficients over 0.30. Many correlation coefficients do not meet this requirement, but almost all of these coefficients were statistically significant at the level $\alpha = 0.05$.

3.2. Statistical Analysis

The results of the research were further processed using statistical analysis. The aim of this analysis was to compare the results with the Industry 4.0 VPi4 index.

Firstly, the VPi4 index distribution within the first wave of the research was compared with the index results in the second wave of the research. Due to the abnormality of the data, it was necessary to use the non-parametric Mann–Whitney–Wilcox test for the independent samples. In this case, we expected the samples to be similar. Working hypotheses, which formed the subject matter of verification at the 5% level of significance, were as follows:

- $H1_0$: The VPi4 indexes based on data from the first research wave and the second research wave are identical populations.
- $H1_A$: The VPi4 indexes based on data from the first research wave and the second research wave are different populations.

Furthermore, the dependence between the subjective perception of Industry 4.0 level and the VPi4 index was tested using Person and Spearman correlation coefficients. The index was expected to correlate to a certain extent with the subjective perception of the situation in the enterprise. Working hypotheses, which formed the subject matter of verification at the 5% level of significance, were as follows:

- $H2_0$: There is no dependency between the perception of Industry 4.0 in enterprises and the VPi4 index.
- $H2_A$: There is a dependency between the perception of Industry 4.0 in enterprises and the VPi4 index.

Furthermore, the hypotheses about the impact of technological intensity of the industry on the level of Industry 4.0 in the enterprises (expressed through the VPi4 index) were tested. The Mann–Whitney test was used for this purpose. It was assumed that the index would reach higher values in the enterprises with higher technological demands. For this purpose, the analysis was carried out separately for high-tech and low-tech enterprises, and the results were then compared. Working hypotheses, which formed the subject matter of verification at the 5% level of significance, were as follows:

- $H3_0$: There is no difference between the level of Industry 4.0 (VPi4) in high-tech and medium high-tech enterprises (HTI) and in low-tech and medium low-tech enterprises (LTI).
- $H3_A$: There is a difference between the level of Industry 4.0 (VPi4) in high-tech and medium high-tech enterprises (HTI) and in low-tech and medium low-tech enterprises (LTI).

Statistical evaluation of tests was performed using Statistica 12 and R software.

4. Results

The results are divided into three sub-sections: factor analysis, index of Industry 4.0, and verification and evaluation of Industry 4.0 index.

4.1. Results of Factor Analysis

Factor analysis was based on the variables the enterprises were asked about in relation to their implementation of Industry 4.0. Several variants of factor analysis were performed with various parameters with different items of the questionnaire.

Firstly, all 17 monitored items from the questionnaire were included in the exploratory factor analysis. The results of the principal component analysis method showed that four factors explained a total of 51% variance. However, the fourth factor contained only two items, of which drones had a negative factor loading of $f4 = -0.45$ and autonomous vehicles reported a factor loading of $f4 = +0.78$. Further rotation and testing did not improve the situation, and these items were, therefore, excluded from the analysis. The highest factor loadings for additive manufacturing $f2 = +0.37$ and nanotechnology $f2 = +0.32$ were very low (<0.4). Items which have a load less than 0.4 on any factor should be removed and the analysis should be re-run [114]. This means that these items did not saturate the factors sufficiently. In addition, they were not used to a great extent in the enterprises surveyed (usage of these variables in our results: nanotechnology 4.0%, drones 0.7%, additive manufacturing only 9.1%, and autonomous vehicles 2.9%). These items were, therefore, also eliminated from the factor analysis.

Finally, 13 variables were selected for the final design. As mentioned in the methodology, the suitability of the factor analysis was verified using the KMO index and the Bartlett test.

4.1.1. Factors Extraction

Factor extraction was performed using the principal component analysis method. This method is based on a large number of variables to find a smaller set of new variables (Table 2 with less redundancy to provide the best possible data representation [115]. The three factors found accounted for a total of 52.8% variance. The first factor explained 34.6% variance. The Eigen value of the second factor was 1.2, and the variance explained by this factor was 9.3%. The third factor then explained 8.9% of the variance (see Table 2). The remaining factors were always less than 5% of the total variance and their Eigen values were less than one. Based on the Kaiser–Guttman criterion, it was, therefore, appropriate to interpret only the first three factors, as they explained more variance than the original variables.

Table 2. Factor extraction using principal component analysis.

Value	Eigen Value	% Total Variance	Cumulative %
1	4.4920	34.5540	34.5540
2	1.2067	9.2826	43.8365
3	1.1689	8.9911	52.8277

4.1.2. Factor Loadings and Rotation

In factor extraction, factor loads were calculated for each item, representing the correlations between the factors and the variables. They could be used to interpret the factors. Thus, by processing the data, three rather consistent factors were extracted (without rotation). Since initial factor extraction usually does not provide interpretable results, it was done using the Varimax method. The primary factor load aggregate variables are marked in bold in Table 3. The values in Table 3 represent the factor loads of the rotating factors. The sign of factor load expresses the opposite relation to the given factor. In addition to the Varimax method, other methods were used, but it was shown that these results are best interpretable.

Table 3. Factor loadings. Primary factor load aggregate variables are marked in bold. IT—information technology; MES—manufacturing execution system; ERP—enterprise resource planning; M2M—machine-to-machine communication; 3D—three-dimensional.

Variable	Factor 1	Factor 2	Factor 3
We collect data	**0.8212**	0.2961	0.0451
Data storage in the cloud	**0.6260**	0.0023	0.3342
We analyze the data	**0.8603**	0.1839	0.0997
We have the right people	**0.6094**	−0.0134	0.1901
IT infrastructure	0.4481	**0.5251**	−0.0306
Information systems MES, ERP	0.1367	**0.7577**	−0.1363
Linked data (M2M)	0.3207	**0.5750**	0.1562
The use of robots	0.1658	**0.5449**	0.4303
Mobile terminal equipment	0.1383	**0.5448**	0.4186
Using sensors	0.3203	**0.5844**	0.3058
Using learning software	0.1950	0.3306	**0.4448**
Sharing data with suppliers	0.2245	−0.0315	**0.6696**
Use of virtual reality	0.0643	0.0797	**0.6842**
Autonomous vehicles		Eliminated	
Additive manufacturing (3D printing)		Eliminated	
Nanotechnology		Eliminated	
Drones		Eliminated	
Variance explained	2.7416	2.3541	1.7719
Total	0.2109	0.1811	0.1363

In terms of interpretation and for model purposes, the factors were identified as levels 1–3 of Industry 4.0 in the enterprises. It is clear from Table 3 that level 1 was primarily saturated with the human capital variable, collecting data, storing data in the cloud, and analyzing data. These variables have in common that they focus on working with data and the availability of human capital, i.e., the need to operate equipment and technology. Level 2, on the other hand, included all the variables related to the core infrastructure of industry 4.0. This means IT infrastructure, the presence of MES and ERP information systems, M2M-based data interconnection, the use of robots and their arms in production, mobile devices, and sensors. Level 3 included a higher level of Industry 4.0 that can be expressed through the use of learning software, data sharing with suppliers, and virtual reality. The items autonomous vehicles, additive manufacturing, nanotechnology, and drones were eliminated in the preliminary factor analysis (see Section 4.1) and not used for this run of factor analysis.

4.2. Index of Industry 4.0

The results of the factor analysis were further used to create an index for the implementation level of Industry 4.0 (VPi4) in the enterprise. Based on these data, it was possible to divide 13 areas into three levels of Industry 4.0 implementation into the enterprise, using factor analysis, where the numbers after each area represent their factor load.

The first level of introducing Industry 4.0 into an enterprise consists of the following areas:

- We have the right people (mechatronics, mounter, technologist)—0.61;
- We collect data—0.82;
- Data storage in the cloud—0.63;
- We analyze data—0.86.

The second level of introducing Industry 4.0 into an enterprise consists of the following areas:

- IT infrastructure (speed, stability)—0.53;
- MES, ERP—0.75;
- We use linked data (M2M)—0.58;
- Use of robots, robotic arms (in production and elsewhere)—0.54;
- Mobile terminals—0.54;
- Use of sensors—0.58.

The third level of introducing Industry 4.0 into an enterprise consists of the following areas:

- Use of learning software—0.44;
- Suppliers can use our data (response options, predictions)—0.67;
- Use of virtual reality (digital twins, simulation)—0.68.

Figure 1a below shows the data distribution in terms of VPi4 percentage; the intervals were created automatically for legibility. The most frequent interval was 39%–52% with a frequency of 37 enterprises, followed by an interval of 26%–39% with a frequency of 36 enterprises. The least represented interval was 78%–91%, where there were seven enterprises.

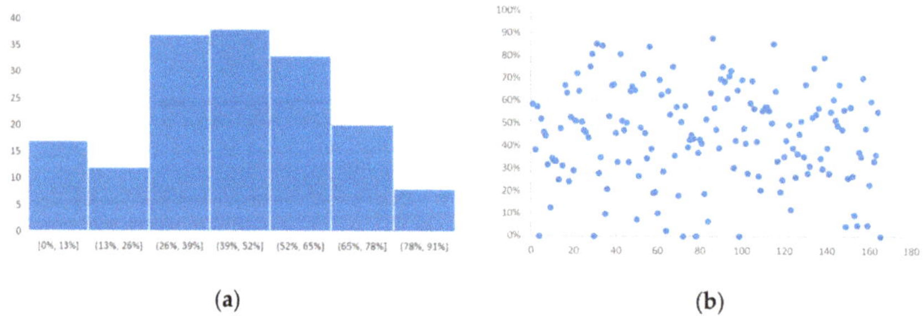

Figure 1. The enterprises by the index of Industry 4.0 (VPi4) percentage: (**a**) distribution of the enterprises by intervals; (**b**) total distribution.

Figure 1b shows the 164 enterprises evaluated under the first wave of the research (x-axis) with their percentage of Vpi4 (y-axis). As seen from the chart, most of the enterprises were between 29% and 60%. Conversely, in the lower quartile, there were 29% of enterprises, while there were 60% of enterprises in the upper quartile.

4.3. Verification and Evaluation of Industry 4.0 Index

The results of the second wave of the research and supplementary questions identifying the subjective perception of the enterprises and the impact of the technological intensity of the industry were also used to assess the results of the Vpi4 index.

On the basis of the results, a scorecard was designed, such that an enterprise is able to determine the level of implementation of Industry 4.0 inside the enterprise based on the answers to the questions. The enterprise finds out the overall score and the fulfilment of different levels of Industry 4.0. At the same time, it can also compare the result with other enterprises in the industry where a set of five icons shows the position compared to other enterprises. Each icon shows a 20% sample distribution (see Figure 2).

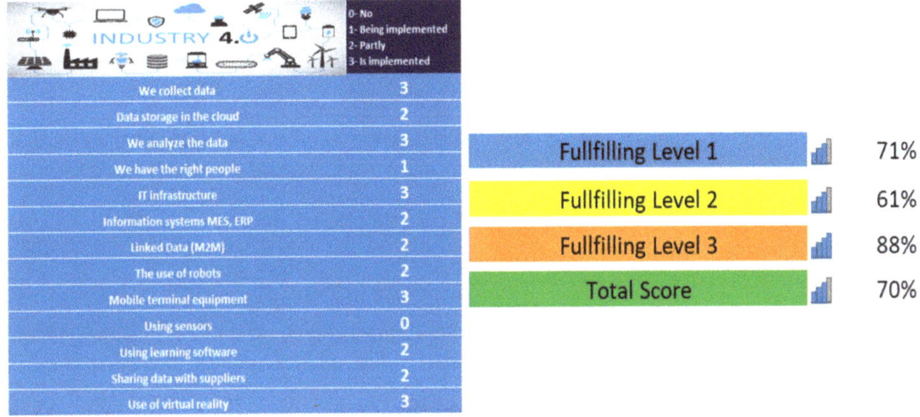

Figure 2. Vpi4 index.

4.3.1. Industry 4.0 Index Distribution

The data of the first and second wave of the research were used to evaluate the data distribution. Figure 3 below shows the enterprises at levels 1–3, which are color-coded (each enterprise is shown three times on the graph), with the y-axis showing the values of each level and the x-axis showing the total Vpi4 as a percentage.

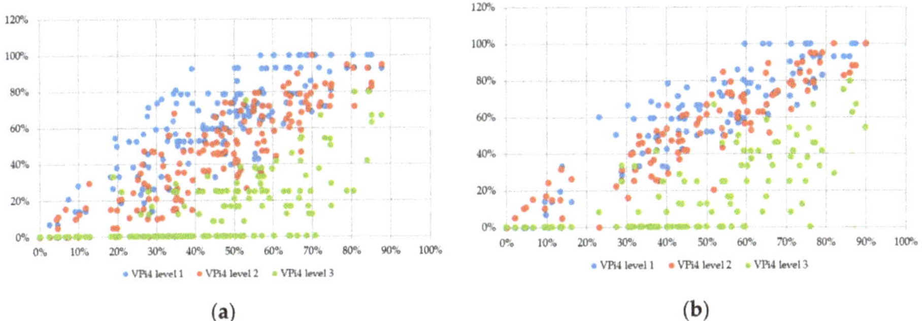

Figure 3. Evaluation of enterprises by VPi4: (**a**) distribution of the enterprises in the first wave of the research; (**b**) distribution of enterprises in the second wave of the research.

Figure 3a shows how the levels overlap; however, most of the enterprises reached the higher level 1, while the second and third levels featured a score pf zero. The distribution of the enterprises in the second wave of Figure 3b was similar. It is interesting to note, for example, that one enterprise achieved a very high overall Vpi4 at 88%, while, at the same time, it had a level 1 score of 93%, level 2 score of 95%, and level 3 score of 67% (three dots to the right of the graph). In total, five companies in the first wave achieved absolutely zero values in Vpi4.

Furthermore, the VPi4 index distribution was statistically compared, using the samples from the first wave and the second wave of the research. For this reason, Mann–Whitney–Wilcox test statistics were used to compare the samples. Table 4 shows that, at all levels of the VPi4 index except the third level, the results of the first and second wave research were identical. Differences were found only in the third level with a p-value = 0.0267. However, the third level of the index is very specific, as higher ranking at this level is often more difficult for enterprises to achieve after the first two levels are met. The enterprises in the second wave achieved a higher level of the VPi4 index at level 3 than in the first wave. The results also show that there was a difference in self-perception and self-assessment of the use of Industry 4.0 for the enterprises in the first and second waves.

Table 4. Industry 4.0 index (VPi4) distribution, using Mann–Whitney–Wilcox test.

Variable	Median w1	Median w2	W	p-Value
VPi4% total	44.39	49.65	8230.0	0.1431
VPi4% level 1	55.33	57.72	8898.0	0.6611
VPi4% level 2	47.48	53.49	8134.0	0.1070
VPi4% level 3	27.30	33.19	7741.0	0.0267
Enterprise perception	2.00	3.00	6771.5	0.0001

The results of the comparison of the index results in the first and second waves of the research show that hypothesis $H1_0$ cannot be rejected, as the results of both surveys showed the same distributions.

4.3.2. Relation of the Index to the Subjective Perception of the Level of Industry 4.0 by the Enterprises

The relation of the index to the subjective perception of the level of Industry 4.0 by the enterprises was carried out in both waves of the research. The correlation between VPi4 index (%) and the scale on which enterprises evaluated themselves in relation to Industry 4.0 from 1–5 was analyzed (1—we do not have Industry 4.0; 5—we fully have Industry 4.0). Pearson and Spearman coefficients were used for testing. Firstly, a coefficient of determination of R^2 = 0.2784 was calculated; thus, the dependence explained 28% of the variability of the number of points. On average, the enterprises rated themselves 2.1 with Industry 4.0, with an average rating of the recalculated VPi4 index being 45% more similar to the score of 3. One-quarter of enterprises had a VPi4 value below 29%, with the upper quarter having a value above 60%. In terms of their own perception, 50% of the enterprises ranged from 1–3 on a five-point scale. As also reported by the minima and maxima, some of the enterprises did not achieve any points in the VPi4 index. The maximum was 88% and 34 points in VPi4.

The normalization of the data of both variables was verified by the Shapiro–Wilk test, with the p-value of VPi4% = 0.09 assuming the normality of the data, as also shown by the histogram. With the perception of Industry 4.0 by the enterprises, the p-value test was close to zero; therefore, the normality of the data was not assumed. The results of the correlation of both variables are shown in Table 5. Here, on the basis of a p-value = 0.0000, the null hypothesis of independence was rejected in favor of the alternative using the Pearson coefficient at the significance level of 5%. We proved the existence of a linear dependence, which was also proven by the positive Pearson correlation coefficient (0.5277). At the same time, in terms of Spearman correlation, where the p-value was very close to zero with R = 0.5147, the null hypothesis was rejected in favor of $H2_A$ on the dependence of both variables.

Table 5. Relation to subjective perception of Industry 4.0, based on Pearson and Spearman coefficients.

		Pearson			Spearman		
Research	Variable	Perception	VPi4	p-Value	Perception	VPi4	p-Value
Wave 1	Perception	1.0000	0.5277	0.0000	1.0000	0.5147	0.0000
	Index VPi4	0.5277	1.0000		0.5147	1.0000	
Wave 2	Perception	1.0000	0.4129	0.0001	1.0000	0.4054	0.0001
	Index VPi4	0.4129	1.0000		0.4054	1.0000	

Similarly, a second questionnaire survey was used for the second wave of the research (Table 5). The coefficient of determination was lower than in the first wave of the research (R^2 = 0.1705). Dependence, therefore, explained only 17% of the variability. In the normality verification by the Shapiro–Wilcox test, the values were low to zero in both cases. Therefore, the normality of data, for the VPi4 index and the perception values of Industry 4.0 by the enterprises, was not considered. It was, therefore, better to compare the dependence of the Spearman coefficient. Its value was 0.4054, i.e., compared to the results in the first wave, the level of dependence was lower. Its value was, however, statistically significant.

Given the proven dependence in both surveys, it was possible to conclude the correct setting of VPi4 by means of factor analysis and the suitability of the questions, as it largely corresponded to the perception of the enterprises in terms of Industry 4.0.

4.3.3. Relation of the Index to Intensity of Technology and Index Weighting

The relation to the intensity of technology in the industry was tested in the first wave of the research only (Table 6). The Mann–Whitney test determined the null hypothesis at the sample significance level of $\alpha = 0.05$, where X = high-technology sector difficulty and Y = low-technology sector difficulty. The hypotheses were tested, providing $H3_0 = x0.50 - y0.50 = 0$ and $H3_A = x0.50 > y0.50$, as viewed from VPi4. As shown in the table below, the null hypothesis of both samples was rejected when the *p*-value was close to zero, and a positive *Z* confirmed the alternative hypothesis, claiming that the higher-tech enterprises have a higher level of Industry 4.0 (VPi4).

Table 6. Intensity of technology levels (HTI—high-tech intensity; LTI—low-tech intensity), based on Mann–Whitney test.

Variable	Sum of HTI	Sum of LTI	U	Z	*p*-Value
VPi4% total	7672.5	5857.5	2617.5	2.4410	0.0146
VPi4% level 1	7178.5	6351.5	3111.5	0.8159	0.4146
VPi4% level 1	7751.5	5778.5	2538.5	2.7009	0.0069
VPi4% level 2	7354.5	6175.5	2935.5	1.3949	0.1631
Enterprise perception	7295.5	6234.5	2994.5	1.2001	0.2298

Interestingly, it was not possible to reject this hypothesis for the perception of the enterprises from the perspective of Industry 4.0, with a *p*-value of 0.2298; therefore, this hypothesis could not be rejected, and we can further assume that the high- and low-tech enterprises saw themselves in a similar way. The hypothesis testing also failed to reject the null $H3_0$ hypothesis at levels 1 and 3 of Industry 4.0 implementation, as the *p*-values were greater than α. On the other hand, for the second phase of Industry 4.0 implementation, it was possible to prove the differences between the two groups. The enterprises with higher technological demands were often more successful.

The VPi4 index was adjusted for comparing enterprises with different intensities of technology. The Mann–Whitney test was used to compare more independent samples where the *p*-value did not indicate statistically significant sectoral differences from the entire sample (Table 7), except for LTS, where a *p*-value (0.0184) indicated a difference at the 0.05 level of significance. For this reason, it was necessary to adjust the index (index obtained by the median difference of 0.0899) for LTS companies, so that their results could be compared with the values of other groups. This fact was logical in terms of the lower use of technologies that were included in VPi4 for enterprises belonging to the low-technology sector. After adjusting the index for LTS enterprises, the Mann–Whitney test was re-conducted, where no significant difference between the whole sample and LTS was found (*p*-value = 0.0967).

Table 7. Intensity of technology sectors, based on Mann–Whitney test. M—medium; HT—high-tech; LT—low-tech; S—sector.

Variable	U	Z	p-Value
HTS vs. data sample	819.5	−1.3833	0.1665
MHTS vs. data sample	5325.5	−1.0364	0.2999
MLTS vs. data sample	1370.0	−1.1984	0.2307
LTS vs. data sample	3907.0	2.3572	0.0184

5. Discussion

In relation to the impact of Industry 4.0, this paper proved that 62% of the enterprises feel influenced by Industry 4.0. According to research (sample of 105 enterprises) of the Confederation of Industry of the Czech Republic [116], 65.7% of companies started implementing Industry 4.0 because it is important for their future. In comparison, the research of Sommer [5] reported that 82% of the enterprises in Germany feel ready for digitalization (a sample of 247 enterprises), and there were 68% of 28 enterprises mentioned by Schulze [117], questioned whether they used technology associated with Industry 4.0. Other influences include the degree of cooperation of SMEs (Small and Medium-sized Enterprises) with universities and research centers mentioned by Sastoque et al. [118].

Firstly, we discuss the structure of the VPi4 index levels. The initial level of index includes the basic requirements of Industry 4.0 such as well-qualified (the right) people [119], and processes of data collecting [120], storage in the cloud [121], and analysis of data [122]. These processes and variables are necessary for higher levels of Industry 4.0 and can be limits for the future development and introduction of Industry 4.0 implementation. The second level of the index consists of necessary infrastructure which is needed to operate with technologies. This level has more parts, such as using sensors for collecting data [121], IT infrastructure including MES [123], ERP information systems [123], linked data via M2M (or IoT), robots [44], and user-end technologies such as mobile terminals [121]. The last level is more advanced in terms of using learning software [122], virtual reality, and simulations including digital twins [97] or sharing data with other parties [124].

Furthermore, we discuss the findings and results of similar models of Industry 4.0. There is still no consensus on which model or index is most relevant to determine the level of Industry 4.0 introduction and implementation. Applications foreseen are not only in the high-technological industry but also in other sectors and branches.

Stefan et al. [125] emphasized considering the same meaning of three dimensions—technology, organization, and personnel—in assessing the level of Industry 4.0, as done by Block [126]. However, these dimensions were classified in more detail into three additional levels, assigning them four to seven characteristics. The characteristics, identical to those used in this paper, include data storage, IT infrastructure, and data evaluation. On the other hand, they emphasized data security, process methodology, and personnel development, in contrast to this paper. They set target value criteria for all these characteristics. In the proposed model, they defined criteria, relations, and dependencies between these dimensions to help the enterprises classify the current state of the implementation of Industry 4.0 and identify opportunities for improvement.

Scremin [127] also divided the Industry 4.0 enterprises into three main dimensions—strategy, readiness, and performance—identifying a number of additional areas (2–3) within each dimension, which they then subdivided into more detailed factors that influence the dimensions. Identical factors within VPi4 include IT infrastructure, data sharing, providing data to suppliers, data analysis, and employees.

Ślusarczyk [124] used secondary data at the level of the United States of America (USA), Germany, Japan, and Poland for his research and concluded that 80% of the enterprises perceive Industry 4.0 as very significant. This can be seen as a similar value to that published in this paper, as, out of 1018, 62% of SMEs responded in the same way. It is important to note that the research of Ślusarczyk [124] was based on data from large enterprises in the US, Germany, Japan, and Poland, which are technologically highly developed countries featuring large enterprises. In this paper, the importance of Industry 4.0 was reported as also increased according to the size of the enterprise, as the medium-sized enterprises reached 74%. It can, therefore, be assumed that large enterprises would reach 80%.

Shumacher, Erol and Sihn [109] conducted an assessment of the readiness to implement Industry 4.0, as well as the maturity of the enterprises in this respect. For this purpose, they created a model evaluating the enterprise in nine dimensions (strategy, leadership, customers, product, processes, culture, people, legislation, technology), and each of these dimensions was divided into other sub-parts, which were evaluated on a five-point scale in their questionnaire (not implemented (1) to fully implemented (5))—this scale also confirmed the accuracy of the four-degree scale in the research (1—we do not use it, 2—being implemented, 3—we use it partially, 4—implemented). From the results of the research, it is evident that the enterprises considered the dimension of the product and people as the most important. The results of this work show that companies mostly deployed IT infrastructure. They then assigned weights to these parts and made readiness calculations. They proposed this model for the enterprises as a means of self-assessment. These authors designed their model very generally, as some dimensions are very difficult to evaluate within subjective perception. For this reason, this paper used a specifically focused indicator, which does not aim to evaluate all the factors, but only the factors related to the technology possible to be evaluated by the enterprise itself.

Frank, Dalenogare, and Ayala [108] conducted a cross-sectional survey among 92 Brazilian manufacturing enterprises, as they identified them as the most affected by Industry 4.0, similar to this paper. They verified that the level of Industry 4.0 implementation depends on the size of the enterprise, as in this work, where it was shown that large enterprises achieve significantly higher VPI4% and Industry 4.0 affects SMEs. They also found enterprises with an advanced level of Industry 4.0 (also divided into three levels). They identified automation, virtualization, and flexibility as the key criteria and barriers to a high level of Industry 4.0.

Durana, Kral, Stehel, Lazaroiu, and Sroka [128], using factor analysis, described a model of quality culture, the fulfilment of which helps the company in the implementation of Industry 4.0, as the most important factor. They found consistency with the research results in terms of the collection of information and emphasis on employees.

Human resources that were identified as the most significant limit in this paper could not be identified in the research of Industry 4.0 technologies, such as implementation patterns in manufacturing companies [108], because the authors did not include them in the questionnaire. They also asked about the sensors, ERP and MES systems, virtual simulation, robot use, and M2M. The main factors affecting the level of Industry 4.0 in the company were equally divided into three levels as in this paper. Interestingly, their allocation of ERP, MES, and sensors to the first level differed from this research. However, robots and M2M were also assigned by Frank [108] to the second level according to cluster analysis.

Other models that summarized the levels of Industry 4.0 implementation in manufacturing enterprises included a model [122] that set six levels of Industry 4.0 in an enterprise. However, these levels were very difficult to measure as they were measured on the basis of general questions. The enterprises that implemented Industry 4.0 throughout the value chain, innovating business processes, reached the highest levels.

6. Conclusions

Industry 4.0 is currently identified as a major factor in the future competitiveness of enterprises. However, the implementation of different technologies varies from one enterprise to another. Based on the performed factor analysis, an Industry 4.0 index (VPi4) was created, which allows the enterprises to determine their current level of Industry 4.0.

The proposed index was statistically verified by supplementary research in the second wave of the research. The consistency of the index was confirmed by the fact that it was not possible to reject the $H1_A$ hypothesis of different sample distributions.

The correctness of the results ($H2_A$) was also shown by the observed dependence between the subjective perception of the enterprises and the results of the index.

Finally, the model was verified due to the intensity of technology in the industry. It was found that companies with a higher intensity more often achieved a higher index level in terms of Industry 4.0 ($H3_A$).

6.1. Managerial Implications

The VPi4 index and its methodology allow enterprises to easily identify their own level of technology readiness within Industry 4.0. The index is a tool for managers to set strategic objectives and formulate strategies in line with the challenges of the Fourth Industrial Revolution. It can also be a criterion in deciding on investment plans in terms of selecting priorities for the further development of an enterprise. The proposed solution allows better assessment of strategic initiatives in terms of their future return. The managers can also help to decide which projects should be implemented in order to ensure greater synergies. The index includes technologies that need to be implemented in the enterprises, as well as the processes that need to be set up, changed, and reintroduced. In this sense, it can also, in addition to project management, help with the management and identification of key processes in the organization.

6.2. Theoretical Implications

Regarding the theory, this paper offers a new way of looking at Industry 4.0 in terms of key processes and technologies. This approach aims to categorize Industry 4.0 components into a clear framework. The proposed index brings a new three-level structure of the Industry 4.0 phenomenon. The main theoretical contribution is, in particular, the determination of the content of the term and the determination of the importance of different factors in the context of the readiness of companies to implement Industry 4.0 concepts. The differences between more technologically and less technologically demanding industries confirm the specifics of different fields in the use of new technologies. This confirms the conclusions of many other researches and the fact that new technologies are largely being introduced, especially in the field of mechanical engineering. The results also indicate that the subjective perception of enterprises of their own level of Industry 4.0 corresponds more or less to the actual situation. The problem, however, is probably the lack of visibility in terms of the current challenges, priorities, and complexity of technology.

6.3. Limitations and Suggestions for Future Research

However, this paper has several limitations that must be considered. The VPi4 index does not include some industry-specific applications of Industry 4.0 technologies, such as drone use, 3D printers, nanotechnology, and autonomous vehicles, due to the lower incidence in the monitored businesses. In the early stages of the index preparation and in the initial factor analysis, these factories were included; however, due to the low factor load, they were subsequently removed from the index. All these technologies fell in the highest (third) level of the index. In the future, the authors assume that, with an increase in their use in enterprises, the index will be supplemented by these specific applications. Alternatively, it was considered to create different variants of the index for different industries.

A certain limitation of the paper is related to the method of data collection. At data collection, the expected return on questionnaires was 15% at a 5% margin of error and 95% confidence level. However, the real rate of return was 12.5%. With a usable 276 questionnaires and 95% confidence level, the margin of error was 5.86%. The error, therefore, slightly exceeded the planned level. Sample size was not representative in the case of proportionality of the Czech Republic enterprise population. The intention of the authors was that the research sample of the enterprises was composed evenly with regard to the size of the enterprises and technological demands of the industry. Therefore, the VPi4 index is not primarily intended to only determine the level of Industry 4.0 in the Czech Republic.

The main questionnaire survey limits are as follows: the limitations in terms of ignorance of the material and terminology by the respondents [129], as well as the fact that the respondents only reported their individual perspective on the situation [130], and that respondents tried portraying the situation (business) in a better light [131]. For this reason, a personal meeting with the representatives of the enterprises was used, who often liked to show off how Industry 4.0 works in the enterprise. An important limit, as mentioned by Roberts and Giddens [129,132], is related to the accuracy of the survey, as there was a small percentage of responses obtained; thus, the research results were often based on only 10% of the original sample; this is a problem faced by every research. The questionnaire also omitted the open questions noted by Saunders [133].

In terms of verification of the resulting VPi4 index, the authors plan to perform a confirmatory factor analysis in combination with the structural model equation method to further refine the adjustment of individual factors within the third wave of the research. It will also include the creation of an Industry 4.0 implementation model. However, recent results from the second wave of the research and comparison presented in the paper suggest that this is unlikely to be a significant intervention in the configuration of the coefficients of different variables and index factors. The authors also plan to analyze the relation of the index to the size of the enterprises.

Another disadvantage of the index could be the fact that the enterprises operating in the Czech Republic only participated in the research in both waves. In the case of large enterprises, however, most of these were foreign-owned enterprises, mostly from the European Union (EU), mostly from Germany. In the third wave of the research, the authors are also planning to do research abroad and include enterprises from developed countries such as Japan, the USA, etc.

Lastly, the proposed VPi4 index is only the first output of the Industry 4.0 project, which deals with the issue more comprehensively. Future research will bring further results.

Author Contributions: Conceptualization, J.V. and M.P.; methodology, J.V. and M.P.; validation, J.V.; formal analysis, M.P.; investigation, J.V. and M.P.; resources, J.V. and M.P.; data curation, M.P.; writing—original draft preparation, J.V. and M.P.; writing—review and editing, M.P.; visualization, J.V.; supervision, J.V.; project administration, J.V.

Funding: This research was funded by "EF-150-GAJU 047/2019/S".

Acknowledgments: The authors thank the enterprises taking part in the research.

Conflicts of Interest: The authors declare no conflicts of interest.

References

1. Weyrich, M.; Schmidt, J.-P.; Ebert, C. Machine-to-Machine Communication. *IEEE Softw.* **2014**, *31*, 19–23. [CrossRef]
2. Kelkar, O. Studie Industrie 4.0—Eine Standortbestimmung der Automobil-und Fertigungsindustrie. Mieschke Hofmann und Partner (MHP), Ludwigsburg, Germany. Available online: https://www.mhp.com/fileadmin/mhp.de/assets/studien/MHP-Studie_Industrie4.0_V1.0.pdf (accessed on 11 November 2019).
3. Pricewaterhouse Coopers PwC-und Strategy&-Studie: Industrie 4.0 hat Hohes Nutzenpotenzial für Deutsche Unternehmen. Available online: https://www.pwc.de/de/digitale-transformation/pwc-studie-industrie-4-0-steht-vor-dem-durchbruch.html (accessed on 8 November 2019).
4. Dörfler, M. Industrie 4.0 ist im Mittelstand Noch Nicht Angekommen. Available online: https://www.marktundmittelstand.de/404/ (accessed on 11 November 2019).
5. Sommer, L. Industrial revolution—Industry 4.0: Are German manufacturing SMEs the first victims of this revolution? *J. Ind. Eng. Manag.* **2015**, *8*, 1512–1532. [CrossRef]
6. Tupa, J.; Simota, J.; Steiner, F. Aspects of Risk Management Implementation for Industry 4.0. *Procedia Manuf.* **2017**, *11*, 1223–1230. [CrossRef]
7. Zhong, R.Y.; Xu, X.; Klotz, E.; Newman, S.T. Intelligent Manufacturing in the Context of Industry 4.0: A Review. *Engineering* **2017**, *3*, 616–630. [CrossRef]
8. Lu, B.H.; Bateman, R.J.; Cheng, K. RFID enabled manufacturing: Fundamentals, methodology and applications. *Int. J. Agile Syst. Manag.* **2006**, *1*, 73. [CrossRef]

9. Mansoor, K.; Ghani, A.; Chaudhry, S.A.; Shamshirband, S.; Ghayyur, S.A.K.; Mosavi, A. Securing IoT-Based RFID Systems: A Robust Authentication Protocol Using Symmetric Cryptography. *Sensors* **2019**, *19*, 4752. [CrossRef]
10. Lucke, D.; Constantinescu, C.; Westkämper, E. Smart Factory—A Step towards the Next Generation of Manufacturing. In *Proceedings of the Manufacturing Systems and Technologies for the New Frontier, Tokyo, Japan, 26–28 May 2008*; Mitsuishi, M., Ueda, K., Kimura, F., Eds.; Springer: London, UK, 2008; pp. 115–118.
11. Tan, P.; Wu, H.; Li, P.; Xu, H. Teaching Management System with Applications of RFID and IoT Technology. *Educ. Sci.* **2018**, *8*, 26. [CrossRef]
12. Reis, M.; Gins, G. Industrial Process Monitoring in the Big Data/Industry 4.0 Era: From Detection, to Diagnosis, to Prognosis. *Processes* **2017**, *5*, 35. [CrossRef]
13. Smajic, H.; Wessel, N. Remote Control of Large Manufacturing Plants Using Core Elements of Industry 4.0. In *Proceedings of the Online Engineering & Internet of Things, Columbia University, New York, USA 15–17 March 2017*; Auer, M.E., Zutin, D.G., Eds.; Springer International Publishing: Cham, Switzerland, 2018; pp. 546–551.
14. McAfee, A.; Brynjolfsson, E. Big Data: The Management Revolution. *Harv. Bus. Rev.* **2012**, *90*, 60–68.
15. Mayer-Schönberger, V.; Cukier, K. *Big Data*; Computer Press: Brno, Czech Republic, 2014; ISBN 978-80-251-4119-9.
16. Boyd, D.; Crawford, K. Critical questions for big data: Provocations for a cultural, technological, and scholarly phenomenon. *Inf. Commun. Soc.* **2012**, *15*, 662–679. [CrossRef]
17. Laney, D. 3D Data Management: Controlling Data Volume, Velocity, and Variety. Available online: http://blogs.gartner.com/doug-laney/files/2012/01/ad949-3D-Data-Management-Controlling-Data-Volume-Velocity-and-Variety.pdf (accessed on 6 February 2001).
18. Berman, J.J. *Principles of Big Data: Preparing, Sharing, and Analyzing Complex Information*; Elsevier, Morgan Kaufmann: Amsterdam, The Netherlands, 2013; ISBN 978-0-12-404576-7.
19. Gantz, J.; Reinsel, D. Extracting value from chaos. *IDC IView* **2011**, *1142*, 1–12.
20. Sivarajah, U.; Kamal, M.M.; Irani, Z.; Weerakkody, V. Critical analysis of Big Data challenges and analytical methods. *J. Bus. Res.* **2017**, *70*, 263–286. [CrossRef]
21. Ge, M.; Bangui, H.; Buhnova, B. Big Data for Internet of Things: A Survey. *Future Gener. Comput. Syst.* **2018**, *87*, 601–614. [CrossRef]
22. Xu, X. From cloud computing to cloud manufacturing. *Robot. Comput. Integr. Manuf.* **2012**, *28*, 75–86. [CrossRef]
23. Tsai, C.-W.; Lai, C.-F.; Chao, H.-C.; Vasilakos, A.V. Big data analytics: A survey. *J. Big Data* **2015**, *2*, 21. [CrossRef]
24. Tjahjono, B.; Esplugues, C.; Ares, E.; Pelaez, G. What does Industry 4.0 mean to Supply Chain? *Procedia Manuf.* **2017**, *13*, 1175–1182. [CrossRef]
25. Ertürk, M.A.; Aydın, M.A.; Büyükakkaşlar, M.T.; Evirgen, H. A Survey on LoRaWAN Architecture, Protocol and Technologies. *Future Internet* **2019**, *11*, 216. [CrossRef]
26. Rao, S.K.; Prasad, R. Impact of 5G Technologies on Industry 4.0. *Wirel. Pers. Commun.* **2018**, *100*, 145–159. [CrossRef]
27. Leber, J. General Electric Pitches an Industrial Internet. Available online: https://www.technologyreview.com/s/507831/general-electric-pitches-an-industrial-internet/ (accessed on 8 November 2019).
28. Floyer, D. Defining and Sizing the Industrial Internet—Wikibon. Available online: http://wikibon.org/wiki/v/Defining_and_Sizing_the_Industrial_Internet#Constraints_on_Adoption_of_the_Industrial_Internet (accessed on 8 November 2019).
29. Fuertes, W.; Reyes, F.; Valladares, P.; Tapia, F.; Toulkeridis, T.; Pérez, E. An Integral Model to Provide Reactive and Proactive Services in an Academic CSIRT Based on Business Intelligence. *Systems* **2017**, *5*, 52. [CrossRef]
30. Petrović, M.; Vučković, M.; Turajlić, N.; Babarogić, S.; Aničić, N.; Marjanović, Z. Automating ETL processes using the domain-specific modeling approach. *Inf. Syst. E Bus. Manag.* **2017**, *15*, 425–460. [CrossRef]
31. Armbrust, M.; Fox, A.; Griffith, R.; Joseph, A.D.; Katz, R.H.; Konwinski, A.; Lee, G.; Patterson, D.A.; Rabkin, A.; Zaharia, M. *Above the Clouds: A Berkeley View of Cloud Computing*; Electrical Engineering and Computer Sciences University of California: Berkley, CA, USA, 2009.
32. Buyya, R. Market-Oriented Cloud Computing: Vision, Hype, and Reality of Delivering Computing as the 5th Utility. In Proceedings of the 9th IEEE/ACM International Symposium on Cluster Computing and the Grid, Shanghai, China, 19–21 May 2009; p. 1.

33. Vaquero, L.M.; Rodero-Merino, I.; Caceres, J.; Lindner, R. A break in the cloud: Towards a cloud definition. *ACM SIGCOMM Comput. Commun. Rev.* **2009**, *39*, 50–55. [CrossRef]
34. McKinsey and Co. *Clearing the Air on Cloud Computing*; McKinsey and Co.: New York, NY, USA, 2009.
35. Jamsa, D.K. *Cloud Computing: SaaS, PaaS, IaaS, Virtualization, Business Models, Mobile, Security and More*, 1st ed.; Jones & Bartlett Learning: Burlington, MA, USA, 2012; ISBN 978-1-4496-4739-1.
36. Mell, P.; Grance, T. *Perspectives on Cloud Computing and Standards*; National Institute of Standards and Technology: Gaithersburg, MD, USA, 2008.
37. Wu, L.; Yang, C. A Solution of Manufacturing Resources Sharing in Cloud Computing Environment. In *Proceedings of the Cooperative Design, Visualization, and Engineering, Hangzhou, China, 21–24 October 2018*; Luo, Y., Ed.; Springer: Berlin/Heidelberg, Germany, 2010; pp. 247–252.
38. Trakadas, P.; Nomikos, N.; Michailidis, E.T.; Zahariadis, T.; Facca, F.M.; Breitgand, D.; Rizou, S.; Masip, X.; Gkonis, P. Hybrid Clouds for Data-Intensive, 5G-Enabled IoT Applications: An Overview, Key Issues and Relevant Architecture. *Sensors* **2019**, *19*, 3591. [CrossRef] [PubMed]
39. Zavoral, P. ICT Revue. Available online: //ictrevue.ihned.cz/ (accessed on 11 November 2019).
40. Parn, E.A.; Edwards, D. Cyber threats confronting the digital built environment: Common data environment vulnerabilities and block chain deterrence. *Eng. Constr. Archit. Manag.* **2019**, *26*, 245–266. [CrossRef]
41. Abad, I.; Cerrada, C.; Cerrada, J.A.; Heradio, R.; Valero, E. Managing RFID Sensors Networks with a General Purpose RFID Middleware. *Sensors* **2012**, *12*, 7719–7737. [CrossRef] [PubMed]
42. Aghenta, L.O.; Iqbal, M.T. Low-Cost, Open Source IoT-Based SCADA System Design Using Thinger.IO and ESP32 Thing. *Electronics* **2019**, *8*, 822. [CrossRef]
43. Wang, C.; Chen, X.; Soliman, A.-H.A.; Zhu, Z. RFID Based Manufacturing Process of Cloud MES. *Future Internet* **2018**, *10*, 104. [CrossRef]
44. Rojko, A. Industry 4.0 Concept: Background and Overview. *Int. J. Interact. Mob. Technol. IJIM* **2017**, *11*, 77–90. [CrossRef]
45. Aruväli, T.; Maass, W.; Otto, T. Digital Object Memory Based Monitoring Solutions in Manufacturing Processes. *Procedia Eng.* **2014**, *69*, 449–458. [CrossRef]
46. Unger, H.; Börner, F.; Müller, E. Context Related Information Provision in Industry 4.0 Environments. *Procedia Manuf.* **2017**, *11*, 796–805. [CrossRef]
47. Haddara, M.; Elragal, A. The Readiness of ERP Systems for the Factory of the Future. *Procedia Comput. Sci.* **2015**, *64*, 721–728. [CrossRef]
48. Schlank, R. *Dynamic Memory: A Theory of Reminding and Learning in Computers and People*; Cambridge University Press: Cambridge, UK, 1982.
49. Wen, J.; Li, S.; Lin, Z.; Hu, Y.; Huang, C. Systematic literature review of machine learning based software development effort estimation models. *Inf. Softw. Technol.* **2012**, *54*, 41–59. [CrossRef]
50. Shekapure, S.; Patil, D.D. Enhanced e-Learning Experience using Case based Reasoning Methodology. *Int. J. Adv. Comput. Sci. Appl.* **2019**, *10*, 236–241. [CrossRef]
51. Lajnef, T.; Chaibi, S.; Ruby, P.; Aguera, P.-E.; Eichenlaub, J.-B.; Samet, M.; Kachouri, A.; Jerbi, K. Learning machines and sleeping brains: Automatic sleep stage classification using decision-tree multi-class support vector machines. *J. Neurosci. Methods* **2015**, *250*, 94–105. [CrossRef]
52. Sun, X.; Chen, C.; Wang, L.; Kang, H.; Shen, Y.; Chen, Q. Hybrid Optimization Algorithm for Bayesian Network Structure Learning. *Information* **2019**, *10*, 294. [CrossRef]
53. Zeng, W.Z.; Zhang, D.Y.; Fang, Y.H.; Wu, J.W.; Huang, J.S. Comparison of partial least square regression, support vector machine, and deep-learning techniques for estimating soil salinity from hyperspectral data. *J. Appl. Remote Sens.* **2018**, *12*, 022204. [CrossRef]
54. Yu, J.; Sun, W.; Huang, H.; Wang, W.; Wang, Y.; Hu, Y. Crack Sensitivity Control of Nickel-Based Laser Coating Based on Genetic Algorithm and Neural Network. *Coatings* **2019**, *9*, 728. [CrossRef]
55. López, R.; González Gurrola, L.; Trujillo, L.; Prieto, O.; Ramírez, G.; Posada, A.; Juárez-Smith, P.; Méndez, L. How Am I Driving? Using Genetic Programming to Generate Scoring Functions for Urban Driving Behavior. *Math. Comput. Appl.* **2018**, *23*, 19. [CrossRef]
56. Álvarez-Díaz, M.; González-Gómez, M.; Otero-Giráldez, M. Forecasting International Tourism Demand Using a Non-Linear Autoregressive Neural Network and Genetic Programming. *Forecasting* **2018**, *1*, 90–106. [CrossRef]

57. Guettiche, A.; Guéguen, P.; Mimoune, M. Seismic vulnerability assessment using association rule learning: Application to the city of Constantine, Algeria. *Nat. Hazards* **2017**, *86*, 1223–1245. [CrossRef]
58. Thabtah, F.; Qabajeh, I.; Chiclana, F. Constrained dynamic rule induction learning. *Expert Syst. Appl.* **2016**, *63*, 74–85. [CrossRef]
59. Sahu, B.K.; Pati, S.; Mohanty, P.K.; Panda, S. Teaching–learning based optimization algorithm based fuzzy-PID controller for automatic generation control of multi-area power system. *Appl. Soft Comput.* **2015**, *27*, 240–249. [CrossRef]
60. Celaschi, F. Advanced design-driven approaches for an Industry 4.0 framework: The human-centred dimension of the digital industrial revolution. *Strateg. Des. Res. J.* **2017**, *10*, 97–104. [CrossRef]
61. Guizzo, E.; Ackerman, E. The rise of the robot worker. *IEEE Spectr.* **2012**, *49*, 34–41. [CrossRef]
62. QB Robotics. Saphari—Safe and Autonomous Physical Human-Aware Robot Interaction. Available online: https://qbrobotics.com/projects/saphari-safe-autonomous-physical-human-aware-robot-interaction/ (accessed on 8 November 2019).
63. Koch, P.J.; van Amstel, M.K.; Dębska, P.; Thormann, M.A.; Tetzlaff, A.J.; Bøgh, S.; Chrysostomou, D. A Skill-based Robot Co-worker for Industrial Maintenance Tasks. *Procedia Manuf.* **2017**, *11*, 83–90. [CrossRef]
64. Kamarul Bahrin, M.A.; Othman, M.F.; Nor Azli, N.H.; Talib, M.F. Industry 4.0: A review on industrial automation and robotic. *J. Teknol.* **2016**, *78*, 137–143. [CrossRef]
65. Arnaiz-González, Á.; Fernández-Valdivielso, A.; Bustillo, A.; López de Lacalle, L.N. Using artificial neural networks for the prediction of dimensional error on inclined surfaces manufactured by ball-end milling. *Int. J. Adv. Manuf. Technol.* **2016**, *83*, 847–859. [CrossRef]
66. Meng, Z.; Wu, Z.; Gray, J. A Collaboration-Oriented M2M Messaging Mechanism for the Collaborative Automation between Machines in Future Industrial Networks. *Sensors* **2017**, *17*, 2694. [CrossRef]
67. Chen, M. Machine-to-Machine Communications: Architectures, Standards and Applications. *KSII Trans. Internet Inf. Syst.* **2012**. [CrossRef]
68. Ali, A.; Shah, G.A.; Arshad, J. Energy Efficient Resource Allocation for M2M Devices in 5G. *Sensors* **2019**, *19*, 1830. [CrossRef]
69. Jacob Taquet, E.; Astorga, J.; Uncilla Galan, J.J.; Huarte, M.; Garcia Conejo, D.; Lopez De La Calle Marcaide, L.N. Hacia una ingraestructura de fabricación flexible, conectada e integrable en redes 5G. *DYNA Ing. E Ind.* **2018**, *93*, 656–662.
70. Seo, D.; Jeon, Y.-B.; Lee, S.-H.; Lee, K.-H. Cloud computing for ubiquitous computing on M2M and IoT environment mobile application. *Clust. Comput.* **2016**, *19*, 1001–1013. [CrossRef]
71. Pilloni, V. How Data Will Transform Industrial Processes: Crowdsensing, Crowdsourcing and Big Data as Pillars of Industry 4.0. *Future Internet* **2018**, *10*, 24. [CrossRef]
72. Veber, J. *Management Inovací*; Management Press: Prague, Czech Republic, 2016; ISBN 978-80-7261-423-3.
73. Burian, P. *Internet Inteligentních Aktivit*; Grada: Praha, Czech Republic, 2014; ISBN 978-80-247-5137-5.
74. Cao, M.; Zhang, Q. Supply chain collaboration: Impact on collaborative advantage and firm performance. *J. Oper. Manag.* **2011**, *29*, 163–180. [CrossRef]
75. Liu, J.; Zhou, H.; Tian, G.; Liu, X.; Jing, X. Digital twin-based process reuse and evaluation approach for smart process planning. *Int. J. Adv. Manuf. Technol.* **2019**, *100*, 1619–1634. [CrossRef]
76. Shen, B.; Chan, H.-L. Forecast Information Sharing for Managing Supply Chains in the Big Data Era: Recent Development and Future Research. *Asia Pac. J. Oper. Res.* **2017**, *34*, 1740001. [CrossRef]
77. Choi, T.-M. Quick response in fashion supply chains with dual information updating. *J. Ind. Manag. Optim.* **2006**, *2*, 255–268. [CrossRef]
78. Du, T.C.; Lai, V.S.; Cheung, W.M.; Cui, X.L. Willingness to share information in a supply chain: A partnership-data-process perspective. *Inf. Manag.* **2012**, *49*, 89–98. [CrossRef]
79. Cachon, G.P.; Lariviere, M.A. Contracting to Assure Supply: How to Share Demand Forecasts in a Supply Chain. *Manag. Sci.* **2001**, *47*, 629–646. [CrossRef]
80. Croson, R.; Donohue, K. Imapct of pos data sharing on supply chain management: An experimental study. *Prod. Oper. Manag.* **2009**, *12*, 1–11. [CrossRef]
81. Christopher, M.; Ryals, L.J. The Supply Chain Becomes the Demand Chain. *J. Bus. Logist.* **2014**, *35*, 29–35. [CrossRef]
82. Juttner, U.; Christopher, M.; Baker, S. Demand chain management-integrating marketing and supply chain management. *Ind. Mark. Manag.* **2007**, *36*, 377–392. [CrossRef]

83. Vezzetti, E. Product lifecycle data sharing and visualisation: Web-based approaches. *Int. J. Adv. Manuf. Technol.* **2009**, *41*, 613–630. [CrossRef]
84. Li, W.D.; Fuh, J.Y.H.; Wong, Y.S. An Internet-enabled integrated system for co-design and concurrent engineering. *Comput. Ind.* **2004**, *55*, 87–103. [CrossRef]
85. Zhao, F.L.; Tso, S.K.; Wu, P.S.Y. A cooperative agent modelling approach for process planning. *Comput. Ind.* **2000**, *41*, 83–97. [CrossRef]
86. Chen, Y.-M.; Liang, M.-W. Design and implementation of a collaborative engineering information system for allied concurrent engineering. *Int. J. Comput. Integr. Manuf.* **2000**, *13*, 11–30. [CrossRef]
87. Nikolakis, N.; Maratos, V.; Makris, S. A cyber physical system (CPS) approach for safe human-robot collaboration in a shared workplace. *Robot. Comput. Integr. Manuf.* **2019**, *56*, 233–243. [CrossRef]
88. Blossey, R. Self-cleaning surfaces—Virtual realities. *Nat. Mater.* **2003**, *2*, 301–306. [CrossRef]
89. Diemer, J. Sichere Industrie-4.0-Plattformen auf Basis von Community-Clouds. In *Handbuch Industrie 4.0: Produktion, Automatisierung und Logistik*; Vogel-Heuser, B., Bauernhansl, T., ten Hompel, M., Eds.; Springer: Berlin/Heidelberg, Germany, 2016; pp. 1–28. ISBN 978-3-662-45537-1.
90. Burdea, C.G.; Coiffet, P. *Virtual Reality Technology*; John Wiley & Sons: New York, NY, USA, 2003.
91. Steuer, J. Defining virtual reality—dimensions determining telepresence. *J. Commun.* **1992**, *42*, 73–93. [CrossRef]
92. Crocetta, T.B.; de Araujo, L.V.; Guarnieri, R.; Massetti, T.; Ferreira, F.; de Abreu, L.C.; Monteiro, C.B.D. Virtual reality software package for implementing motor learning and rehabilitation experiments. *Virtual Real.* **2018**, *22*, 199–209. [CrossRef]
93. Yin, X.; Fan, X.; Yang, X.; Qiu, S.; Zhang, Z. An Automatic Marker—Object Offset Calibration Method for Precise 3D Augmented Reality Registration in Industrial Applications. *Appl. Sci.* **2019**, *9*, 4464. [CrossRef]
94. Jurášková, O. *Velký Slovník Marketingových Komunikací*; Grada Publishing Inc.: Prague, Czech Republic, 2012; ISBN 978-80-247-4354-7.
95. Soderberg, R.; Warmefjord, K.; Carlson, J.S.; Lindkvist, L. Toward a Digital Twin for real-time geometry assurance in individualized production. *Cirp Ann. Manuf. Technol.* **2017**, *66*, 137–140. [CrossRef]
96. Rosen, R.; von Wichert, G.; Lo, G.; Bettenhausen, K.D. About the Importance of Autonomy and Digital Twins for the Future of Manufacturing. *IFAC Pap. OnLine* **2015**, *48*, 567–572. [CrossRef]
97. El Saddik, A. Digital Twins the Convergence of Multimedia Technologies. *IEEE Multimed.* **2018**, *25*, 87–92. [CrossRef]
98. Batty, M. Digital twins. *Environ. Plan. B Urban Anal. City Sci.* **2018**, *45*, 817–819. [CrossRef]
99. Lutters, E. Pilot production environments driven by digital twins. *S. Afr. J. Ind. Eng.* **2018**, *29*, 40–53. [CrossRef]
100. Zhuang, C.; Liu, J.; Xiong, H. Digital twin-based smart production management and control framework for the complex product assembly shop-floor. *Int. J. Adv. Manuf. Technol.* **2018**, *96*, 1149–1163. [CrossRef]
101. Atorf, L.; Schorn, C.; Rossmann, J.; Schlette, C. A framework for simulation-based optimization demonstrated on reconfigurable robot workcells. In Proceedings of the 2017 IEEE International Systems Engineering Symposium (ISSE), Vienna, Austria, 11–13 October 2017; pp. 1–6.
102. Czech Statistical Office High-Tech Sektor. Available online: https://www.czso.cz/csu/czso/high_tech_sektor (accessed on 25 November 2019).
103. European Commission. *Recommendation of 6 May 2003 Concerning the Definition of Micro, Small and Medium-Sized Enterprises*; Text with EEA Relevance, Notified Under Document Number C (2003) 1422; European Commission: Brussels, Belgium, 2003.
104. Burt, C.; Burt, C.L. *The Factors of the Mind: An Introduction to Factor-Analysis in Psychology*; University of London Press: London, UK, 1940.
105. McDonald, R.P. *Factor Analysis and Related Methods*; Lawrence Erlbaum Associates: Hillsdale, NJ, USA, 1985; ISBN 978-0-89859-388-4.
106. Gorsuch, R.L. *Factor Analysis*, Classic ed.; Routledge: New York, NY, USA; Taylor & Francis Group: London, UK, 2015; ISBN 978-1-138-83198-8.
107. Veza, I.; Mladineo, M.; Peko, I. Analysis of the current state of croatian manufacturing industry with regard to industry 4.0. In Proceedings of the 15th International Scientific Conference on Production Engineering - CIM'2015: Computer Integrated Manufacturing and High Speed Machining, Vodice, Croatia, 10–13 June 2015; pp. 249–254.

108. Frank, A.G.; Dalenogare, L.S.; Ayala, N.F. Industry 4.0 technologies: Implementation patterns in manufacturing companies. *Int. J. Prod. Econ.* **2019**, *210*, 15–26. [CrossRef]
109. Schumacher, A.; Erol, S.; Sihn, W. A Maturity Model for Assessing Industry 4.0 Readiness and Maturity of Manufacturing Enterprises. *Procedia CIRP* **2016**, *52*, 161–166. [CrossRef]
110. Pett, M.A.; Lackey, N.R.; Sullivan, J.J. *Making Sense of Factor Analysis: The use of Factor Analysis for Instrument Development in Health Care Research*; Sage Publications: Thousand Oaks, CA, USA, 2003; ISBN 978-0-7619-1949-0.
111. Hutcheson, G.; Sofroniou, N. *The Multivariate Social Scientist: Introductory Statistics Using Generalized Linear Models*; Sage Publications: London, UK; Thousand Oaks, CA, USA, 1999; ISBN 978-0-7619-5200-8.
112. Cohen, J. *Statistical Power Analysis for the Behavioral Sciences*, 2nd ed.; Lawrence Erlbaum Associates: Hillsdale, NJ, USA, 1988; ISBN 978-0-8058-0283-2.
113. Tabachnick, B.G.; Fidell, L.S.; Ullman, J.B. *Using Multivariate Statistics*, 7th ed.; Pearson: New York, NY, USA, 2019; ISBN 978-0-13-479054-1.
114. Samuels, P. *Advice on Exploratory Factor Analysis*; Birmingham City University: Birmingham, UK, 2016.
115. Řezanková, H.; Húsek, D.; Snášel, V. *Shluková Analýza Dat*; Professional Publishing: Prague, Czech Republic, 2009; ISBN 978-80-86946-81-8.
116. Rolinek, L.; Kopta, D.; Plevny, M.; Rost, M.; Kubecova, J.; Vrchota, J.; Marikova, M. Level of process management implementation in SMEs and some related implications. *Transform. Bus. Econ.* **2015**, *14*, 360–377.
117. Schulze, A. Industrie 4.0 steht noch ganz am Anfang. *FLYACTS—Digit. Innov. Fact.* **2014**. Available online: https://www.flyacts.com/industrie-4-0-steht-noch-ganz-am-anfang (accessed on 25 November 2019).
118. Sastoque Pinilla, L.; Llorente Rodríguez, R.; Toledo Gandarias, N.; López de Lacalle, L.N.; Ramezani Farokhad, M. TRLs 5–7 Advanced Manufacturing Centres, Practical Model to Boost Technology Transfer in Manufacturing. *Sustainability* **2019**, *11*, 4890. [CrossRef]
119. Scremin, L.; Armellini, F.; Brun, A.; Solar-Pelletier, L.; Beaudry, C. Towards a framework for assessing the maturity of manufacturing companies in industry 4.0 adoption. In *Analyzing the Impacts of Industry 4.0 in Modern Business Environments*; Hershey: Derry Township, PA, USA, 2018; pp. 224–254.
120. Oesterreich, T.D.; Teuteberg, F. Understanding the implications of digitisation and automation in the context of Industry 4.0: A triangulation approach and elements of a research agenda for the construction industry. *Comput. Ind.* **2016**, *83*, 121–139. [CrossRef]
121. Akdil, K.Y.; Ustundag, A.; Cevikcan, E. Maturity and Readiness Model for Industry 4.0 Strategy. In *Industry 4.0: Managing the Digital Transformation*; Ustundag, A., Cevikcan, E., Eds.; Springer International Publishing: Cham, Switzerland, 2018; pp. 61–94. ISBN 978-3-319-57870-5.
122. Gökalp, E.; Şener, U.; Eren, P.E. Development of an Assessment Model for Industry 4.0: Industry 4.0-MM. In *Software Process Improvement and Capability Determination*; Mas, A., Mesquida, A., O'Connor, R.V., Rout, T., Dorling, A., Eds.; Springer International Publishing: Cham, Switzerland, 2017; Volume 770, pp. 128–142. ISBN 978-3-319-67382-0.
123. Lee, J.; Jun, S.; Chang, T.-W.; Park, J. A Smartness Assessment Framework for Smart Factories Using Analytic Network Process. *Sustainability* **2017**, *9*, 794. [CrossRef]
124. Ślusarczyk, B. Industry 4.0—Are we ready? *Pol. J. Manag. Stud.* **2018**, *17*, 232–248. [CrossRef]
125. Stefan, L.; Thom, W.; Dominik, L.; Dieter, K.; Bernd, K. Concept for an evolutionary maturity based Industrie 4.0 migration model. *Procedia CIRP* **2018**, *72*, 404–409. [CrossRef]
126. Block, C.; Freith, S.; Kreggenfeld, N.; Morlock, F.; Prinz, C.; Kreimeier, D.; Kuhlenkötter, B. Industrie 4.0 als soziotechnisches Spannungsfeld: Ganzheitliche Betrachtung von Technik, Organisation und Personal. *ZWF Z. Für Wirtsch. Fabr.* **2015**, *110*, 657–660. [CrossRef]
127. Brunet-Thornton, R.; Martinez, F. (Eds.) *Analyzing the Impacts of Industry 4.0 in Modern Business Environments*; Advances in Business Information Systems and Analytics; IGI Global: Hershey, PA, USA, 2018; ISBN 978-1-5225-3468-6.
128. Durana, P.; Kral, P.; Stehel, V.; Lazaroiu, G.; Sroka, W. Quality Culture of Manufacturing Enterprises: A Possible Way to Adaptation to Industry 4.0. *Soc. Sci.* **2019**, *8*, 124. [CrossRef]
129. Giddens, A.; Sutton, P.W. *Sociologie*; Argo: Prague, Czech Republic, 2013; ISBN 978-80-257-0807-1.
130. Hair, J.F.; Money, A.H.; Samouel, P.; Page, M. Research Methods for Business. *Educ. Train.* **2007**, *49*, 336–337. [CrossRef]

131. Greener, S. *Business Research Methods;* Ventus Publishing: London, UK, 2008; ISBN 978-87-7681-421-2.
132. Roberts, F.S. The questionnaire method. In *Structure of Decision: The Cognitive Maps of Political Elites;* Princeton University Press: Princeton, NJ, USA, 2015; pp. 333–342.
133. Saunders, M.N.K.; Lewis, P.; Thornhill, A. *Research Methods for Business Students*, 5th ed.; Prentice Hall: New York, NY, USA, 2009; ISBN 978-0-273-71686-0.

Review

Review and Development Trend of Digital Hydraulic Technology

Qiwei Zhang [1], Xiangdong Kong [1,2,*], Bin Yu [1], Kaixian Ba [1], Zhengguo Jin [1] and Yan Kang [1]

1 School of Mechanical Engineering, Yanshan University, Qinhuangdao 066004, China; zhangqiwei@stumail.ysu.edu.cn (Q.Z.); yb@ysu.edu.cn (B.Y.); bkx@ysu.edu.cn (K.B.); jzg@stumail.ysu.edu.cn (Z.J.); KY18733508996@163.com (Y.K.)

2 School of Mechanical Engineering, Nanjing Institute of Technology, Nanjing 211167, China

* Correspondence: xdkong@ysu.edu.cn; Tel.: +86-0335-8051166

Received: 8 December 2019; Accepted: 9 January 2020; Published: 13 January 2020

Abstract: Since the emergence of digital hydraulic technology, it has achieved good results in intelligence, integration, energy saving, etc. After decades of development, and it has also attracted wide attention in the industry. However, for many years, the definition of digital hydraulic technology has differed between researchers, and there is no uniform definition. Such a situation affects the development of it to a certain extent. Therefore, this paper gives the exact definition of digital hydraulic technology based on a large number of researches on it. At the same time, the paper analyzes the research status and developmental process of the such a technology, and we forecast the development trend of it.

Keywords: digital hydraulic technology; digital hydraulic components; digital hydraulic system

1. Introduction

Although the foundation and development of hydraulic technology can be traced back to the middle of the 17th century, the rapid development and application of it really began 100 years ago. The development of electro-hydraulic servo-control technology began in the 20th century based on the continuous improvement of control theory and its engineering practice. After that, with the support of microelectronics technology, the hydraulic control unit became able to integrate with microprocessor, electronic power amplifier and sensor. So, the dynamic control accuracy, intelligence, reliability and robustness of the hydraulic control unit were improved [1–3]. When industry entered the 21st century, the tremendous cost of labor and energy forced industry to develop towards energy conservation and intelligence. However, the high cost of hydraulic components does not correspond to the direction of industrial development. Achten P statistic shows that the cost of a hydraulic transmission component which is to 40–80 pounds/kg, is more than three times that of a mechanical transmission component [4]. Low energy efficiency is also a key factor limiting the further application and development of hydraulic technology. Generally, more than half of the output power of hydraulic pumps or motors is dissipated to throttle or overflow. Take the excavator hydraulic system as an example; up to 80% of energy dissipation occurs in such hydraulic systems [5]. In addition, after Germany put forward the definition of "Industry 4.0" in 2013, the industrial system put forward higher requirements for the intelligent hydraulic system and its application in the "Intelligent factory." So, it can be said that if the hydraulic technology wants to survive from the intense market competition, high energy efficiency and low cost will be the inevitable direction of its development. And digital hydraulic technology is providing a feasible way to fulfil that purpose.

After referring to the concept of "digital" in electronic technology, digital hydraulic technology gradually develops and takes shape; it accelerates the pace of intelligent hydraulic technology, and because of its huge advantages compared with traditional proportional servo control technology, it has attracted extensive attention from researchers. However, the definition of digital hydraulic technology has always been in dispute. There are two typical definitions of digital hydraulic technology. The one comes from M. Linjama; he says that "digital fluid power refers to hydraulic and pneumatic systems that use discrete value component to actively control output of the systems," but from his further explanation, it can be found that his definition only includes the parallel technology and high-speed switching technology that are based on switching valve [6]. The other comes from Yang Huayong; he presented a definition of digital hydraulic which included control signal discretization and fluid flow discretization [7]; there is also the content of digital signal indirect control (proportional control and servo control).

The definition of digital hydraulic technology is not unified, which limits its development and application to some extent. With the intelligent and green development of global industry, only the technology meeting social needs can survive, and digital hydraulic technology is providing a feasible innovative development path for traditional hydraulic technology. Therefore, on the basis of a lot of research on digital hydraulic technology, this paper gives the exact definition of such a technology, and expounds its developmental course and trend, so as to make more researchers understand and promote the further development of it.

2. Review of Digital Hydraulic Technology

2.1. Definition of Digital Hydraulic Technology

The definition of digital hydraulic technology has not been unified yet. At present, all the mainstream digital hydraulic definitions can reflect the characteristics of digital hydraulic partly, but they all have an inappropriate coverage, which leads the definition of digital hydraulic technology to be ambiguous. Among them, the definition proposed by M. Linjama points out two characteristics of digital hydraulic technology; namely, discrete and active control. The former is the inherent property of digital hydraulic technology, because the signal controlling the digital hydraulic components is a discrete digital signal. The latter gives the essential characteristics of digital hydraulic technology only partly, because active control is not equal to intelligent control. For example, the proportion-integration-differentiation control (PID control), which is common in hydraulic systems, can realize the active control of system's output. But it cannot be called intelligent control for the reason that it cannot perform intelligent behaviors related to human intelligence, such as judgment, reasoning, perception, communication, etc. Therefore, M. Linjama's definition does not completely reflect the essential characteristics of digital hydraulics. Another mainstream definition of digital hydraulic was put forward by Yang Huayong, which mainly highlights the discrete characteristics of digital hydraulic technology. It has two aspects, control signal discretization and fluid flow discretization, but it ignores the essential characteristic of digital hydraulics—the intelligent control. For example, traditional proportional control and simple switch control also have discrete characteristics, but they will never be considered digital hydraulic technology. Therefore, Yang Huayong's definition is not perfect either.

On that basis, this paper gives the definition of digital hydraulic technology based on the viewpoints of researchers all over the world. We define the digital hydraulic technology as a system which controls a discrete fluid with a modulated, discrete, digital signal directly to realize active and intelligent control of the system output. The hydraulic components with such technical characteristics can be defined as digital hydraulic components. The system composed by digital hydraulic components can also be defined as digital hydraulic system. What is more, the essential feature of digital hydraulic technology is intelligent control; technology which can only realize on/off control cannot be classified as digital hydraulic technology.

2.2. Classification of Digital Hydraulic Technology

As defined in Section 2.1, digital hydraulic technology can be divided into three main categories.

2.2.1. Parallel Digital Hydraulic Technology

The parallel digital hydraulic technology requires all the components to be connected in parallel, and the composite states of all the components are controlled by modulated discrete digital signals. Different composite states give different discrete fluid flows which can be used to realize intelligent control of system output. The parallel digital hydraulic system has a fixed number of discrete outputs which depend on the composite state of components, and it does not need frequent on/off switching of components.

2.2.2. High-Speed Switching Digital Hydraulic Technology

In order to realize intelligent control of system output, high-speed switching digital hydraulic technology put forward high requirements for high-speed switching components which can switch quickly and continuously to output fluid flow with different discrete values. Theoretically, the output of high-speed switching digital hydraulic system can be any value within a certain range, but it is still a discrete quantity due to the switching frequency of components. However, if the switching frequency is very high or the switching quantity is fine enough, the pulsation caused by the discrete quantity to the system can be acceptable in the control system. And if the pulse width modulation (PWM) signal is used to control the high-speed switching components, the output is proportional to the width of pulse.

2.2.3. Stepping Digital Hydraulic Technology

The stepping digital hydraulic technology relies on precise stepping motor which controlled by modulated discrete digital signals. The rotation of stepping motor can be used to control the movement of the spool though a mechanical structure. So, the discrete fluid flow can be obtained to realize intelligent control of the system output.

The key of stepping digital hydraulic technology is the accurate conversion between the rotation of stepping motor and the movement of the spool.

3. Digital Hydraulic Components

As with the traditional hydraulic components, digital ones also contain three main categories: digital control components (digital hydraulic valve), digital power components (digital hydraulic pump, transformer and power control system) and digital actuators (digital hydraulic cylinder and motor). And all kinds of components with different classifications of digital hydraulic technology also have different products.

3.1. Digital Hydraulic Valve

According to different working principles, digital hydraulic valve can be divided into three different types.

3.1.1. Parallel Digital Hydraulic Valves

The parallel digital hydraulic valve also can be called a digital flow control unit (DFCU). It can realize the accurate flow control through encoding control of multiple switching valves connected in parallel. Figure 1 shows the working principle of a typical parallel digital two-way valve.

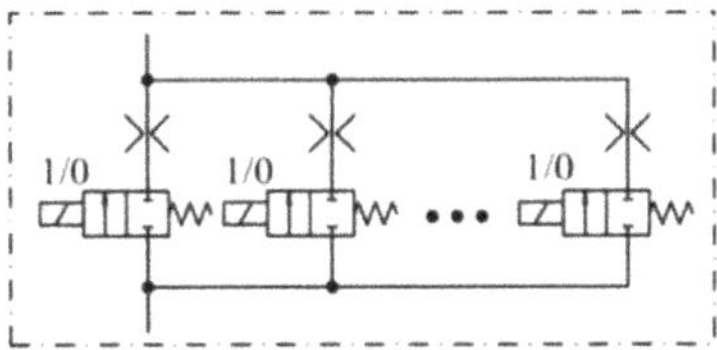

Figure 1. Parallel digital two-way valve.

The flow rate of a DFCU is the sum of all the switching valves' flow which are set to the "on" state. DFCU's steady-state characteristics are affected by two factors: the number of switching valves (N) and the encoding mode of switching valves' control signal. The commonly used encoding methods are binary encoding, Fibonacci encoding and PNM (pulse number modulation) encoding. There are 2^N combinations of switching valves, which can be named the states of DFCU. So, there are 2^N kinds of flow output under different states of DFCU. The essential difference between DFCU and a high-speed switching digital hydraulic valve is that the former does not require the frequent switching of one single valve between on and off to obtain continuous system output. And the state switching of a valve is only used to adjust the state of DFCU [8].

Similarly, if DFCU is used for independent metering control, the function of three-position four-way valve (which is shown in Figure 2a) can also be realized. The working principle is shown in Figure 2b. And each metering of this valve is independently and precisely controlled by DFCU, which is composed of five switching valves [9].

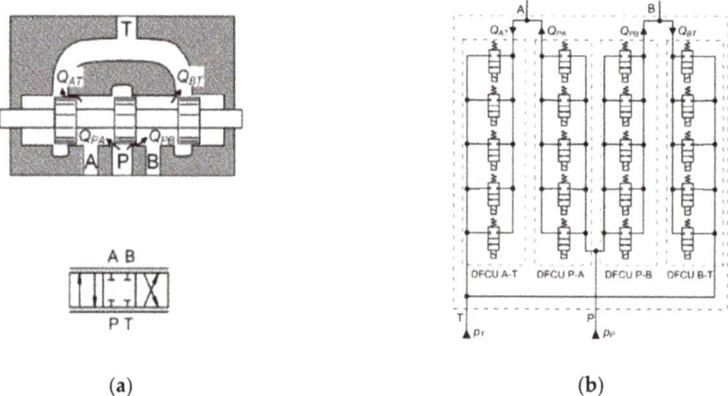

Figure 2. Two kinds of four-way valve: (**a**) traditional three-position four-way valve; (**b**) digital flow control unit (DFCU) four-way valve.

3.1.2. High-Speed Switching Digital Hydraulic Valve

The high-speed switching digital hydraulic valve is also called pulse modulation switching digital hydraulic valve. Figure 3 shows the working principle of a typical high-speed switching two-way valve. The on/off switching of a valve is controlled by the pulse signal with high/low electrical level. And the average flow of a valve is controlled by the digital signal of high frequency modulation [10]. Among the signal modulation modes, pulse width modulation (PWM) is one of the most commonly used. Theoretically, the controllable flow rate of the valve can be set to any value, but the ratio of the maximum to minimum flow rate can only change in a limited range due to the dynamic characteristics of the valve itself. The control performance of a high-speed switching valve is directly related to the switching frequency. Low frequency control performance is better but will cause obvious pressure pulsation and noise.

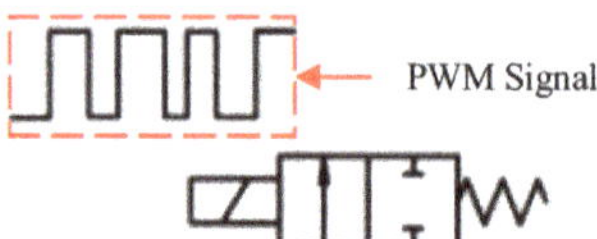

Figure 3. High-speed switching two-way valve controlled by pulse width modulation (PWM).

If a high-speed switching valve is used for independent metering control, the function of four-way valve, as shown in Figure 2a, can be achieved, and its working principle is shown in Figure 4.

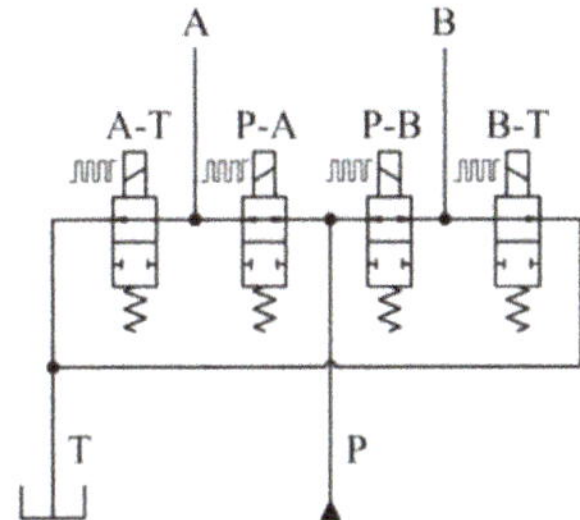

Figure 4. High-speed switching four-way valve.

3.1.3. Stepping Digital Hydraulic Valve

A stepping digital hydraulic valve controls the duty cycle of input pulse signal (which can be represented by the rotation angle and rotation speed of stepping motor) through PWM encoding, so as to realize the active intelligent position control of the spool. Because the stepping motor has no accumulated error and almost no hysteresis, the stepping digital hydraulic valve has a higher positional accuracy of the spool. Four typical forms of stepping digital hydraulic valves are shown in Figure 5 [11].

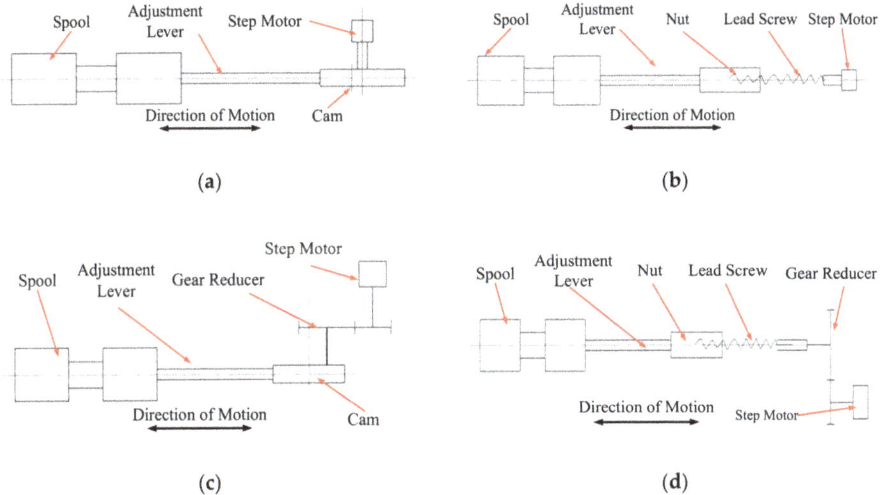

Figure 5. Four typical forms of stepping digital hydraulic valves: (**a**) cam type; (**b**) screw nut type; (**c**) gear reducer-cam type; (**d**) gear reducer-screw nut type.

However, the stepping motor outputs rotational motion, which needs to be transformed into linear motion to drive the spool. So, the conversion mechanisms, such as cam and ball screws, are indispensable. But all of the conversion mechanisms have great friction and inertia, which affects the frequency response characteristics of a stepping digital hydraulic valve. In addition, the stepping motor is prone to being out-of-step at high frequency. These problems limit the application of stepping digital hydraulic valves.

3.2. Digital Hydraulic Pump

The function of the digital hydraulic pump can be realized through the digital control of its output variation. The research on it mainly includes the variable output control of the quantitative pump and the variable mode control of the variable pump.

3.2.1. Variable Output Control of Quantitative Pump

Figure 6 shows the principle of high-speed switching digital hydraulic pump, which is composed of the quantitative hydraulic pump and high-speed switching valve.

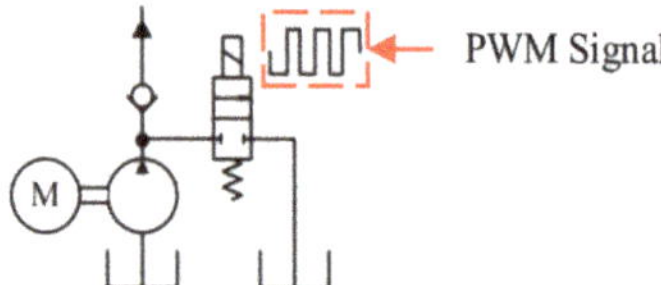

Figure 6. High-speed switching digital hydraulic pump.

High-speed switching digital hydraulic pump regulates the inlet-flow of the system through the high-speed switching valve at the outlet of the quantitative pump. And its control performance is directly related to the switching frequency of the valve.

The parallel digital hydraulic pump consists of several coaxial quantitative pumps in parallel; its working principle is shown in Figure 7.

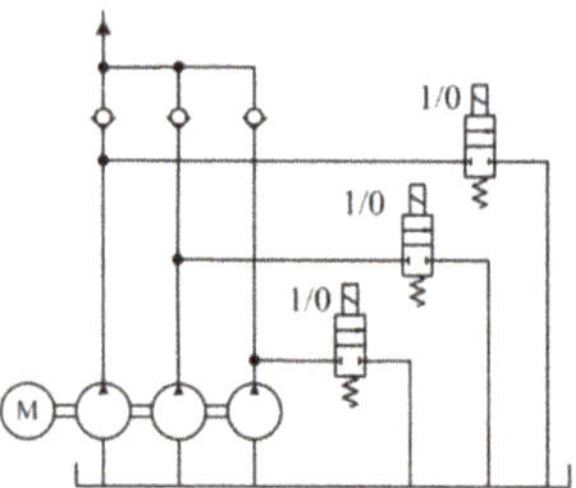

Figure 7. Parallel digital hydraulic pump.

Each quantitative pump in a parallel digital hydraulic pump system is independently controlled by switching valve at outlet of each quantitative pump. The maximum displacement is the sum of the displacements of all quantitative pumps in parallel; the minimum displacement is the displacement of minimum quantitative pump; the displacement between maximum and minimum displacement depends on the encoding mode of the switching valves. So, a parallel digital hydraulic pump has 2^N kinds of displacement (N is the number of quantitative pumps in parallel). The essential difference between parallel digital hydraulic pump and high-speed switching digital hydraulic pump is that the former does not need to control the output of the system through one single switching valve's frequently state change between on and off. The state changing of switching valves is only used to adjust the combined form of parallel pumps.

The piston chamber independent control digital hydraulic pump can realize the active and intelligent control of displacement commendably. The working principle of such a pump is shown in Figure 8a. And it can be seen that each piston chamber of the digital hydraulic pump can be switched between working state and no-load state independently under the control of switching valves. The average displacement of the pump depends on the ratio of working piston chamber number and the no-load piston chamber number. In addition, with switching valves applied to pumps or hydraulic motors, it is possible to use a part of the stroke of each cylinder as well, in order to obtain more flexible system output.

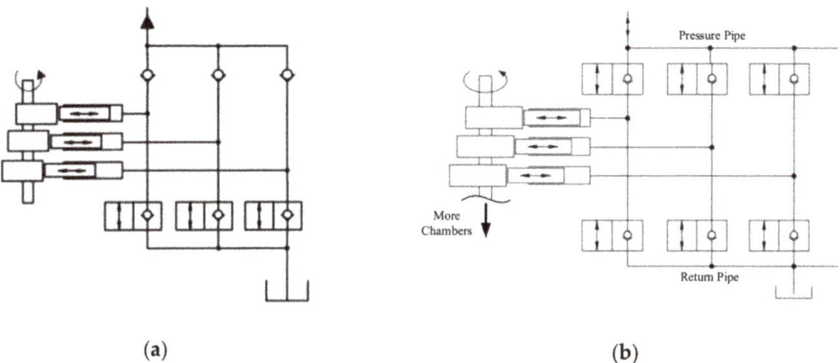

Figure 8. The piston chamber's independent control digital hydraulic components: (**a**) the digital hydraulic pump; (**b**) the digital hydraulic pump-motor.

Figure 8b shows the piston chamber independent control digital hydraulic pump-motor. The working principle is the same as that of the piston chamber independent control digital hydraulic pump, but it can be switched to be a motor. When it turns to be a digital motor, its rotation speed can be controlled by turning the switching valve on and off at high frequencies [12].

3.2.2. Digital Control of a Variable Pump

The working principle of combination cylinder control digital hydraulic pump is shown in Figure 9. When the controller outputs different code under the control of digital input signal, the number of opened switching valves varies, as does the input flow of the combined cylinder, which makes the extension length of piston rod vary. And the piston rod can control the inclination of the swash plate to realize the intelligent control of pump displacement. This kind of digital hydraulic pump can obtain different pressures and flow rates according to different encoding combinations. When the combination cylinder has N level combinations, there are 2^N kinds of pump displacement [13]. But this is just a theoretical possibility, because the forces on the swash plate are very much variable actually, and a force control on the swash plate is not enough to obtain a displacement control.

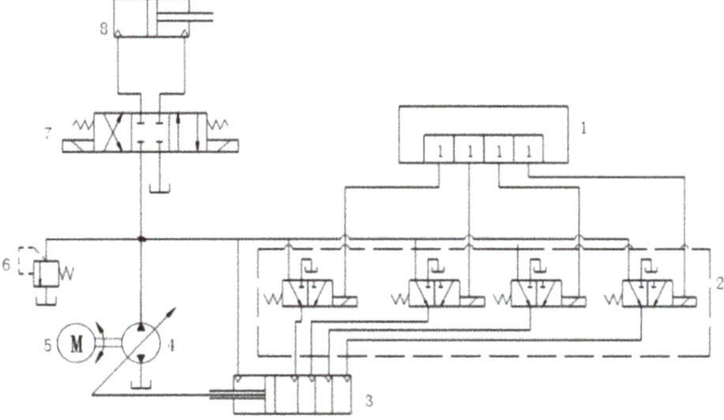

Figure 9. The digital hydraulic pump controlled by the combination cylinder.

Stepping motor control digital hydraulic pump takes the stepping motor as the driver of variable mechanism to realize the intelligent displacement control. As shown in Figure 10, the stepping motor receives the input digital signal, and converts its rotary motion to the linear motion of the spool through the conversion mechanism (such as cam and ball screw). Finally, the spool drives the variable piston rod to change the inclination of the swash plate to realize the hydraulic pump displacement control [14].

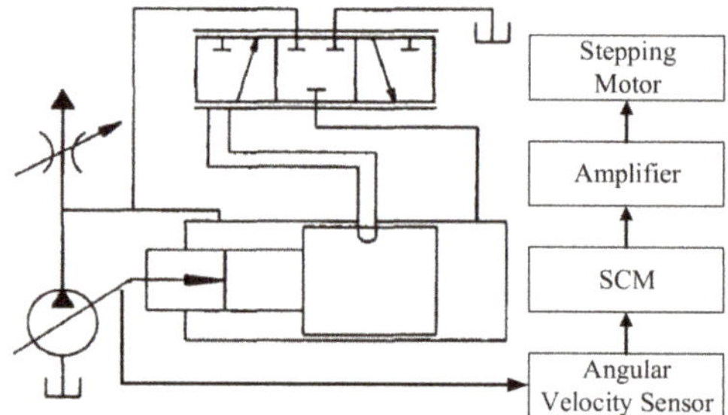

Figure 10. Digital hydraulic pump controlled by the stepping motor.

The high-speed switching valve control digital hydraulic pump adjusts the extension length of piston rod by switching the state of the high-speed switching valve between on and off at high frequency. And the extension length of piston rod can be used to control the inclination of the swash plate to realize the intelligent control of pump displacement. The working principle of such a pump is shown in Figure 11 [15].

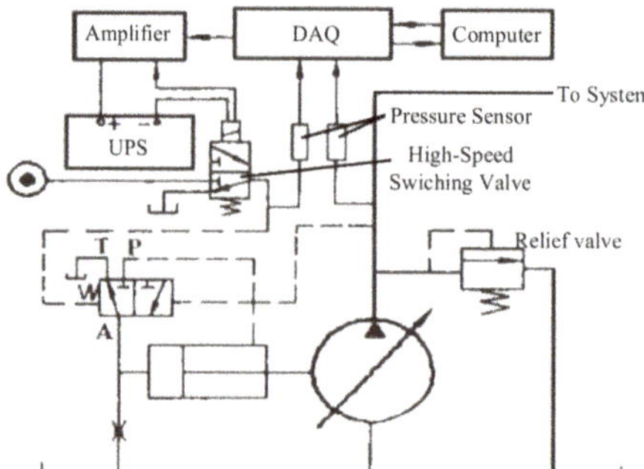

Figure 11. Digital hydraulic pump controlled by high-speed switching valve.

3.3. Digital Hydraulic Transformer

The hydraulic transformer is a new type of hydraulic component based on the constant pressure network secondary regulation system. It can adjust the system pressure to any value within the pressure variation range without throttling loss, and such a process is reversible, which means that the system can output energy to the load or recover energy from the load to the accumulator. However, the low-speed operation stability, the robustness and vibration noise of hydraulic transformer limit its application [16]. Digital hydraulic technology provides two possible solutions. The one is high-speed switching digital hydraulic transformer, which is composed of high-speed switching valves with appropriate hydraulic impedance (Figure 12a); the other is parallel linear digital hydraulic transformer, which is composed of hydraulic cylinders with multi-piston chambers (Figure 12b) [8].

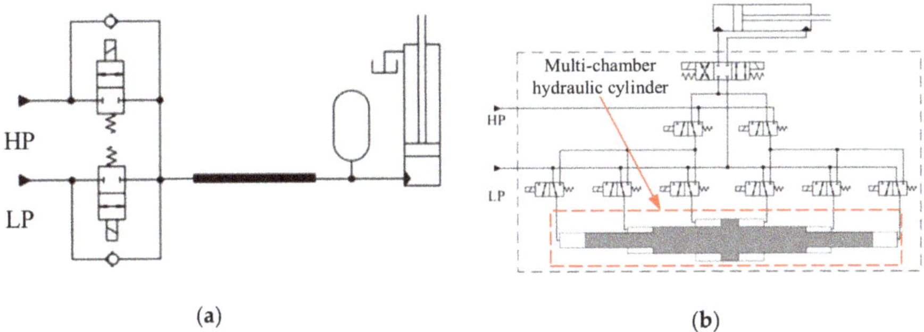

(a) (b)

Figure 12. Digital hydraulic transformers: (**a**) high-speed switching digital hydraulic transformer; (**b**) parallel linear digital hydraulic transformer.

3.4. Digital Hydraulic Power Control System

The digital hydraulic power control system (DHPMS) is an integrated volumetric component which can provide multiple independent outputs [8]. Its working pressure can be adjusted actively and adaptively according to working conditions. Figure 13a,b shows two forms of the DHPMS.

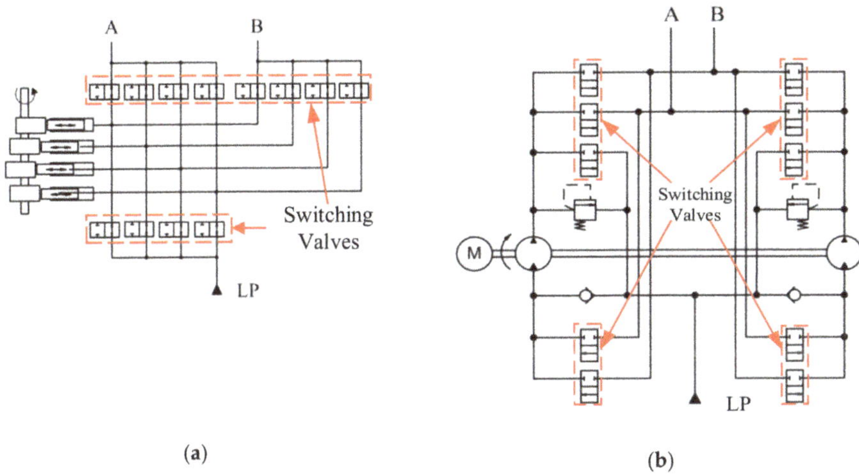

(a) (b)

Figure 13. Digital hydraulic power control system: (**a**) piston type digital hydraulic power control system (DHPMS); (**b**) quantitative pump-motor type DHPMS.

The piston type DHPMS (Figure 13a) is derived from the digital pump-motor shown in Figure 8b. The switching valves allow any piston chamber to be switched between working state and no-load state. The quantitative pump-motor type DHPMS is composed of two coaxial pump-motors, as shown as Figure 13b, and its input/output is controlled by the switching valves. A certain DHPMS with a different structure has the common, interesting feature of every independent outlet behaving like a digital pump-motor. That means the hydraulic power from the load lowering can be recovered to the accumulator, even if accumulator pressure is higher than load pressure. Thus, the whole energy storing capacity of the accumulator can be utilized [8].

3.5. Digital Hydraulic Actuator

3.5.1. Digital Hydraulic Cylinder

According to the different control methods, there are three kinds of digital hydraulic cylinder: the stepping digital hydraulic cylinder, the high-speed switching valve control digital hydraulic cylinder and the parallel digital hydraulic cylinder.

The stepping digital hydraulic cylinder (which is shown as Figure 14) controls the rotation angle and rotation speed of the stepping motor by PWM encoding of digital signal. The stepping motor's rotation can be converted into the position change of the spool. So, the flow rate of oil coming in or out of the hydraulic cylinder can be controlled intelligently and digitally and the position control of piston can be realized.

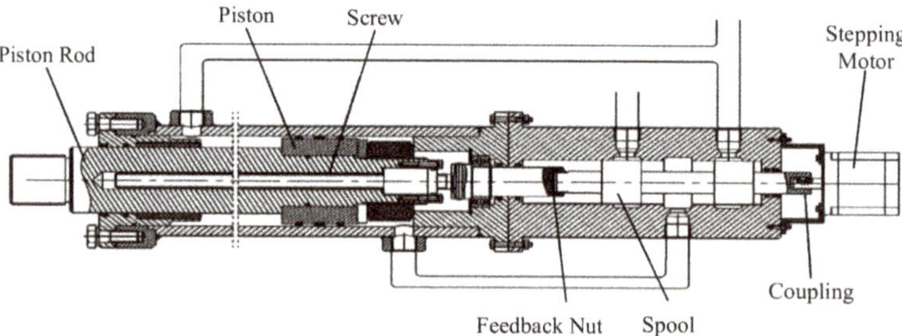

Figure 14. Stepping digital hydraulic cylinder.

The working principle of the high-speed switching valve control digital hydraulic cylinder is similar to that of the high-speed switching valve control hydraulic motor, which is shown in Figure 15a. The flow rate coming in or out of the cylinder (the displacement of the piston rod) can be adjusted through the duty cycle of PWM signal, which is used to control the high-speed switching valve. However, because the flow rate of the current high-speed switching valve is generally small, this kind of hydraulic cylinder is difficult to adapt to the conditions with high pressure and large flow.

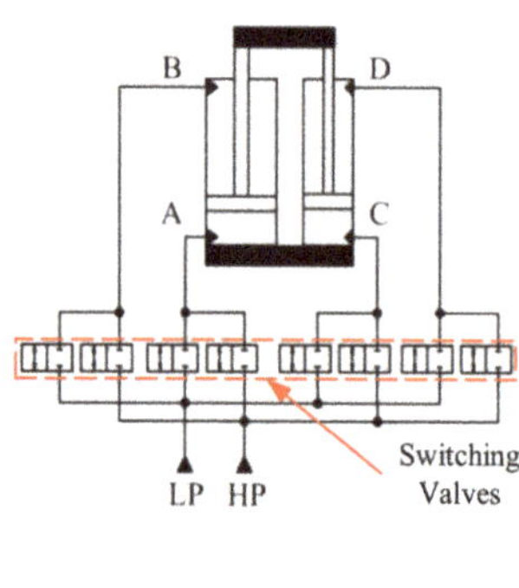

(**a**)

Figure 15. *Cont.*

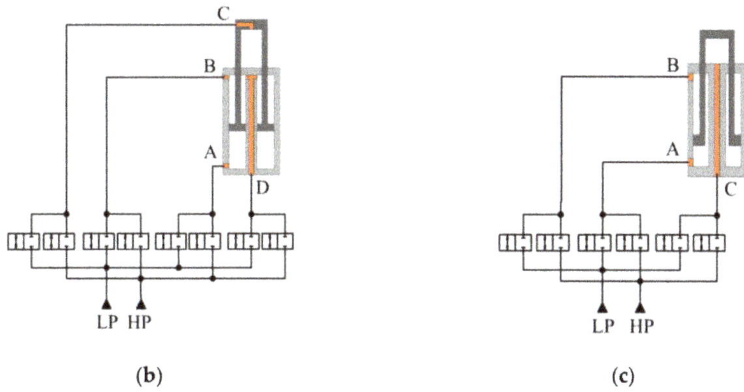

Figure 15. Parallel digital hydraulic cylinders: (**a**) multiple cylinder parallel connection mode; (**b**) parallel digital hydraulic cylinder with multiple piston cavities (1); (**c**) parallel digital hydraulic cylinder with multiple piston cavities (2).

Parallel digital hydraulic cylinder can be realized in a variety of ways, the simplest of which is to connect multiple hydraulic cylinders in parallel. The working principle is shown in Figure 15a; it controls the flow rate come in or out of the cylinder (the displacement of the piston rod) through the switching valves. However, the integration of such a system is poor. So, it needs a large installation space [8].

To avoid the deficiency of multiple cylinders in parallel connection, researchers designed some parallel digital hydraulic cylinders with integrated multiple piston cavities, whose principles are shown in Figure 15b,c. At present, this kind of parallel digital hydraulic cylinder can be integrated with up to four piston cavities, and it can provide 16 kinds of discrete output forces under different combinations of the switching valves. And this kind of cylinder can also obtain more kinds of output force by increasing the number of discrete pressure sources. When N is the number of discrete pressure sources and M is the number of piston cavities, N^M kinds of force can be output. Similar to other parallel digital hydraulic components, the parallel digital hydraulic cylinder has different dynamic/static characteristics with different encoding modes of the switch valves. The weak point is that continuous switching between control modes is required in order to obtain quasi-steady velocity. The situation is not so demanding as in the switching systems because there are much more force values available [8].

3.5.2. Digital Hydraulic Motor

There are two main kinds of digital hydraulic motors: high-speed switching valve control hydraulic motors and parallel digital hydraulic motors.

A high-speed switching valve control hydraulic motor can be realized in two ways, the principles of which are shown in Figure 16a,b respectively. The high-speed switching digital hydraulic motor shown in Figure 16a controls the duty cycle of PWM signal to adjust the flow rate or pressure of the hydraulic motor. So, the rotation speed or torque of the motor can be controlled digitally [8]. Figure 16b shows the digital flow distribution hydraulic motor with low speed and high torque; it uses five sets of high-speed switching valves (two valves for each set) to control the flow distribution of the five piston chambers [17]. So, the output torque of the motor can be controlled digitally and intelligently.

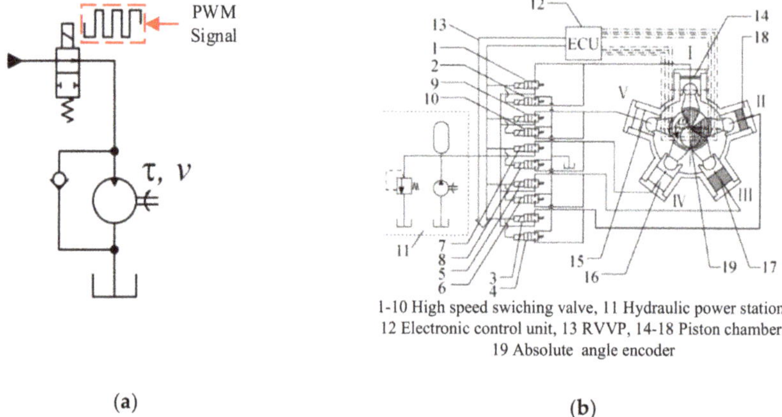

(a) (b)

Figure 16. Hydraulic motors each controlled by a high-speed switching valve: (**a**) high-speed switching digital hydraulic motor; (**b**) flow distribution digital hydraulic motor with low speed and high torque.

The principle of the parallel digital hydraulic motor is shown in Figure 17; it is composed of multiple coaxial hydraulic motors in parallel, in which each motor is independently controlled by the switching valve [8]. The maximum torque is the sum torque of all the motors in parallel; the minimum torque is the torque of the minimum motor. And the torque between the maximum torque and minimum torque depends on the different encoding mode of the switch valves.

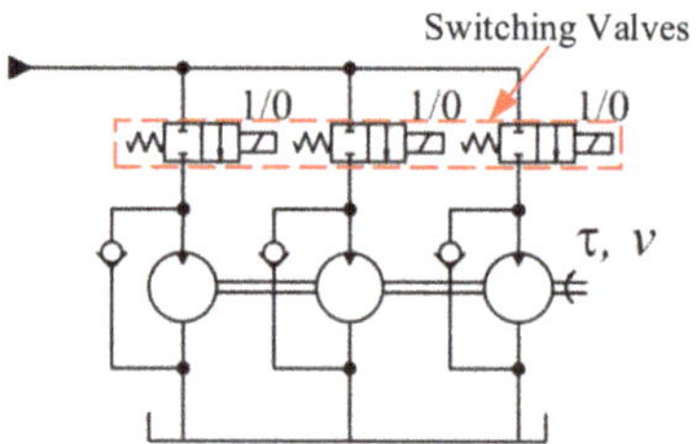

Figure 17. Parallel digital hydraulic motor.

4. Features and Advantages of Digital Hydraulic Technology

In 2013, the concept of "Industry 4.0" was officially launched at Hanover Industrial Expo, aiming to improve the intelligent level of the manufacturing industry. Industry believes that the concept of "Industry 4.0" is the fourth industrial revolution or revolutionary production method led by intelligent manufacturing. "Industry 4.0" mainly includes three aspects; namely, intelligent factories, intelligent production and intelligent logistics. As one of the important transmission technologies in industrial systems, the hydraulic system is required to be more energy efficient, more accurate and more reliable. However, as mentioned before, the low energy efficiency and high cost of conventional hydraulic technology are against such requirements. As the inheritor and innovator of traditional hydraulic technology, digital hydraulic technology is in line with "Industry 4.0." Its unique technical features and advantages will provide technical support for the development of manufacturing industry in the developmental trend of "Industry 4.0."

4.1. Features of Digital Hydraulic Technology

4.1.1. Discrete Output

Discrete output is the most basic feature of a parallel digital hydraulic system. When the number of components in parallel is N, the system has 2^N kinds of combinations of components (this is also called the state of the system). In theory, each system state generates a discrete output, but in practice, the actual number of output values depends on the coding method and the relative size of components in parallel. This section will introduce two extreme cases of output values. The minimum number of output values is achieved by using components with the same size. And the control signal is coded by pulse number modulation (PNM), which means the control signal is a certain number of pulses with the same width and amplitude. Because the components in parallel are the same size, it would not make sense for a single component in the system to receive a pulse signal, and what makes sense is the number of components receiving the pulse signal. That means the number of output values is same as the number of components receiving the pulse signal, and under such a condition the number of outputs values is $N+1$ (one is the condition that no component receives the pulse signal). The maximum number of output values is achieved by using components with entirely different sizes and the encoding of components adopting binary code. In this case, the system discrete output number is 2^N. And the outputs of a DFCU system with such two extreme cases are shown in Figure 18.

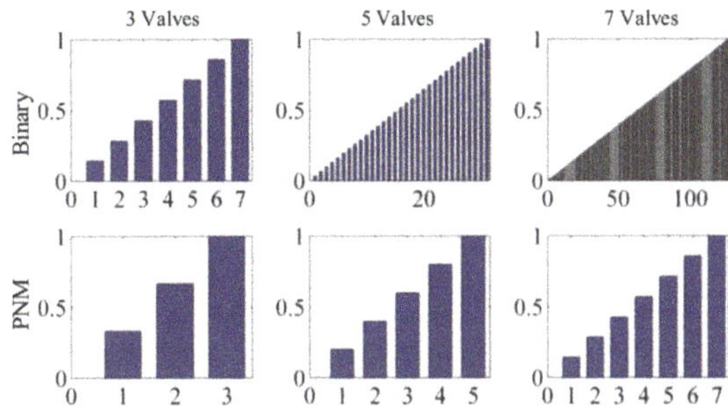

Figure 18. The output of the system with different encoding modes.

It can be seen that binary encoding can achieve a more linear output when the number of valves in parallel is the same.

4.1.2. Fast Response Time Independent of Amplitude

Since each component of the parallel digital hydraulic system works independently, DFCU can realize a direct flow rate mutation from 0% to 100% by opening all the components in parallel at the same time. That means the DFCU has a 2 ms full amplitude response time, which is the same as the switching valve. Conversely, the full amplitude response time of 3 ms is only a dream for traditional pump technology. Such a feature is especially important in the energy efficient cylinder control, because it can improve the production efficiency, which is one of the key features of smart factory in "Industry 4.0."

4.2. Advantages of Digital Hydraulic Technology

4.2.1. Fault-Tolerance Performance

The intelligent factory in "Industry 4.0" puts forward higher requirements for the fault-tolerant performance of hydraulic system, because it is an important factor to ensure production efficiency, and it is also one of the shortcomings of conventional hydraulic technology. But for the parallel digital hydraulic systems, fault-tolerance is an inherent feature. The components in parallel are independent of each other, and the system can maintain the original performance to a large extent when individual components fail. It should be noted that the fault-tolerance performance of parallel digital hydraulic system is closely related to the encoding mode of components in parallel. The fault-tolerance performance is best when the PNM code used to control the state of the system, and it is the opposite when using the binary code. Figure 19 shows fault-tolerance performance of the system when 5-bit binary code and 31-bit PNM code are adopted in case of "valve fails to turn on" [8]. But it has to be said that the "valve does not close" is a more difficult situation. And this problem also plagues researchers very much, and a large proportion of current research on control algorithms is about it.

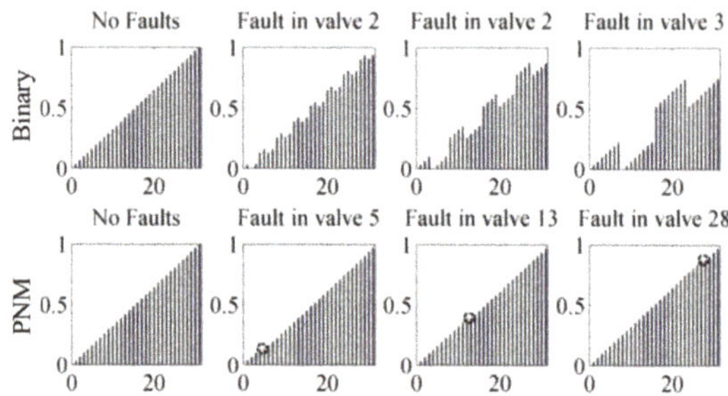

Figure 19. Fault freedom of DFCUs with different encoding modes.

4.2.2. Precise Lossless Control

Intelligent logistics is one of the three themes of "Industry 4.0," and it put forward higher requirements for the mobile outdoor robotic applications, wherein high power density, ruggedness and reliability are key features. So, the low efficiency of conventional proportional control can be a limitation. However, because the high-speed switching valves can turn on/off at high frequencies, the system does not cause excessive oil supply, so the accurate control with high energy efficiency has become the most prominent feature of high-speed switching digital hydraulic system. Take the hydraulic buck converter (HBC) as an example. It is controlled by the switching valve, and its schematic and prototype are shown in Figure 20 [18].

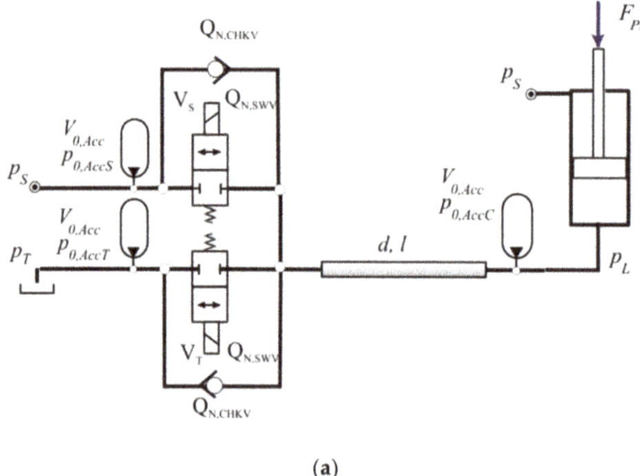

(a)

Figure 20. *Cont.*

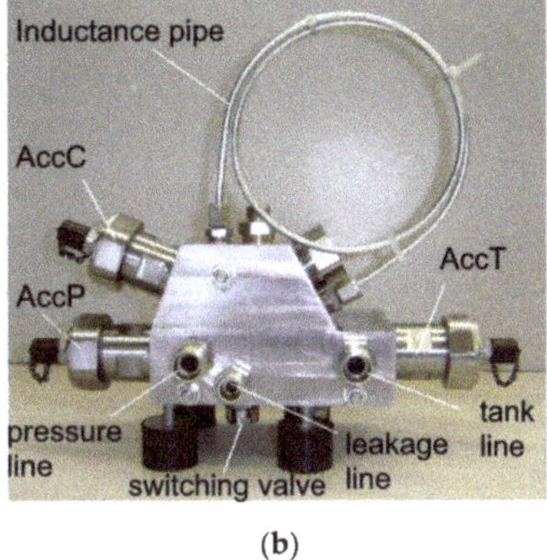

(b)

Figure 20. Hydraulic buck converter (HBC): (a) schematic of an HBC controlled linear drive; (b) prototype of a compact HBC.

Figure 21 shows the graphs of the measured pressures and efficiencies and the efficiency improvement to a proportional drive that provides same pressure and flow rate and works with the same supply pressure p_S. It shows that the measured efficiency of the high-speed switching system reaches 70–85%. At the same time, the HBC has quite a constant efficiency profile in a large operating range, and the improvements over resistance control are higher for low operating pressures when resistance control has high pressure losses [18]. In addition, we can see from Figure 22 that the HBC has the same control accuracy as the hydraulic proportional control system (p_{LHPD}).

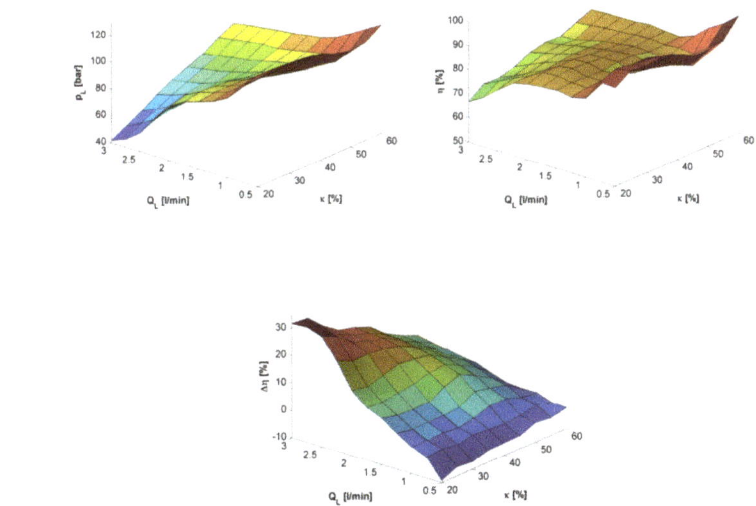

Figure 21. Measured load pressures p_L, efficiencies η, and efficiency improvements over resistance control $\Delta\eta$ of an HBC for different duty cycles κ and flow rates Q_L for a switching frequency 100 Hz.

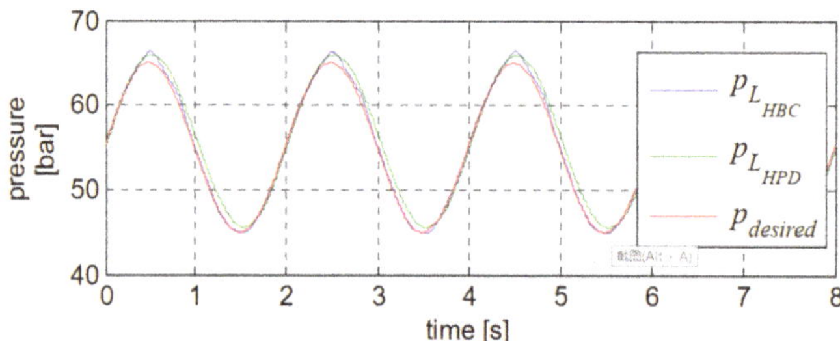

Figure 22. Measured accumulator pressures of an HBC and hydraulic proportional control for the periodic charging and discharging of a hydraulic accumulator.

4.2.3. High Accuracy Sensorless Control

Precise position control is an important actuation function in intelligent manufacturing. The conventional approach uses a cylinder with a precise position sensor and a closed loop control via a proportional or servo valve. Although its control accuracy is very high, the costs and low reliability caused by sensor, cabling, connectors and controller input modules are the main factors limiting the conventional proportional or servo valve system's application. So, the avoidance of such sensors is an effective measure to improve system reliability and make it survive and develop in "Industry 4.0." And digital hydraulic technology is providing possible approaches of high accuracy sensorless control.

A hydraulic stepping actuator is presented by Andreas Plöckinger, and its schematic is shown in Figure 23.

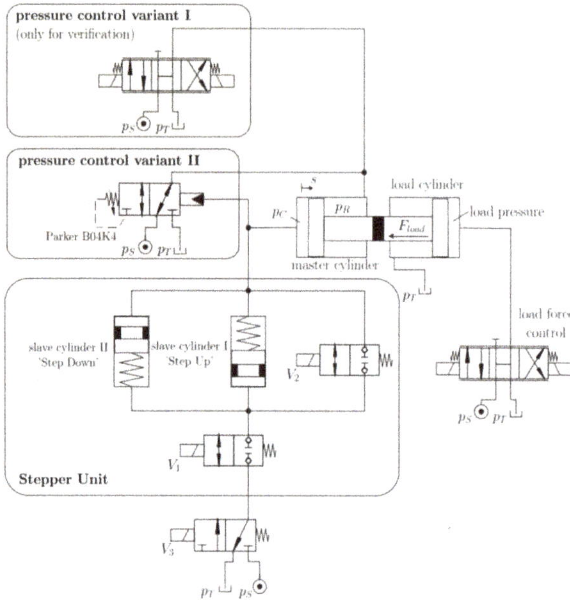

Figure 23. Schematic of the hydraulic circuit for two variants of the proposed pressure compensation definition. Variant I with an active proportional valve. Variant II with a passive pressure regulator valve.

The basic idea is to use a digital stepping unit to control the load force instead of pressure sensor. So, the load force is automatically fed back via the valve. Thus, the system would be a real "sensorless" hydraulic stepping actuator [19]. The mechanical design of the test rig is shown in Figure 24a.

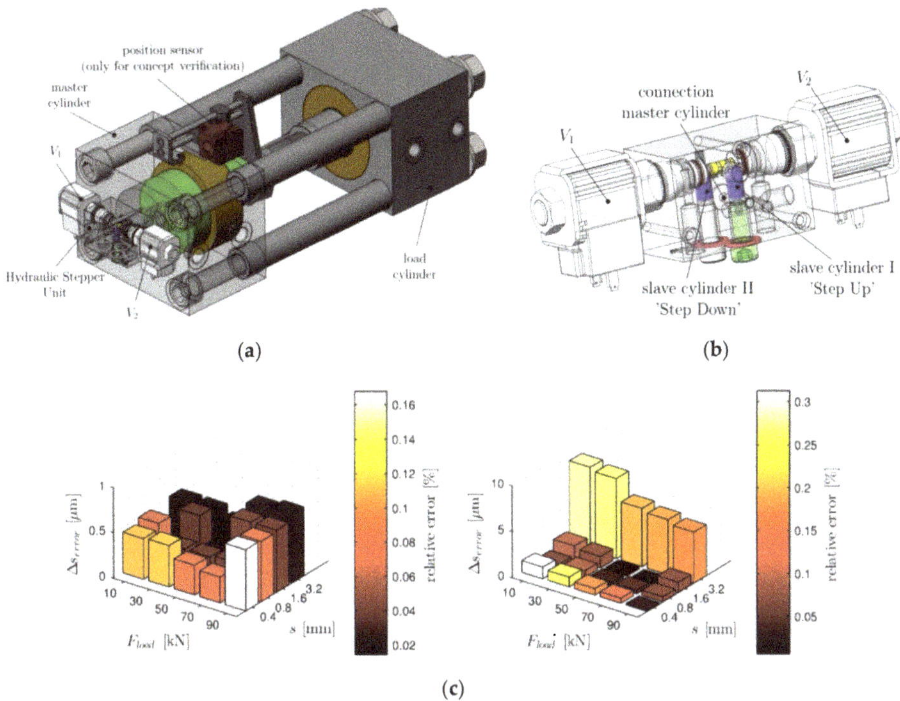

Figure 24. Hydraulic stepping actuator: (**a**) Mechanical design of the test rig; (**b**) Detail view of the Hydraulic Stepper Unit; (**c**) Position repeatability for various strokes and load forces (**left**); position accuracy of the sensorless control (**right**).

And we can see from Figure 24c that the position error is with ±0.8 µm and the relative error is less than 0.17%. For that the error after 50, 100, 200, 300, 400, 600 steps and at different forces from 10 kN to 90 kN were measured and calculated. The maximum absolute error is less than 16 µm and the relative error less than 0.55%.

4.3. Advantages and Challenges of Digital Hydraulic Technology

By integrating the technical features of parallel digital hydraulic system and high-speed switching digital hydraulic system, it can be seen that digital hydraulic technology has unique advantages over traditional hydraulic technology.

- Digital hydraulic technology directly adopts the digital signal to control without D/A conversion, which simplifies the control mode; makes the signal data storage, processing and transmission more convenient; and has stronger disturbance rejection ability, which is helpful for improving the robustness of the system.
- The digital hydraulic system has a better integration and programmability which can improve the application and maintenance performance of the system. It also facilitates networking of the system.
- The performance of the system depends on the control of the combination state of components rather than the performance of individual components. Therefore, simple and reliable components can be widely used to improve the robustness and fault-tolerance performance of the system.
- The digital hydraulic system avoids the use of proportional and servo components, and it improves the anti-pollution performance.

- It is easier to realize independent metering control. And because of the switching control mode, the system can reduce the throttle loss and improve efficiency.
- Digital hydraulic technology has obvious advantages, but there are also challenges to limit its application.
- The high frequency switching of the components can cause noise and pressure impact.
- The durability of high-speed switching hydraulic technology severely limits its application at present.
- The parallel digital hydraulic technology needs to use a large number of components, and such a situation would cause a dramatic increase in size and cost.
- A complex, unconventional control strategy would also bring difficulties to the application of digital hydraulic technology.

5. Developments and the Current Situation of Digital Hydraulic Technology

5.1. Parallel Digital Hydraulic Technology

The idea of using multiple hydraulic valves in parallel has existed since the birth of the hydraulic valves. In the document that can be found at present, as early as 1930, Rickenberg applied a patent regarding using three electromagnetic valves with different flow rates in parallel [20]. Murphy [21] also applied for a patent of a four-way valve for which the load-side is independently controlled by using DFCU. Virvalo [22] achieved the application of DFCU in the velocity control of hydraulic cylinder in 1978. However, due to the basic computer technology at that time, the parallel technology was difficult to be applied in practice. With the development of the computer technology, research and applications of the parallel technology are gradually becoming abundant.

5.1.1. Parallel Digital Hydraulic Components

At present, parallel digital hydraulic valve technology is one of the most significant research directions among the parallel digital hydraulic technology. Represented by Tampere University of Technology, the University of Aalborg and the Federal University of the State of Santa Catarina have conducted in-depth research on the parallel digital hydraulic valve technology. Among them, Linjama of Tampere University of Technology and his team put forward the basic definition of DFCU based on ordinary commercial valve earlier, and they tested the dynamic and static characteristics of DFCU [23–25].

After that, the team carried out a series aimed at the control strategy [26,27], pressure peak [28], fault detection [29] and energy saving [30,31] of DFCU. And they successfully applied the DFCU to a paper cutting machine; it is superior to the traditional hydraulic system in terms of cost, control performance and energy saving [32,33]. Now, a new generation DFCU that can integrate more switching valves is being developed [34]. Some researchers have carried out research on improving DFCU energy efficiency and fault-tolerant performance, and they also achieved some results [9,35].

The use of the parallel digital hydraulic pump (which is shown in Figure 7) can be dated back to the London water supply system in 1883 [36], and it was applied in many industries due to its excellent control and energy saving performance. However, research about the parallel digital hydraulic pump is hardly seen; only in the 1980s were there some studies on its control and energy saving [37,38]. There are more studies aimed at the digital hydraulic pump-motor (which is shown in Figure 9b); Artemis started research on the piston hydraulic pump-motor in the 1980s, but the results were not published until the 1990s [39–41]. At present, the six-piston digital pump-motor developed by Artemis can realize the independent control of each piston chamber. At the same time, Tampere University and Purdue University also carried out some research on piston-type digital hydraulic pump-motor [42–44].

The research of the parallel digital hydraulic actuator mainly centers on the hydraulic cylinder. Tampere university has studied the impedance control of a three-piston parallel digital hydraulic cylinder, and the experimental results show that the energy loss of the parallel digital hydraulic cylinder in the system with constant input pressure is reduced by 30%–60% compared with the traditional hydraulic cylinder [45]. At the same time, Linjama also found that when adopting the secondary regulation hydraulic source without throttling, the parallel digital hydraulic cylinder can achieve better energy saving effect, and this conclusion was verified by the experiment of a four-piston parallel digital hydraulic cylinder [46]. Furthermore, De Gier carried out abundant research on application of the parallel digital hydraulic cylinders with multiple pistons in high-speed and high-pressure stamping machines. He increased the movement speed of the piston rod by reducing the piston area, and he increased the output force of the hydraulic cylinder by increasing the piston area [47].

5.1.2. Application of the Parallel Digital Hydraulic Technology

The parallel digital hydraulic technology is mainly used in two aspects: direct parallel control of components and applications of parallel digital valve.

Direct parallel control of components is realized by multiple components with the same/different specifications connected in parallel, and the number of components in the accessed system (which can be considered as the output/input of the system) is intelligently controlled by combination of switching valves. Direct parallel control of components mainly contains the parallel digital hydraulic pump shown in Figure 7 and the parallel digital hydraulic motor shown in Figure 17.

The application of the parallel digital valve is the main research direction of parallel digital hydraulic technology. Bishop E. has developed a parallel linear hydraulic transformer with unique advantages over traditional hydraulic transformers in terms of pressure ratio, response time and transfer efficiency [48]. The team of Tampere University of Technology creatively has put forward the definition of digital hydraulic momentum control system (DHPMS), and published its simulation analysis results for the first time in 2009 [49,50], and their experimental analysis results were published in the following year. The results show that DHPMS has obvious advantages in terms of control performance and energy efficiency [51]. At present, the research and development of the second generation DHPMS (which is shown in Figure 25) have also made breakthroughs [52].

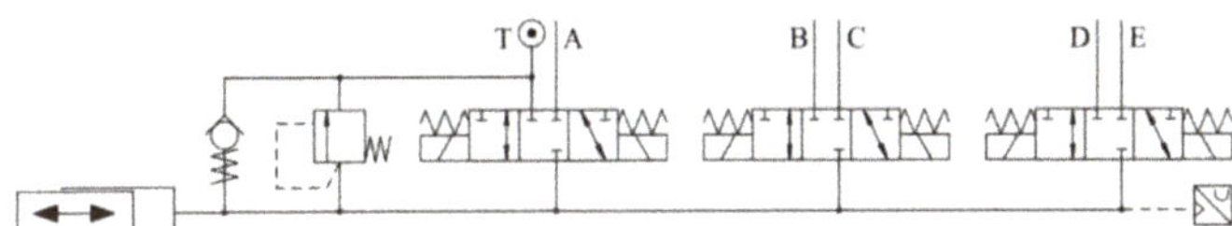

Figure 25. Schematic diagram of second generation DHPMS.

Johan Ersfolk optimized the parallel digital valve control of hydraulic cylinders by using a modern embedded graphics processing unit (GPU), and the results showed that the large-scale parallel characteristics brought by the GPU enabled the controller to obtain better control performance than conventional controllers [53].

5.2. High-Speed Switching Digital Hydraulic Technology

The automobile industry has greatly promoted the development of high-speed switching digital hydraulic technology. The high-speed switching valves which can turn on and off more than 1000 times per second are the first and the most widely applied in the ABS braking system [54,55].

The high-pressure fuel injection technology which emerged around the year of 2000 put forward a series of technical requirements for high-speed switching valve. Since the working pressure is 200 Mpa, it requires that the valve can realize five times on/off switching for each combustion. And the valve failures are not allowed during vehicle service. Although the high-speed switching solenoid valve in the automobile industry generally has a small flow rate and is not suitable for hydraulic system, its successful promotion and application have proved that the high-speed switching technology is feasible and reliable. And that is the reason why some valves in the automobile industry are mentioned below when the high-speed switching digital hydraulic valve is introduced.

5.2.1. High-Speed Switching Digital Hydraulic Valve

The research of high-speed switching hydraulic technology focuses on the R&D of high-speed switching valves, which has two aspects. The one is the research and development of new high-speed switching valves, and the other is the control method of high-speed switching valves.

The development of the high-speed switching valve can be traced back to the end of the 1970s. The company Lucas in Britain developed the high-speed solenoid switching valve using two special shape electromagnets: spiral tube type and taper type, which were called the "Helenoid valve" and "Colenoid valve." They overcame the problem that the acceleration of armature is inversely proportional to electromagnetic force.

But due to their complicated structures and difficulty in processing and manufacturing, they are not widely used [56,57]. Since then, many companies and research institutions launched various types of high-speed switching valves. BKM company launched a three-way spherical, cartridge type, high-speed electromagnetic switching valve with a response time of 2–3 ms, and its working pressure is 10 MPa [58]. But this valve can only be used for direct digital control of electronic unit injectors because of its small flow rate. Yukio Tanaka developed a two-way high-speed switching valve and a three-way high-speed switching valve in the 1980s; their response times are all around 3 ms [59].

In the late 1980s, Masahiko Miyamoto developed an ultra-high-pressure high-speed switching valve with working pressure of 120 MPa and response time of 0.4 ms [60–62]. However, because these valves do not overcome the small flow rate, their applications are limited to the field of fuel injection. Bosch company also successfully developed a high-speed electromagnetic switching valve suitable for an ultra-high-pressure environment, and its response time is between 0.3 and 0.65 ms [63]. Linz center developed a high frequency switching ball valve which is shown in Figure 26a, and the spool is controlled by current feedback. The spool position of this valve is only 5 mm, and its frequency response is up to 1 kHz; its flow rate can reach 14 L/min [64]. But for now, the exemplary application for such a valve could be the actuation of automotive wet clutches; in particular, those of dual clutch systems. Tampere optimized the surface material and the heat-treatment process of the cone valve, and they obtained a high-speed switching water hydraulic valve with high reliability and long service life (which is shown in Figure 26b) [65]. Minnesota university developed a rotary high-speed switching valve with a special structure pilot spool which is shown in Figure 26c; this valve's maximum flow rate is 40 L/min, and its frequency response is 100 Hz [66]. Zhejiang University designed a PZT piezoceramic high-speed switching valve with temperature compensation, which is shown in Figure 26d. This valve's working pressure can reach 20 MPa, and its frequency response and flow rate are, respectively, 200Hz and 10 L/min [67]. Guizhou Honglin machinery factory cooperated with BKM company developed a thread cartridge HSV high-speed electromagnetic switching valve, whose opening time is 3 ms, and closing time is 2 ms. This valve 's highest working pressure is 20 MPa, and its flow rate can reach 2–9 L/min [68].

Zhejiang University of Technology developed a high-speed switching valve with a high frequency and large flow rate. This valve's working pressure is 21 Mpa, and its flow rate is 450 L/min. The spool position of the valve reaches 6 mm, but the response time is only about 8 ms [69].

The control of high-speed switching valve has a crucial effect on its performance, but few commercial controllers can be used for high-speed switching valve at present [70,71]. To solve such a problem, Linjama developed a set of controllers with response time of 8–12 ms that can directly control the cartridge valve [59]. Zhejiang University put forward an intelligent voltage control method based on current feedback. The self-adaptability of coil resistance makes the excitation time of each voltage segment vary adaptively with the change of coil resistance. And such a control method keeps the dynamic characteristics of the valve at a high level during its working procedure [72]. Wuhan University proposed a zero-flow switching control method, which enables the spool of high-speed switching valve to always switch at the zero-flow point; it avoids the pressure impact and energy loss caused by high-frequency switching [73].

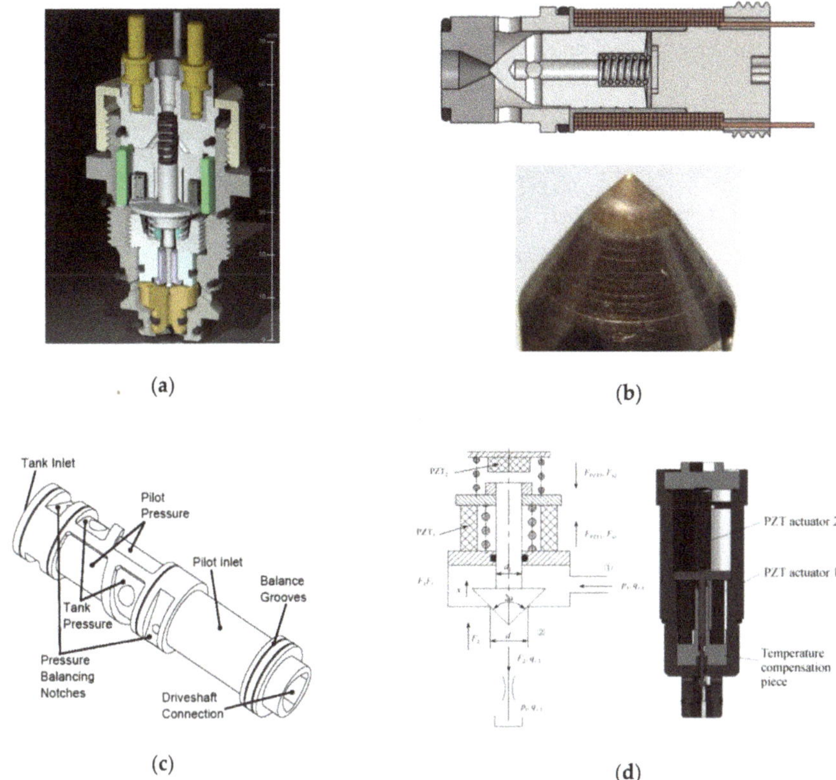

Figure 26. New high-speed switching valves: (**a**) Linz high frequency switching valve; (**b**) Tampere hydraulic water valve; (**c**) rotation pilot spool; (**d**) PZT piezoceramic high-speed switching valve.

5.2.2. Applications of High-Speed Switching Hydraulic Technology

Because the current high-speed switching valves generally have a small flow rate, they are mostly used as the pilot control parts of other hydraulic components to achieve hydraulic components' intelligent digital. Among them, the pilot control of a proportional valve emerged in the 1990s as a successful example [74–76].

With the continuous development of high-speed switch technology, its application is gradually widespread. Minnesota University used high-speed switching valves to control the input/output of each piston chamber of a low-speed radial hydraulic motor with large torque (which is shown in Figure 27a). And so far, such a digital motor is used as part of a 7 MW wind turbine drive train (which is shown in Figure 27b). The application of high-speed switching valves improves the energy efficiency of the hydraulic motor [77]. Tyler Helmus also adopted a similar method to control the hydraulic pump-motor, and he also achieved excellent control effect of the pump-motor [78].

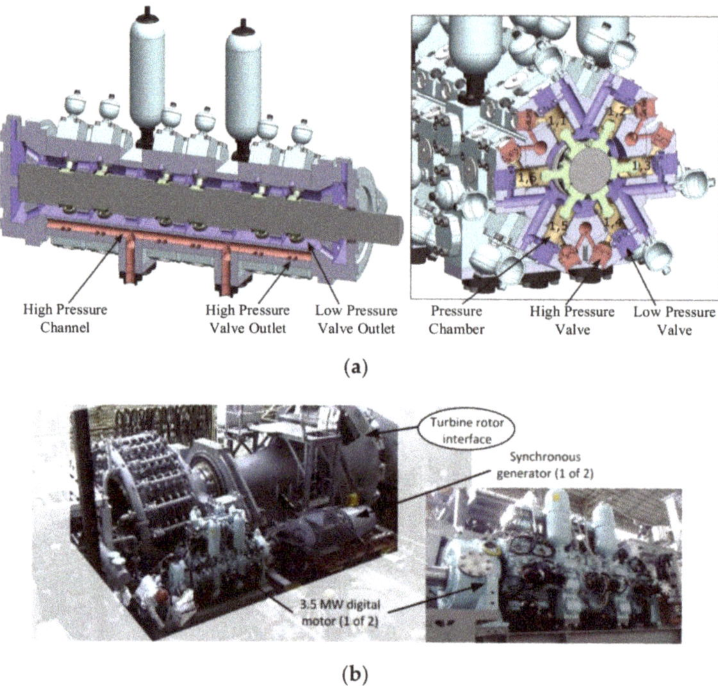

Figure 27. (**a**) High-speed switching valve controlled hydraulic motor; (**b**) 7 MW wind turbine drive train using a large scale digital hydraulic pump and two digital motors.

Yang huayong proposed a variable axial piston pump control method (which is shown in Figure 28) using high-speed switching valve for pilot control, and the experimental results proved that this method also has excellent control effect [79].

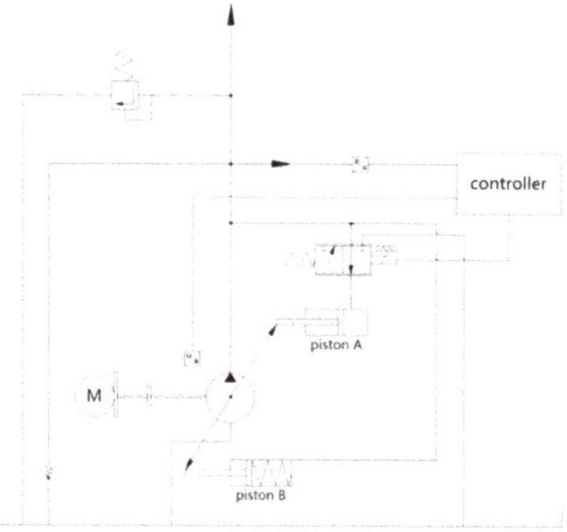

Figure 28. High-speed switching valve controlled variable axial piston pump.

Rainer Haas used different high-speed switching valve layouts to carry out position control of the hydraulic cylinder (which is shown in Figure 29). And he analyzed the position and speed response of the hydraulic cylinder with different control strategies [80].

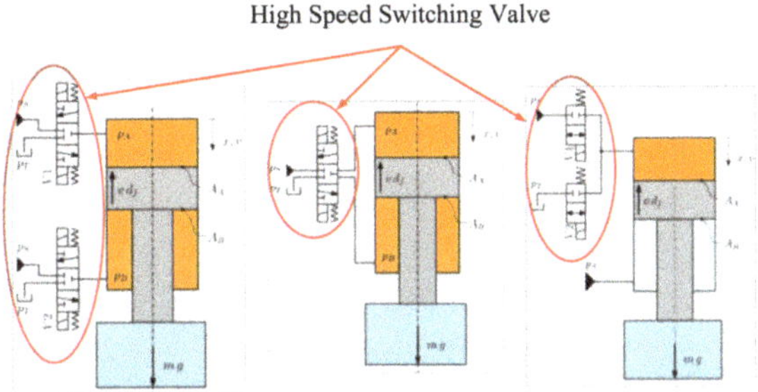

Figure 29. Position and speed control of the hydraulic cylinder.

Scheidl and his research team carried out a series of studies on the control method and energy-saving of high-speed switching hydraulic torque converters. Among them, one small torque converter obtains an average efficiency of 80% when the maximum output power is 1 kW [18,81,82]. Marcos Paulo Nostrani used the high-speed switching technology to control the hydraulic transformer, and they reduced its energy consumption effectively [83]. Shi Guanglin realized precise control of the pneumatic robot by using high-speed switching valves [84].

5.3. Stepping Digital Hydraulic Technology

The development of the stepping digital hydraulic technology is thanks to the mature stepping motor technology. Especially after the 1980s, the stepping motor control method was more flexible and diverse because the cheap microcomputer with multiple functions had appeared in industry. Therefore, the stepping motor gradually met the functional requirements of controlling hydraulic components [85]. But there are still some problems, such as the lower rotation speed, the small torque, falling out-of-step under high frequency, etc. So, the development and popularization of stepping digital hydraulic technology are still limited. What is more, with the development and perfection of parallel digital hydraulic technology and high-speed switching hydraulic technology, the research on the stepping digital hydraulic technology are gradually reduced.

5.3.1. Stepping Digital Hydraulic Components

The R&D of the stepping digital hydraulic valve is more advanced in Japan, and relevant studies and applications have also been carried out in France, Britain, Canada and other countries [86–88]. Among them, the stepping-type digital flow valve and pressure valve of Tokyo Keiki formed a complete product line. The pressure of the valves can reach 210 Mpa, and the flow rate is 1–500 L/min. And the repeatability accuracy and hysteresis accuracy are less than 0.1% [89]. In addition, the companies Yuken, Toyooki, Uchida, Sperry, Vickers, Danfos, Beijing Aemetec Digital Hydraulic Ltd., etc. produced stepping digital hydraulic valve products.

Among them, a 2D digital hydraulic valve developed by Zhejiang University of Technology is the most distinctive. The spool of this valve has dual degrees of freedom; one is the rotation around the axis, and the other is the linear motion along the axis. Stepping motor drives the spool to rotate in a certain angle range through the transmission mechanism to realize the function of the pilot valve. And there is a spiral groove on the inner surface of the valve sleeve, the linear motion of the spool is driven by the area difference between the low-pressure hole, the high-pressure hole and the spiral groove to realize the function of the main valve. Its working principle is shown in Figure 30 [90]. At present, 2D digital hydraulic valve has formed a relatively complete product series, and they have already been put into service in the aviation industry.

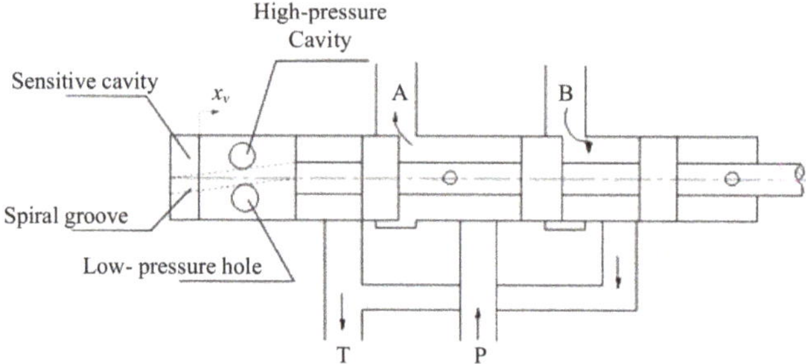

Figure 30. Working principle of 2D digital hydraulic valve.

Research on stepping digital hydraulic cylinder began in the 1960s and 1970s; the world's first digital hydraulic cylinder was exhibited in Olympia Hall by Germany. Since then, the company of Rexroth, Tokyo Keiki, Beijing Aemetec Digital Hydraulic Ltd., etc. have developed a variety of stepping digital hydraulic cylinders with various structures [91].

5.3.2. Applications of Stepping Digital Hydraulic Technology

The most important applications of stepping digital hydraulic technology are the use of customized digital hydraulic cylinders/valves to achieve accurate position/speed control in extreme environments. At the same time, accurate synchronization control of multiple hydraulic cylinders has also been widely applicated. Among them, the most representative digital cylinder/valve products come from Beijing Aemetec Digital Hydraulic Ltd. (Beijing, China); their products have been successfully applied in a number of military engineering, large water conservancy engineering and metallurgical engineering applications [92–94]. For example, a stepping digital hydraulic cylinder is used for the crystallizer liquid level automatic control system (which is shown in Figure 31).

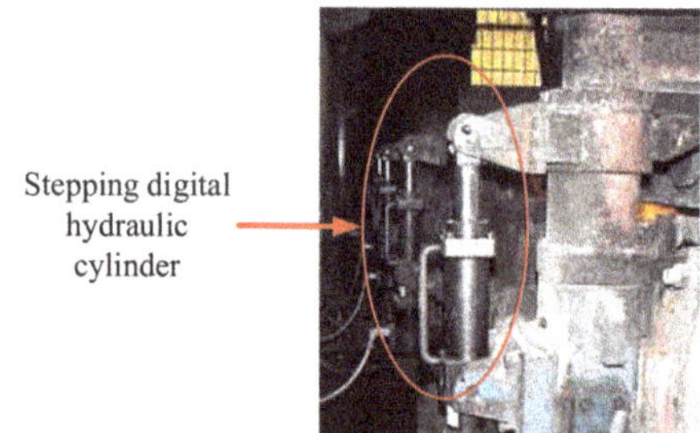

Figure 31. Practical application of digital hydraulic cylinder.

6. Developing Trend of the Digital Hydraulic Technology

After decades of research, digital hydraulic components have formed a series of products with complete functions and various categories. Some products have been used in aerospace, construction machinery, shipbuilding industry and other industries; its technology has become increasingly mature. However, under the development trend of high speed, high pressure and high power to weight ratio of hydraulic system, the problems such as lower flow rate, lower allowable pressure and serious dispersion of system have also become more prominent. To solve such problems, researchers have also invested a lot of energy into optimizing the existing digital hydraulic products. Some of the research results have not only been applied in engineering, but also evolved into some developmental direction of digital hydraulic technology.

6.1. Development of a New Valve Prototype

High-speed switching digital hydraulic technology puts forward the technical index of switching frequency 500 Hz (that means the opening time is 2 ms) for high-speed switching valve. However, the flow rate of the current valve which can reach such a technical index is generally small. So, the valves cannot satisfy the need of continuous switching control under large flow. Therefore, many research institutions and universities have invested a lot of resources in the R&D of new valves, and they had put forward different valve prototypes, but few valve products can be put into industrial applications. It can be concluded that the R&D of new high-speed switching valves with high working pressure and large flow rate will become an inevitable trend of digital hydraulic technology.

6.2. Integration of Digital Hydraulic Technology

Parallel digital hydraulic technology and high-speed switching digital hydraulic technology, are two main branches of current digital hydraulic technology; they have their own unique technological advantages and face different challenges. Therefore, it is also a research direction of digital hydraulic technology to combine the two main branches of current digital hydraulic technology and take their respective advantages. For example, Huova replaced the smallest switching valve unit in DFCU with high-speed switching valve, which successfully reduced the pressure impact of single high-speed switching control. At the same time, they also obtain accurate speed control of hydraulic cylinders [95].

6.3. Improvement of Energy Efficiency

One of the important characteristics of digital hydraulic technology which is different from traditional hydraulic technology, is that it can realize intelligent hydraulic energy supply based on the system requirements. So, digital hydraulic components, such as the digital hydraulic pump-motor, digital hydraulic transformer, DHPMS etc., all have high energy efficiency in theory, but that still needs a lot of experimental verification and optimization. Therefore, improving energy efficiency is also an important development trend of digital hydraulic technology.

7. Conclusions

This paper explains the mainstream definition of digital hydraulics, and it gives a more accurate definition of digital hydraulic technology. Meanwhile, this paper presents a review of developmental works on digital hydraulic components and digital hydraulic technology. The main outcomes of this review work are as follows:

- With the continuous promotion of "Industry 4.0" in the world, traditional hydraulic technology has been marginalized because of its low energy efficiency and lack of intelligence. The digital hydraulic technology will be able to make up for the defects of the traditional hydraulic technology, and play a greater role in intelligent factories and intelligent manufacturing.
- With the advantages of great fault-tolerance and fast response performance, parallel digital hydraulic technology has become one of the mainstream research directions in digital hydraulic technology. But it still needs to solve the problems of high cost and large volume after a large number of switching components are connected in parallel. At the same time, the lack of accurate and suitable control algorithms of parallel systems has also become an important factor hindering the development of parallel digital hydraulic technology.
- High-speed switching digital hydraulic technology is also one of the main research directions in digital hydraulic technology. It can achieve precise lossless control performance, and its response time can also reach considerable millisecond level. However, the development and application of high-speed switching digital hydraulic technology are restricted by the vibration, noise, pulsation

and other problems caused by the frequent opening and closing of high-speed switching valve, and the lives of high-speed switch components themselves. At the same time, the problem of high-speed switching valves' insufficient service lives also needs to be solved.

- The development of stepping digital hydraulic technology started earlier. And it is famous for its high accuracy, sensorless control performance which can greatly simplify a system and improve the system's usability and maintainability. However, the application and development of digital hydraulic technology are limited because the stepping motor is prone to being out-of-step at high frequency. In recent years, with the continuous development and improvement of parallel technology and high-speed switching technology, the stepping digital hydraulic technology has faded out of the mainstream research direction of digital hydraulic technology.
- The "Workshop on Digital Fluid Power" held by Tampere university is the most famous academic conference on digital hydraulic technology in the world. Looking at the papers published in the conference in recent years, it can be seen that the main research directions of researchers on digital hydraulic technology are focused on control algorithms, new valve prototypes, digital pump motors, etc. The purpose of researchers is to further improve the energy efficiency, reliability and practicability of digital hydraulic technology, and lay a foundation for promoting the practical application of digital hydraulic technology.

The definition of Industry 4.0 promotes the progress of the whole society and industrial system. Only the technology that meets the needs of society can survive and develop. And digital hydraulic technology provides a feasible way for the traditional hydraulic industry to develop in the direction of intelligence and greenness. At the same time, as the innovation of and successor to traditional hydraulic technology, digital hydraulic technology will certainly make fluid power technology developing in line with "Industry 4.0."

With continuous in-depth research, digital hydraulic technology will continue to innovate. More mature digital hydraulic components will also have more extensive engineering application prospects.

Author Contributions: Q.Z. is responsible for article writing and research on digital hydraulic technology. X.K. is responsible for the general idea of this article, and he gives the defination of digital hydraulic technology. B.Y. investigates the digital hydraulic components. K.B. investigates the features and advantages of digital hydraulic technology. Z.J. investigates the developments and the current situation of digital hydraulic technology. Y.K. investigates the developing trend of digital hydraulic technology. All authors have read and agreed to the published version of the manuscript.

Funding: This research was funded by NATIONAL KEY RESEARCH AND DEVELOPMENT PROGRAM, grant number 2018YFB2000701, and the NATIONAL NATUAL SCIENCE FOUNDATION OF CHINA, grant number 51975506.

Conflicts of Interest: The authors declare no conflict of interest.

References

1. Lu, Y. History progress and prospects of fluid power transmission and control. *Chin. J. Mech. Eng.* **2010**, *46*, 1–9.
2. Ba, K.; Yu, B.; Zhu, Q.; Gao, Z.; Kong, X. The position-based impedance control combined with compliance-eliminated and feedforward compensation for HDU of legged robot. *J. Frankl. Inst.* **2019**, *356*, 9232–9253. [CrossRef]
3. Ba, K.; Yu, B.; Gao, Z.; Ma, G.; Kong, X. An improved force-based impedance control method for the legged robot HDU. *Isa Trans.* **2019**, *84*, 187–205. [CrossRef] [PubMed]
4. Achten, P. Convicted to innovation in fluid power. *Proc. Inst. Mech. Eng. Part I* **2010**, *224*, 619–621. [CrossRef]
5. Kagoshima, M.; Komiyama, M.; Nanjo, T.; Tsutsui, A. Development of new kind of hybrid excavator. *Res. Dev. Kobe Steel Eng. Rep.* **2007**, *57*, 66–69.
6. Scheidl, R. Discussion: Is the future of fluid power digital? *Proc. Inst. Mech. Eng. Part I* **2012**, *226*, 724–727.
7. Yang, H. *Progress and Trend of Construction Machinery Intelligence*; Construction Machinery Technology and Management: Beijing, China, 2017; pp. 19–21.

8. Linjama, M. Digital Fluid Power—State of the Art. In Proceedings of the Twelfth Scandinavian International Conference on Fluid Power, Tampere, Finland, 18–20 May 2011.
9. Linjama, M. On The Numerical Solution of Steady-State Equations of Digital Hydraulic Valve Actuator System. In Proceedings of the Eighth Workshop on Digital Fluid Power, Tampere, Finland, 24–25 May 2016.
10. Zhao, X. Research on the Theory and Application of HGDV pulse Modulation Switching Digital Hydraulic Valve. Master's Thesis, Lanzhou University of Technology, Lanzhou, China, 2005.
11. Yang, H. Development direction of digital hydraulic valve technology. In Proceedings of the Shanghai: 9th FPTC-2016, Shanghai, China, 21–24 November 2016.
12. Breidi, F.; Helmus, T.; Lumkes, J. The Impact of Peak-and-Hold and Reverse Current Solenoid Driving Strategies on the Dynamic Performance of Commercial Cartridge Valves in a Digital Pump/Motor. *Int. J. Fluid Power* **2015**, *17*, 37–47. [CrossRef]
13. Ding, X. Research on Digital Hydraulic Valve Controlling Axial Piston Pump. Master's Thesis, Taiyuan University of Science and Technology, Taiyuan, China, 2015.
14. Wang, Q. Study of the Digital Control Variable Axial Piston Pump and its SCM Control. Master's Thesis, Shenyang University of Technology, Shenyang, China, 2015.
15. Liu, Z. Research on Electro-Hydraulic Digital Control of Constant Pressure Variable Pump System. *Mach. Tool Hydraul.* **2001**, *2*, 29–31.
16. Yang, H.; Ouyang, X.; Xu, B. Development status of hydraulic transformer. *J. Mech. Eng.* **2003**, *39*, 1–5. [CrossRef]
17. Shi, G.; Yu, L.; Qi, L. Simulation of Radial Piston Constant Flow Pump with Digital Distribution under Random Low Speed Driving. In Proceedings of the Eighth Workshop on Digital Fluid Power, Tampere, Finland, 24–25 May 2016.
18. Kogler, H.; Scheidl, R.; Ehrentraut, M.; Guglielmino, E.; Semini, C.; Caldwell, D.G. A Compact Hydraulic Switching Converter for Robotic Applications. In Proceedings of the Fluid Power and Motion Control (FPMC2010), Bath, UK, 15–17 September 2010; Johnston, D.N., Plummer, A., Eds.; Hadleys Ltd.: Theale, UK, 2010; pp. 55–68.
19. Plöckinger, A.; Grad, C.; Scheid, R. High Accuracy Sensorless Hydraulic Stepping Actuator. In Proceedings of the Eighth Workshop on Digital Fluid Power, Tampere, Finland, 24–25 May 2016.
20. Rickenberg, F. Valve. U.S. Patent No. 1757059, 30 April 1930.
21. Murphy, R.; Weil, J. Hydraulic Control System. U.S. Patent No. 3038449, 20 June 1962.
22. Virvalo, T. Cylinder Speed Synchronization. *Hydraul. Pneum.* **1978**, *31*, 55–57.
23. Linjama, M.; Koskinen, K.T.; Vilenius, M. Pseudo-Proportional Position Control of Water Hydraulic Cylinder Using On/Off Valves. In Proceedings of the Fifth JFPS International Symposium on Fluid Power, Nara, Japan, 12–15 November 2002; pp. 155–160.
24. Laamanen, A.; Linjama, M.; Tammisto, J.; Koskinen, K.T.; Vilenius, M. Velocity Control of Water Hydraulic Motor. In Proceedings of the Fifth JFPS International Symposium on Fluid Power, Nara, Japan, 12–15 November 2002; pp. 167–172.
25. Laamanen, A.; Linjama, M.; Vilenius, M. Characteristics of a Digital Flow Control Unit with PCM Control. In Proceedings of the Seventh Triennial International Symposium on Fluid Control, Measurement and Visualization, Sorrento, Italy, 25–28 August 2003. ISBN 0-9533991-4-1.
26. Linjama, M.; Koskinen, K.T.; Vilenius, M. Accurate Trajectory Tracking Control of Water Hydraulic Cylinder with Non-Ideal on/Off Valves. *Int. J. Fluid Power* **2002**, *4*, 7–16. [CrossRef]
27. Linjama, M.; Vilenius, M. Improved Digital Hydraulic Tracking Control of Water Hydraulic Cylinder Drive. *Int. J. Fluid Power* **2005**, *6*, 29–39. [CrossRef]
28. Siivonen, L.; Linjama, M.; Huova, M.; Vilenius, M. Pressure based fault detection and diagnosis of a digital valve system. In Proceedings of the Power Transmission and Motion Control (PTMC 2007), Bath, UK, 12–14 September 2007; Johnston, D.N., Plummer, A., Eds.; Hadleys Ltd.: Theale, UK, 2007; pp. 67–79.
29. Siivonen, L.; Linjama, M.; Huova, M.; Vilenius, M. Jammed On/Off Valve Fault Compensation with Distributed Digital Valve System. *Int. J. Fluid Power* **2009**, *10*, 73–82. [CrossRef]
30. Linjama, M.; Huova, M.; Vilenius, M. Online Minimization of Power Losses in Distributed Digital Hydraulic Valve System. In Proceedings of the 6th International Fluid Power Conference Dresden, Dresden, Germany, 1–2 April 2008; Volume 1, pp. 157–171.

31. Huova, M.; Karvonen, M.; Ahola, V.; Linjama, M.; Vilenius, M. Energy Efficient Control of Multiactuator Digital Hydraulic Mobile Machine. In Proceedings of the 7th International Fluid Power Conference, Aachen, Germany, 22–24 March 2010; Volume 1, pp. 25–36.
32. Linjama, M.; Hopponen, V.; Ikonen, A.; Rintamäki, P.; Vilenius, M.; Pietola, M. Design and Implementation of Digital Hydraulic Synchronization and Force Control System. In Proceedings of the 11th Scandinavian International Conference on Fluid Power SICFP'09, Linköping, Sweden, 2–4 June 2009. 13p.
33. Hopponen, V.; Linjama, M.; Ikonen, A.; Rintamäki, P.; Pietola, M.; Vilenius, M. Energy Efficient Digital Hydraulic Force Control. In Proceedings of the 11th Scandinavian International Conference on Fluid Power SICFP'09, Linköping, Sweden, 2–4 June 2009. 11p.
34. Linjama, M.; Karvonen, M. Digital Microhydraulics. In Proceedings of the First Workshop on Digital Fluid Power, Tampere, Finland, 3 October 2008; Linjama, M., Laamanen, A., Eds.; pp. 141–152.
35. Linjama, M.; Huova, M.; Karhu, O.; Huhtala, K. Energy Efficient Tracking Control of a Mobile Machine Boom Mockup. In Proceedings of the Eighth Workshop on Digital Fluid Power, Tampere, Finland, 24–25 May 2016.
36. Pugh, B. *The Hydraulic Age—Public Power Supplies before Electricity*; Mechanical Engineering Publications Ltd.: London, UK, 1980; 176p.
37. Lambeck, R.P. *Hydraulic Pumps and Motors: Selection and Application for Hydraulic Power Control Systems*; Dekker: New York, NY, USA, 1983; 154p.
38. Moorhead, J.R. Saving Energy with "Digital" Pump Systems. *Mach. Des.* **1984**, *56*, 40–44.
39. Rampen, W.H.S.; Salter, S.H. The Digital Displacement Hydraulic Piston Pump. In Proceedings of the 9th International Symposium on Fluid Power, Cambridge, UK, 25–27 April 1990; BHR Group: Cambridge, UK, 1990; pp. 33–46.
40. Ehsan, M.; Rampen, W.H.S.; Salter, S.H. Modeling of Digital-Displacement Pump-Motors and Their Application as Hydraulic Drives for Nonuniform Loads. *Asme J. Dyn. Syst. Meas. Control* **2000**, *122*, 210–215. [CrossRef]
41. Payne, G.S.; Kiprakis, A.E.; Ehsan, M.; Rampen, W.H.S.; Chick, J.P.; Wallace, A.R. Efficiency and Dynamic Performance of Digital Displacement^{TM} Hydraulic Transmission in Tidal Current Energy Converters. *Proc. Inst. Mech. Eng. Part A* **2007**, *221*, 207–218. [CrossRef]
42. Tammisto, J.; Huova, M.; Heikkilä, M.; Linjama, M.; Huhtala, K. Measured Characteristics of an In-Line Pump with Independently Controlled Pistons. In Proceedings of the 7th International Fluid Power Conference, Aachen, Germany, 22–24 March 2010; Volume 1, pp. 361–372.
43. Lumkes, J.; Batdorff, M.; Mahrenholz, J. Characterization of Losses in Virtually Variable Displacement Pumps. *Int. J. Fluid Power* **2009**, *10*, 17–27. [CrossRef]
44. Merril, K.J.; Lumkes, J.H., Jr. Operating Strategies and Valve Requirements for Digital Pump/Motors. In Proceedings of the 6th FPNI—PhD Symposium, West Lafayette, IN, USA, 15–19 June 2010; pp. 249–258.
45. Huova, M.; Laamanen, A. Control of Three-Chamber Cylinder with Digital Valve System. In Proceedings of the Second Workshop on Digital Fluid Power, Linz, Austria, 12–13 November 2009; Scheidl, R., Winkler, B., Eds.; pp. 94–105.
46. Linjama, M.; Vihtanen, H.-P.; Sipola, A.; Vilenius, M. Secondary Controlled Multi-Chamber Hydraulic Cylinder. In Proceedings of the 11th Scandinavian International Conference on Fluid Power SICFP'09, Linköping, Sweden, 2–4 June 2009. 15p.
47. De Gier, G. Hydraulic Cylinder for Use in a Hydraulic Tool. Patent EP1580437, 15 September 2004.
48. Bishop, E.D. Digital Hydraulic Transformer—Approaching Theoretical Perfection in Hydraulic Drive Efficiency. In Proceedings of the Ninth Scandinavian International Conference on Fluid Power, Linköping, Sweden, 2–4 June 2009. 19p.
49. Linjama, M.; Huhtala, K. Digital pump-motor with independent outlets. In Proceedings of the 11th Scandinavian International Conference on Fluid Power SICFP'09, Linköping, Sweden, 2–4 June 2009. 16p.
50. Linjama, M.; Tammisto, J. New Alternative for Digital Pump-Motor Transformer. In Proceedings of the Second Workshop on Digital Fluid Power, Linz, Austria, 12–13 November 2009; Scheidl, R., Winkler, B., Eds.; pp. 49–61.
51. Heikkilä, M.; Tammisto, J.; Huova, M.; Huhtala, K.; Linjama, M. Experimental Evaluation of a Piston-Type Digital Pump-Motor-Transformer with Two Independent Outlets. In Proceedings of the Fluid Power and Motion Control (FPMC 2010), Bath, UK, 15–17 September 2010; Johnston, D.N., Plummer, A., Eds.; pp. 83–97.

52. Heikkilä, M.; Tammisto, J.; Linjama, M.; Huhtala, K. Digital Hydraulic Power Management System—Measured Characteristics of a Second Prototype. In Proceedings of the Eighth Workshop on Digital Fluid Power, Tampere, Finland, 24–25 May 2016.
53. Ersfolk, J.; Boström, P.; Timonen1, V.; Westerholm1, J.; Wiik, J.; Karhu, O.; Linjama, M.; Waldén, M. Optimal Digital Valve Control Using Embedded, GPU. In Proceedings of the Eighth Workshop on Digital Fluid Power, Tampere, Finland, 24–25 May 2016.
54. Ballard, R.L. System for Minimizing Skidding. U.S. Patent No 3528708, 18 February 1968.
55. Wennmacher, G. Untersuchung und Anwendung Schnellschaltender Elektrohydraulischer Ventile für den Einsatz in Kraftfahrzeugen. Ph.D. Thesis, RWTH Aachen University, Aachen, Germany, 1995.
56. Seilly, A.H. *Helenoid Actuators—A New Definition in Extremely Fast Acting Solenoids. SAE Paper, 790119*; SAE International: Warrendale, PA, USA, 1979.
57. Seilly, A.H. *Colenoid Actuators-Further Developments in Extremely Fast Acting Solenoids. SAE Paper, 810462*; SAE International: Warrendale, PA, USA, 1979.
58. Beck, N.J.; Barkhimer, R.L.; Calkins, M.A.; Johnson, W.P.; Weseloh, W.E. *Direct Digital Control of Electronic Unit Injectors, SAE 840273*; SAE International: Warrendale, PA, USA, 1979; pp. 21332–21340.
59. Tanaka, H. *Digital Control and Application of Hydraulic and Pneumatic*; Chongqing University Press: Chongqing, China, 1992.
60. Tanaka, H. Research on high speed electromagnetic on-off valve. *Transaclions Jsme* **1984**, *50*, 1594–1601. [CrossRef]
61. Tanaka, H. Digital control and its application. *Oil Air Compression Design.* **1984**, *22*, 16–23.
62. Tanaka, H.; Tanaka, H.; Araki, K. Digital control of three-way high-speed solenoid valve. *Trans. Jsme (B)* **1984**, *50*, 2663–2666. [CrossRef]
63. Zou, Z. Research on 2D Digital Valve and Electromechanical Converter. Master's Thesis, Zhejaing University of Technology, Hangzhou, China, 2010.
64. Florian, M.; Rudolf, S. Development and Experimental Results of a Small Fast Switching Valve Derived from Fuel Injection Technology. In Proceedings of the Eighth Workshop on Digital Fluid Power, Tampere, Finland, 24–25 May 2016.
65. Paloniitty, M.; Matti, L.; Huhtala, K. Durability Study on High Speed Water Hydraulic Miniature On/Off-Valve. In Proceedings of the Eighth Workshop on Digital Fluid Power, Tampere, Finland, 24–25 May 2016.
66. Rannow, M.B.; Li, P.Y.; Chase, T.R. Discrete Piston Pump/Motor Using a Mechanical Rotary Valve Control Mechanism. In Proceedings of the Eighth Workshop on Digital Fluid Power, Tampere, Finland, 24–25 May 2016.
67. Ouyang, X.; Yang, H.Y.; Jiang, H.; Xu, B. Simulation of the Piezoelectric High-speed on/off Valve. *Chin. Sci. Bull.* **2008**, *53*, 2706–2711. [CrossRef]
68. Liu, P.; Fan, L.; Hayat, Q.; Xu, D.; Ma, X.; Song, E. Research on Key Factors and Their Interaction Effects of Electromagnetic Force of High-Speed Solenoid Valve. *Sci. World J.* **2014**, *2014*, 567242. [CrossRef] [PubMed]
69. Zhang, B.; Ruan, J.; Nie, W. Dynamic response analysis of high-speed locking valve. *J. Mech. Electr. Eng.* **2008**, *25*, 69–72. [CrossRef]
70. Plöckinger, A.; Scheidl, R.; Winkler, B. Performance, Durability and Applications of a Fast Switching Valve. In Proceedings of the Second Workshop on Digital Fluid Power, Linz, Austria, 12–13 November 2009; Scheidl, R., Winkler, B., Eds.; pp. 129–143.
71. Zöppig, V.; Neumann, K. Switching Magnetic Valve Electronics. In Proceedings of the 7th International Fluid Power Conference (7th IFK), Aachen, Germany, 22–24 March 2010; Volume 2, pp. 407–418.
72. Zhong, Q.; Zhang, B.; Yang, H.; Ma, J.; Fung, R. Performance analysis of a high-speed on/off valve based on an intelligent pulse-width modulation control. *Adv. Mech. Eng.* **2017**, *9*. [CrossRef]
73. Peng, S. The Definition of a Zero-Flowrate-Switching Controller. In Proceedings of the Eighth Workshop on Digital Fluid Power, Tampere, Finland, 24–25 May 2016.
74. Sauer-Danfoss. PVE Series 4 for PVG 32, PVG 100 and PVG 120, Technical Information. In *Sauer-Danfoss Brochure No 520L0553 Rev EA*; Sauer-Danfoss: Ames, IA, USA, 2010; 32p.
75. Becker, U. The Behavior of a Position Controlled Actuator with Switching Valves. In Proceedings of the Fourth Scandinavian International Conference on Fluid Power, Tampere, Finland, 26–29 September 1995; pp. 160–167.

76. Muto, T.; Yamada, H.; Tsuchiya, S. A Precision Driving System Composed of a Hydraulic Cylinder and High-Speed ON/OFF Valves. In Proceedings of the 49th National Conference on Fluid Power, Las Vegas, NV, USA, 19–21 March 2002; pp. 627–638.
77. Roemer, D.B.; Norgaard, C.; Bech, M.M.; Johansen, P. Valve and Manifold Considerations for Efficient Digital Hydraulic Machines. In Proceedings of the Eighth Workshop on Digital Fluid Power, Tampere, Finland, 24–25 May 2016.
78. Helmus, T.; Breidi, F.J., Jr. Simulation of a Variable Displacement Mechanically Actuated Digital Pump Unit. In Proceedings of the Eighth Workshop on Digital Fluid Power, Tampere, Finland, 24–25 May 2016.
79. Zhang, B.; Hong, H.; Zhong, Q.; Guan, R.; Yang, H. A Pilot Control Method for a Variable Displacement Axial Piston Pump Using Switching Technology. In Proceedings of the Eighth Workshop on Digital Fluid Power, Tampere, Finland, 24–25 May 2016.
80. Haas, R.; Hinterbichler, C.; Lukachev, E.; Schoberl, M. Optimal Digital Hydraulic Feed-forward Control Applied to Simple Cylinder Drives. In Proceedings of the Eighth Workshop on Digital Fluid Power, Tampere, Finland, 24–25 May 2016.
81. Scheidl, R.; Riha, G. Energy Efficient Switching Control by a Hydraulic Resonance-Converter. In Proceedings of the Workshop on Power Transmission and Motion Control (PTMC 1999), Bath, UK, 8–11 September 1999; Burrows, C.R., Edge, K.A., Eds.; pp. 267–273.
82. Scheidl, R.; Mikota, G. The Role of Resonance in Elementary Hydraulic Switching Control. *Proc. Inst. Mech. Eng. Part I* **2003**, *217*, 469–480. [CrossRef]
83. Nostrani, M.P.; Galloni, A.; Raduenz, H.; De Negri, V.J. Theoretical and Experimental Analysis of a Hydraulic Step-Down Switching Converterfor Position and Speed Control. In Proceedings of the Eighth Workshop on Digital Fluid Power, Tampere, Finland, 24–25 May 2016.
84. Shi, G.; Lee, B.; Yang, L. On the Control Strategy for Pneumatic Robot Driven by High Speed Solenoid On/OFF Valves above Rough Ground. In Proceedings of the Eighth Workshop on Digital Fluid Power, Tampere, Finland, 24–25 May 2016.
85. Liu, B. The Study of Exactly Congtrol Stepping Motor. Master's Thesis, Shandong University, Jinan, China, 2010.
86. Trostmann, E. *Water Hydraulics Control Technology*; Danfoss: Nordborg, Denmark, 1996.
87. Koskinen, K.T.; Vilenius, M.J.; Virvalo, T. Water as a pressure medium in position servo systems. In Proceedings of the Forth Scandinavian International Conference on Fluid Power, Tampere, Finland, 26–29 September 1995.
88. Urata, E.; Miyakawa, S.; Yamashina, C.; Nakao, Y.; Usami, Y.; Shinoda, M. Development of a water hydraulic servovalve. *Jsme Int. J. Ser. B* **1998**, *41*, 286–294. [CrossRef]
89. Zhang, Q. Research on the Static and Dynamic Characteristics of the 2D Digital Valve and Compensation of the Dead Zone Nonlinear. Master's Thesis, Zhejiang University of Technology, Hangzhou, China, 2011.
90. Li, S.; Ruan, J.; Meng, B. Dither Compensation Technology for Hysteresis of 2D Digital Valve. *Trans. Chin. Soc. Agric. Mach.* **2012**, *42*, 208–218.
91. Katakura, H.; Yamane, R.; Takenka, T. Fundamental research on digital positioning by several hydraulic cylinder and a microcomputer. *J. Jpn. Hydraul. Pneum. Soc.* **1991**, *22*, 63–70.
92. Yang, S. Unusual Hydraulic Synchronization and By-talking Experience on Innovation. *Hydraul. Pneum. Seals* **2015**. [CrossRef]
93. Yang, T. Talking about the Digital Hydraulic by YIMEIBO. *Hydraul. Pneum. Seals* **2017**, *37*, 16–19.
94. Aemetec Co. Ltd. [EB/OL]. Available online: http://www.china-hydraulic.com/ (accessed on 17 October 2019).
95. Huova, M.; Plöckinger, A. Improving Resolution of Digital Hydraulic Valve System by Utilizing Fast Switching Valves. In Proceedings of the Third Workshop on Digital Fluid Power, Tampere, Finland, 13–14 October 2010; Laamanen, A., Linjama, M., Eds.; pp. 79–92.

MDPI
St. Alban-Anlage 66
4052 Basel
Switzerland
Tel. +41 61 683 77 34
Fax +41 61 302 89 18
www.mdpi.com

Applied Sciences Editorial Office
E-mail: applsci@mdpi.com
www.mdpi.com/journal/applsci

www.ingramcontent.com/pod-product-compliance
Lightning Source LLC
LaVergne TN
LVHW070143170726
843515LV00007B/1854

* 9 7 8 3 0 3 9 4 3 5 8 3 8 *